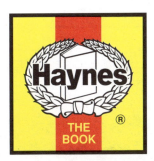

Honda CBR400RR
Service and Repair Manual

by Matthew Coombs

Models covered

Honda CBR400RR-J and K (NC23) Tri-Arm. 399 cc. 1988-on
Honda CBR400RR-L, N and R (NC29) Gull-Arm/FireBlade. 399 cc. 1990-on

Note: This manual does not include the CBR400R Aero, CB-1 or CB400 Super Four.

(3552-5Y1-264)

© Haynes Publishing 2000

ABCDE
FGHIJ
KLMNO

Printed in the USA

A book in the **Haynes Service and Repair Manual Series**

All rights reserved. No part of this book may be reproduced or transmitted in any form or by any means, electronic or mechanical, including photocopying, recording or by any information storage or retrieval system, without permission in writing from the copyright holder.

ISBN 1 85960 552 4

British Library Cataloguing in Publication Data
A catalogue record for this book is available from the British Library

Haynes Publishing
Sparkford, Yeovil, Somerset BA22 7JJ, England

Haynes North America, Inc
861 Lawrence Drive, Newbury Park, California 91320, USA

Editions Haynes
4, Rue de l'Abreuvoir
92415 COURBEVOIE CEDEX, France

Haynes Publishing Nordiska AB
Box 1504, 751 45 UPPSALA, Sweden

Contents

LIVING WITH YOUR HONDA CBR

Introduction
The Birth of a Dream	Page 0•4
Acknowledgements	Page 0•7
About this manual	Page 0•7
Identification numbers	Page 0•8
Buying spare parts	Page 0•8
Unofficial (grey) imports	Page 0•8
Safety first!	Page 0•10

Daily (pre-ride checks)
Engine/transmission oil level check	Page 0•11
Brake fluid level checks	Page 0•12
Coolant level check	Page 0•13
Suspension, steering and final drive checks	Page 0•13
Tyre checks	Page 0•14
Legal and safety checks	Page 0•14

MAINTENANCE

Routine maintenance and servicing
Specifications	Page 1•1
Recommended lubricants and fluids	Page 1•2
Maintenance schedule	Page 1•3
Component locations	Page 1•4
Maintenance procedures	Page 1•6

Contents

REPAIRS AND OVERHAUL

Engine, transmission and associated systems

Engine, clutch and transmission	Page	**2•1**
Cooling system	Page	**3•1**
Fuel and exhaust systems	Page	**4•1**
Ignition system	Page	**5•1**

Chassis and bodywork components

Frame, suspension and final drive	Page	**6•1**
Brakes	Page	**7•1**
Wheels	Page	**7•14**
Tyres	Page	**7•20**
Fairing and bodywork	Page	**8•1**

Electrical system

	Page	**9•1**

Wiring diagrams

	Page	**9•30**

REFERENCE

Dimensions and Weights	Page	**REF•1**
Tools and Workshop Tips	Page	**REF•2**
Conversion factors	Page	**REF•20**
Motorcycle Chemicals and Lubricants	Page	**REF•21**
MOT Test Checks	Page	**REF•22**
Storage	Page	**REF•27**
Fault Finding	Page	**REF•30**
Fault Finding Equipment	Page	**REF•39**
Technical Terms Explained	Page	**REF•43**

Index

	Page	**REF•47**

The Birth of a Dream

by Julian Ryder

There is no better example of the Japanese post-War industrial miracle than Honda. Like other companies which have become household names, it started with one man's vision. In this case the man was the 40-year old Soichiro Honda who had sold his piston-ring manufacturing business to Toyota in 1945 and was happily spending the proceeds on prolonged parties for his friends. However, the difficulties of getting around in the chaos of post-War Japan irked Honda, so when he came across a job lot of generator engines he realised that here was a way of getting people mobile again at low cost.

A 12 by 18-foot shack in Hamamatsu became his first bike factory, fitting the generator motors into pushbikes. Before long he'd used up all 500 generator motors and started manufacturing his own engine, known as the 'chimney', either because of the elongated cylinder head or the smoky exhaust or perhaps both. The chimney made all of half a horsepower from its 50 cc engine but it was a major success and became the Honda A-type. Less than two years after he'd set up in Hamamatsu, Soichiro Honda founded the Honda Motor Company in September 1948. By then, the A-type had been developed into the 90 cc B-type engine, which Mr Honda decided deserved its own chassis not a bicycle frame. Honda was about to become Japan's first post-War manufacturer of complete motorcycles. In August 1949 the first prototype was ready. With an output of three horsepower, the 98 cc D-type was still a simple two-stroke but it had a two-speed transmission and most importantly a pressed steel frame with telescopic forks and hard tail rear end. The frame was almost triangular in profile with the top rail going in a straight line from the massively braced steering head to the rear axle. Legend has it that after the D-type's first tests the entire workforce went for a drink to celebrate and try and think of a name for the bike. One man broke one of those silences you get when people are thinking, exclaiming 'This is like a dream!' 'That's it!' shouted Honda, and so the Honda Dream was christened.

> 'This is like a dream!'
> 'That's it'
> shouted Honda

Mr Honda was a brilliant, intuitive engineer and designer but he did not bother himself with the marketing side of his business. With hindsight, it is possible to see that employing Takeo Fujisawa who would both sort out the home market and plan the eventual expansion into overseas markets was a masterstroke. He arrived in October 1949 and in 1950 was made Sales Director. Another vital new name was Kiyoshi Kawashima, who along with Honda himself, designed the company's first four-stroke after Kawashima had told them that the four-stroke opposition to Honda's two-strokes sounded nicer and therefore sold better. The result of that statement was the overhead-valve 148 cc E-type which first ran in July 1951 just two months after the first drawings were made. Kawashima was made a director of the Honda Company at 34 years old.

The E-type was a massive success, over 32,000 were made in 1953 alone, but Honda's lifelong pursuit of technical innovation sometimes distracted him from commercial reality. Fujisawa pointed out that they were in danger of ignoring their core business, the motorised bicycles that still formed Japan's main means of transport. In May 1952 the F-type Cub appeared, another two-stroke despite the top men's reservations. You could buy a complete machine or just the motor to attach to your own bicycle. The result was certainly distinctive, a white fuel tank with a circular profile went just below and behind the saddle on the left of the bike, and the motor with its horizontal cylinder and bright red cover just below the rear axle on the same side of the bike. This was the machine that turned Honda into the biggest bike maker in Japan

Honda C70 and C90 OHV-engined models

Introduction

The CB250N Super Dream became a favorite with UK learner riders of the late seventies and early eighties

with 70% of the market for bolt-on bicycle motors, the F-type was also the first Honda to be exported. Next came the machine that would turn Honda into the biggest motorcycle manufacturer in the world.

The C100 Super Cub was a typically audacious piece of Honda engineering and marketing. For the first time, but not the last, Honda invented a completely new type of motorcycle, although the term 'scooterette' was coined to describe the new bike which had many of the characteristics of a scooter but the large wheels, and therefore stability, of a motorcycle. The first one was sold in August 1958, fifteen years later over nine-million of them were on the roads of the world. If ever a machine can be said to have brought mobility to the masses it is the Super Cub. If you add in the electric starter that was added for the C102 model of 1961, the design of the Super Cub has remained substantially unchanged ever since, testament to how right Honda got it first time. The Super Cub made Honda the world's biggest manufacturer after just two years of production.

Honda's export drive started in earnest in 1957 when Britain and Holland got their first bikes, America got just two bikes the next year. By 1962 Honda had half the American market with 65,000 sales. But Soichiro Honda had already travelled abroad to Europe and the USA, making a special point of going to the Isle of Man TT, then the most important race in the GP calendar. He realised that no matter how advanced his products were, only racing success would convince overseas markets for whom 'Made in Japan' still meant cheap and nasty. It took five years from Soichiro Honda's first visit to the Island before his bikes were ready for the TT. In 1959 the factory entered five riders in the 125. They did not have a massive impact on the event being benevolently regarded as a curiosity, but sixth, seventh and eighth were good enough for the team prize. The bikes were off the pace but they were well engineered and very reliable.

The TT was the only time the West saw the Hondas in '59, but they came back for more the following year with the first of a generation of bikes which shaped the future of motorcycling - the double-overhead-cam four-cylinder 250. It was fast and reliable - it revved to 14,000 rpm - but didn't handle anywhere near as well as the opposition. However, Honda had now signed up non-Japanese riders to lead their challenge. The first win didn't come until 1962 (Aussie Tom Phillis in the Spanish 125 GP) and was followed up with

The GL1000 introduced in 1975, was the first in Honda's line of Goldwings

Introduction

a world-shaking performance at the TT. Twenty-one year old Mike Hailwood won both 125 and 250 cc TTs and Hondas filled the top five positions in both races. Soichiro Honda's master plan was starting to come to fruition, Hailwood and Honda won the 1961 250 cc World Championship. Next year Honda won three titles. The other Japanese factories fought back and inspired Honda to produce some of the most fascinating racers ever seen: the awesome six-cylinder 250, the five-cylinder 125, and the 500 four with which the immortal Hailwood battled Agostini and the MV Agusta.

When Honda pulled out of racing in '67 they had won sixteen rider's titles, eighteen manufacturer's titles, and 137 GPs, including 18 TTs, and introduced the concept of the modern works team to motorcycle racing. Sales success followed racing victory as Soichiro Honda had predicted, but only because the products advanced as rapidly as the racing machinery. The Hondas that came to Britain in the early '60s were incredibly sophisticated. They had overhead cams where the British bikes had pushrods, they had electric starters when the Brits relied on the kickstart, they had 12V electrics when even the biggest British bike used a 6V system. There seemed no end to the technical wizardry and when in 1968 the first four-cylinder CB750 road bike arrived the world changed for ever. They even had to invent a new word for it: superbike. Honda raced again with the CB750 at Daytona and won the World Endurance title with a prototype DOHC version that became the CB900 roadster. There was the six-cylinder CBX, the first turbocharged production bike, they invented the full-dress tourer with the Goldwing and came back to GPs with the revolutionary oval-pistoned NR500 four-stroke, a much-misunderstood bike that was more rolling experiment than racer. It was true, though, that Mr Honda was not keen on two-strokes - early motocross engines had to be explained away to him as lawnmower motors! However, in 1982 Honda raced the NS500, an agile three-cylinder lightweight against the big four-cylinder opposition in 500 GPs. The bike won in the first year and in '83 took the world title for Freddie Spencer. In four-stroke racing the V4 layout took over from the straight four, dominating TT, F1 and Endurance championships and when Superbike arrived Honda were ready with the RC30. On the roads the VFR V4 became an instant classic while the CBR600 invented another new class of bike on its way to becoming a best-seller.

And then there was the NR750. This limited-edition technological tour-de-force embodied many of Soichiro Honda's ideals. It used the latest techniques and materials in every component, from the oval-piston, 32-valve V4 motor to the titanium coating on the windscreen, it was - as Mr Honda would have wanted - the best it could possibly be. A fitting memorial to the man who has shaped the motorcycle industry and motorcycles as we know them today.

Carl Fogarty in action at the Suzuka 8 Hour on the RC45

An early CB750 Four

Precision Miniatures

When the grey imports phenomenon took off commercially, it bought to the UK two different types of bike. To understand the difference you must first understand a few basic facts about the Japanese home market, where the vast majority of these machines came from. Firstly, Japan's draconian licensing system effectively restricted riders to 400 cc four-strokes and 250 cc two-strokes. Secondly, the Japanese market is fashion conscious in the extreme. If a manufacturer gets it right then it will sell thousands of units. Get it wrong and your bikes will languish in the showrooms for ever, or until some budding entrepreneur buys them up cheaply and ships them to Europe.

So when the greys started landing, UK motorcyclists were not very surprised to discover that there was a 400 cc model in the CBR range beside the 1000 cc sports tourer and the best-selling CBR600. These were used bikes which those same restrictive laws effectively took off the roads of Japan once they were a few years old. The second type tended to be weird models that had failed Japanese youth's style test and arrived over here hardly used and sometimes even brand new.

Right at the head of the first category was the CBR400RR that first hit Japan's roads in 1988 and instantly stormed to the top of the sales charts. And no wonder. Although the motor was a development of the earlier CBR400 Aero that looked just like an early 'jellymould' CBR, the double-R had gear-driven cams and an aluminium frame as well as a 17-inch front wheel and mouth-watering styling complete with twin-headlamp fairing. Bear in mind that the FireBlade was four years away and the 600 cc CBR had camshafts driven by chains and a tubular steel chassis and you can see why the double-R was such a success. In fact in terms of technology, it was nearer to the RC30 that was launched for 1988 and its 400 cc sibling the NC30, or

The CBR400RR Gull-Arm

VFR400R, as shown by the fact that the 400 cc in-line four also had a factory code: NC23. In fact it was better known as the Tri-Arm, a reference to the sticker on the side of the fairing meant to draw attention to the heavily triangulated swinging arm that looked like it had just been unbolted from a factory racer.

The commercial domination of the Tri-Arm was continued in 1990 by the next generation of CBR400RR, the L-model, the first of the Gull-Arms or NC29. Despite the superficial similarities of the two bikes, the Gull-Arm was a completely new machine whose cycle parts had much in common with the VFR400R including 17-inch wheels front and rear. The motor's cylinders and upper crankcase half were cast in one piece, again before the FireBlade and the CBR600 M-model used the same layout. The bike's name came from the new swinging arm design, again derived from current race-track practice. Instead of the straight, rectangular cross-sectional members, each side of the arm was a single, massive fabrication. On the right side, it had an elbow bend in it to accommodate the exhaust pipe without compromising ground clearance. Again this was copied from Grand Prix practice (despite the real racers being two-strokes) just a year after the design was first seen on the track, and it was this feature that gave the bike the Gull-Arm name.

The smaller bike has regularly been given the same colour schemes as the bigger Blade and has basically remained otherwise unchanged. There have been minor adjustments and the claimed power output has even gone down a fraction and the weight up by a kilogram - just like on the 400 cc V4s.

When the 900 cc FireBlade hit world markets and revolutionised the sports bike market sector in 1992 the Gull-Arm got the same paintwork and even a FireBlade sticker on the fairing. Even though the 400 was the earlier design, it was immediately (or rather lately) christened the Baby Blade and as far as anyone can tell became the best-selling grey import machine in the UK.

Acknowledgements

Our thanks are due to Elliott Motorcycles of Swindon who supplied the machines featured in the illustrations throughout this manual and provided technical literature. We would also like to thank NGK Spark Plugs (UK) Ltd for supplying the colour spark plug condition photos and the Avon Rubber Company for supplying information on tyre fitting.

The introduction 'The Birth of a Dream' was written by Julian Ryder.

About this Manual

The aim of this manual is to help you get the best value from your motorcycle. It can do so in several ways. It can help you decide what work must be done, even if you choose to have it done by a dealer; it provides information and procedures for routine maintenance and servicing; and it offers diagnostic and repair procedures to follow when trouble occurs.

We hope you use the manual to tackle the work yourself. For many simpler jobs, doing it yourself may be quicker than arranging an appointment to get the motorcycle into a dealer and making the trips to leave it and pick it up. More importantly, a lot of money can be saved by avoiding the expense the shop must pass on to you to cover its labour and overhead costs. An added benefit is the sense of satisfaction and accomplishment that you feel after doing the job yourself.

References to the left or right side of the motorcycle assume you are sitting on the seat, facing forward.

We take great pride in the accuracy of information given in this manual, but motorcycle manufacturers make alterations and design changes during the production run of a particular motorcycle of which they do not inform us. No liability can be accepted by the authors or publishers for loss, damage or injury caused by any errors in, or omissions from, the information given.

Identification numbers

The frame serial number is stamped into the right side of the steering head. The engine number is stamped into the right upper side of the crankcase, directly above the clutch unit. Both of these numbers should be recorded and kept in a safe place so they can be furnished to law enforcement officials in the event of a theft. The carburettor number is stamped into the back of each carburettor.

The frame serial number, engine serial number and carburettor identification number should also be kept in a handy place (such as with your driver's licence) so they are always available when purchasing or ordering parts for your machine.

Identifying model years

The procedures in this manual identify the bikes by model code. The model code (e.g. CBR400RR-**L**) is printed on the colour code label, which is located on the top of the rear mudguard under the passenger seat. The model code and production year can also be determined from the engine and frame serial numbers in the accompanying table. **Note:** *Do not identify your bike using the date of registration; in some cases, especially where a new bike has been imported into the UK, the registration date will differ considerably from the model code year.*

Model, code and production year	Frame number	Engine number	Carburettor number
CBR400RR-J (1988)	NC23-1020001 to 1036454	NC23E-1020001 to1036510	VG04A
CBR400RR-K (1989)	NC23 1090001 to 1098116	NC23E 1090001 to1098123	VG04B
CBR400RR-L (1990 and 91)	NC29-1000001 to 1010598	NC23E 1300001 to1310636	VP01A
CBR400RR-N (1992 and 93)	NC29-1050001 on	NC23E-1420001 on	VP01A
CBR400RR-R (1994-on)	NC29-1100001 on	NC23E-1500001 on	VP01B

Buying spare parts

Once you have found all the identification numbers, record them for reference when buying parts. Since the manufacturers change specifications, parts and vendors (companies that manufacture various components on the machine), providing the ID numbers is the only way to be reasonably sure that you are buying the correct parts.

Whenever possible, take the worn part to the dealer so direct comparison with the new component can be made. Along the trail from the manufacturer to the parts shelf, there are numerous places that the part can end up with the wrong number or be listed incorrectly.

The two places to purchase new parts for your motorcycle – the accessory shop and the motorcycle dealer – differ in the type of parts they carry. While dealers can obtain virtually every part for your motorcycle, the accessory shop is usually limited to normal high wear items such as shock absorbers, tune-up parts, various engine gaskets, cables, chains, brake parts, etc. Rarely will an accessory outlet have major suspension components, cylinders, transmission gears, or cases.

Used parts can be obtained for roughly half the price of new ones, but you can't always be sure of what you're getting. Once again, take your worn part to the breaker for direct comparison.

Whether buying new, used or rebuilt parts, the best course is to deal directly with someone who specialises in parts for your particular make.

Unofficial (grey) imports

All CBR400RR models in the UK are unofficial (grey) imports from Japan. The majority are second-hand machines and are allocated age-related licence plates for UK use (the licence plate letter reflects the production year in Japan), although new CBRs are allocated current year UK registration letter plates.

Common changes made prior to sale in the UK are the disabling of the rev-limiter device (or more correctly 'speed-limiter' device), which is fitted to comply with Japanese market regulations. The device is located in the speedometer head and is linked to the ignition control unit to cut the ignition when 180 kmh (112 mph) is reached. The importers have several methods of disabling the device, either fitting a plug-in unit at the speedometer head or ignition control unit, or by modifying the limiter mechanism.

Speedometers calibrated in kilometres (kmh) must have a miles per hour (mph) scale applied. This can be done simply by applying a suitable overlay to the speedometer lens or a more professional approach is to fit one of the replacement dial faces to the speedometer itself. In each case it is important that the correct size overlay or dial face is used.

The ratings of certain bulbs (headlight, sidelight, brake/tail light and turn signal lights) differ from those normally used on UK market machines. Of these, the brake/tail light bulbs will most likely have been replaced with the regulation 21/5W UK fitment.

Note that restrictor kits can be fitted to the CBR engines to reduce their power output to 33 bhp (25 kw) to comply with the UK full standard category A licence. Kits can be obtained from and fitted by grey importers.

Identification numbers 0•9

The colour code label is under the passenger seat

The engine number is stamped on the top of the crankcase on the right-hand side of the engine

The frame number is stamped on the right-hand side of the steering head

Safety first!

Professional mechanics are trained in safe working procedures. However enthusiastic you may be about getting on with the job at hand, take the time to ensure that your safety is not put at risk. A moment's lack of attention can result in an accident, as can failure to observe simple precautions.

There will always be new ways of having accidents, and the following is not a comprehensive list of all dangers; it is intended rather to make you aware of the risks and to encourage a safe approach to all work you carry out on your bike.

Asbestos

● Certain friction, insulating, sealing and other products - such as brake pads, clutch linings, gaskets, etc. - contain asbestos. Extreme care must be taken to avoid inhalation of dust from such products since it is hazardous to health. If in doubt, assume that they do contain asbestos.

Fire

● Remember at all times that petrol is highly flammable. Never smoke or have any kind of naked flame around, when working on the vehicle. But the risk does not end there - a spark caused by an electrical short-circuit, by two metal surfaces contacting each other, by careless use of tools, or even by static electricity built up in your body under certain conditions, can ignite petrol vapour, which in a confined space is highly explosive. Never use petrol as a cleaning solvent. Use an approved safety solvent.

● Always disconnect the battery earth terminal before working on any part of the fuel or electrical system, and never risk spilling fuel on to a hot engine or exhaust.

● It is recommended that a fire extinguisher of a type suitable for fuel and electrical fires is kept handy in the garage or workplace at all times. Never try to extinguish a fuel or electrical fire with water.

Fumes

● Certain fumes are highly toxic and can quickly cause unconsciousness and even death if inhaled to any extent. Petrol vapour comes into this category, as do the vapours from certain solvents such as trichloro-ethylene. Any draining or pouring of such volatile fluids should be done in a well ventilated area.

● When using cleaning fluids and solvents, read the instructions carefully. Never use materials from unmarked containers - they may give off poisonous vapours.

● Never run the engine of a motor vehicle in an enclosed space such as a garage. Exhaust fumes contain carbon monoxide which is extremely poisonous; if you need to run the engine, always do so in the open air or at least have the rear of the vehicle outside the workplace.

The battery

● Never cause a spark, or allow a naked light near the vehicle's battery. It will normally be giving off a certain amount of hydrogen gas, which is highly explosive.

● Always disconnect the battery ground (earth) terminal before working on the fuel or electrical systems (except where noted).

● If possible, loosen the filler plugs or cover when charging the battery from an external source. Do not charge at an excessive rate or the battery may burst.

● Take care when topping up, cleaning or carrying the battery. The acid electrolyte, evenwhen diluted, is very corrosive and should not be allowed to contact the eyes or skin. Always wear rubber gloves and goggles or a face shield. If you ever need to prepare electrolyte yourself, always add the acid slowly to the water; never add the water to the acid.

Electricity

● When using an electric power tool, inspection light etc., always ensure that the appliance is correctly connected to its plug and that, where necessary, it is properly grounded (earthed). Do not use such appliances in damp conditions and, again, beware of creating a spark or applying excessive heat in the vicinity of fuel or fuel vapour. Also ensure that the appliances meet national safety standards.

● A severe electric shock can result from touching certain parts of the electrical system, such as the spark plug wires (HT leads), when the engine is running or being cranked, particularly if components are damp or the insulation is defective. Where an electronic ignition system is used, the secondary (HT) voltage is much higher and could prove fatal.

Remember...

✗ **Don't** start the engine without first ascertaining that the transmission is in neutral.

✗ **Don't** suddenly remove the pressure cap from a hot cooling system - cover it with a cloth and release the pressure gradually first, or you may get scalded by escaping coolant.

✗ **Don't** attempt to drain oil until you are sure it has cooled sufficiently to avoid scalding you.

✗ **Don't** grasp any part of the engine or exhaust system without first ascertaining that it is cool enough not to burn you.

✗ **Don't** allow brake fluid or antifreeze to contact the machine's paintwork or plastic components.

✗ **Don't** siphon toxic liquids such as fuel, hydraulic fluid or antifreeze by mouth, or allow them to remain on your skin.

✗ **Don't** inhale dust - it may be injurious to health (see Asbestos heading).

✗ **Don't** allow any spilled oil or grease to remain on the floor - wipe it up right away, before someone slips on it.

✗ **Don't** use ill-fitting spanners or other tools which may slip and cause injury.

✗ **Don't** lift a heavy component which may be beyond your capability - get assistance.

✗ **Don't** rush to finish a job or take unverified short cuts.

✗ **Don't** allow children or animals in or around an unattended vehicle.

✗ **Don't** inflate a tyre above the recommended pressure. Apart from overstressing the carcass, in extreme cases the tyre may blow off forcibly.

✓ **Do** ensure that the machine is supported securely at all times. This is especially important when the machine is blocked up to aid wheel or fork removal.

✓ **Do** take care when attempting to loosen a stubborn nut or bolt. It is generally better to pull on a spanner, rather than push, so that if you slip, you fall away from the machine rather than onto it.

✓ **Do** wear eye protection when using power tools such as drill, sander, bench grinder etc.

✓ **Do** use a barrier cream on your hands prior to undertaking dirty jobs - it will protect your skin from infection as well as making the dirt easier to remove afterwards; but make sure your hands aren't left slippery. Note that long-term contact with used engine oil can be a health hazard.

✓ **Do** keep loose clothing (cuffs, ties etc. and long hair) well out of the way of moving mechanical parts.

✓ **Do** remove rings, wristwatch etc., before working on the vehicle - especially the electrical system.

✓ **Do** keep your work area tidy - it is only too easy to fall over articles left lying around.

✓ **Do** exercise caution when compressing springs for removal or installation. Ensure that the tension is applied and released in a controlled manner, using suitable tools which preclude the possibility of the spring escaping violently.

✓ **Do** ensure that any lifting tackle used has a safe working load rating adequate for the job.

✓ **Do** get someone to check periodically that all is well, when working alone on the vehicle.

✓ **Do** carry out work in a logical sequence and check that everything is correctly assembled and tightened afterwards.

✓ **Do** remember that your vehicle's safety affects that of yourself and others. If in doubt on any point, get professional advice.

● If in spite of following these precautions, you are unfortunate enough to injure yourself, seek medical attention as soon as possible.

Daily (pre-ride) checks

Note: *The daily (pre-ride) checks outlined in the owner's manual covers those items which should be inspected on a daily basis.*

1 Engine/transmission oil level check

Before you start:

✔ Start the engine and allow it to reach normal operating temperature.
Caution: Do not run the engine in an enclosed space such as a garage or workshop.
✔ Stop the engine and support the motorcycle in an upright position, using an auxiliary stand if required. Allow it to stand undisturbed for a few minutes to allow the oil level to stabilise. Make sure the motorcycle is on level ground.

Bike care:

● If you have to add oil frequently, you should check whether you have any oil leaks. If there is no sign of oil leakage from the joints and gaskets the engine could be burning oil (see *Fault Finding*).

The correct oil

● Modern, high-revving engines place great demands on their oil. It is very important that the correct oil for your bike is used.
● Always top up with a good quality oil of the specified type and viscosity and do not overfill the engine.

Oil type	API grade SE, SF or SG
Oil viscosity	SAE 10W40

1 Unscrew the oil filler cap (arrowed) from the right-hand crankcase cover. The dipstick is integral with the oil filler cap, and is used to check the engine oil level.

2 Using a clean rag or paper towel, wipe off all the oil from the dipstick. Insert the clean dipstick back into the engine, but **do not** screw it in.

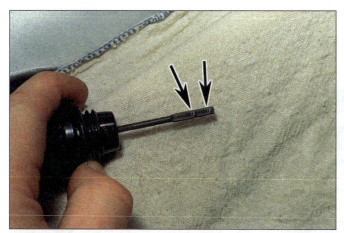

3 Remove the dipstick and observe the level of the oil, which should be somewhere in between the upper and lower level lines (arrowed).

4 If the level is below the lower line, top the engine up with the recommended grade and type of oil, to bring the level up to the upper line on the dipstick. Do not overfill.

0•12 Daily (pre-ride) checks

2 Brake fluid level checks

Warning: *Brake hydraulic fluid can harm your eyes and damage painted surfaces, so use extreme caution when handling and pouring it and cover surrounding surfaces with rag. Do not use fluid that has been standing open for some time, as it absorbs moisture from the air which can cause a dangerous loss of braking effectiveness*

Before you start:

✔ Support the motorcycle in an upright position, using an auxiliary stand if required, and turn the handlebars until the top of the front master cylinder is as level as possible. The rear master cylinder reservoir is located behind the seat cowling on the right-hand side of the machine.
✔ Make sure you have the correct hydraulic fluid. DOT 4 is recommended.
✔ Wrap a rag around the reservoir being worked on to ensure that any spillage does not come into contact with painted surfaces.
✔ Access to the front reservoir cap screws is restricted by the windshield. Use a short or angled screwdriver to access the screws.

Bike care:

● The fluid in the front and rear brake master cylinder reservoirs will drop slightly as the brake pads wear down.
● If any fluid reservoir requires repeated topping-up this is an indication of an hydraulic leak somewhere in the system, which should be investigated immediately.
● Check for signs of fluid leakage from the hydraulic hoses and components – if found, rectify immediately.
● Check the operation of both brakes before taking the machine on the road; if there is evidence of air in the system (spongy feel to lever or pedal), it must be bled as described in Chapter 7.

1 The front brake fluid level is visible through the reservoir body – it must be between the UPPER and LOWER level lines (arrowed).

2 If the level is below the LOWER level line, remove the two reservoir cap screws and remove the cap, the diaphragm plate and the diaphragm.

3 Top up with new DOT 4 hydraulic fluid, until the level is just below the UPPER level line. Do not overfill the reservoir, and take care to avoid spills (see **Warning** above).

4 Ensure that the diaphragm is correctly seated before installing the plate and cap. Tighten the cap screws securely.

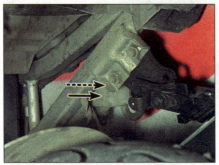

5 On J and K models, the rear brake fluid level is visible by looking across the top of the rear wheel at the window in the rear corner of the reservoir body – the fluid level must be between the UPPER and LOWER level lines (arrowed).

6 If the lines aren't visible, remove the seat cowling (see Chapter 8). The lines are also marked on the outer corner (arrows).

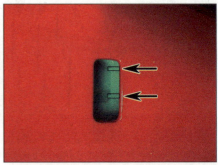

7 On L, N and R models, the rear brake fluid level is visible by looking through the aperture in the seat cowling at the window in the reservoir body – the fluid level must be between the UPPER and LOWER level lines (arrowed).

Daily (pre-ride) checks 0•13

8 If the level is below the LOWER level line, remove the seat cowling (see Chapter 8). Unscrew the reservoir cover screws and remove the cover, diaphragm plate and diaphragm.

9 Top up with new DOT 4 hydraulic fluid, until the level is just below the UPPER level line. Do not overfill the reservoir, and take care to avoid spills (see **Warning** above).

10 Ensure that the diaphragm is correctly seated before installing the plate and cover. Tighten the cover screws securely, then install the seat cowling (see Chapter 8).

3 Coolant level check

> ⚠ **Warning: DO NOT remove the radiator pressure cap to add coolant. Topping up is done via the coolant reservoir tank filler. DO NOT leave open containers of coolant about, as it is poisonous.**

Before you start:
✔ Make sure you have a supply of coolant available (a mixture of 50% distilled water and 50% corrosion inhibited ethylene glycol anti-freeze is needed).
✔ Always check the coolant level when the engine is at normal working temperature. Start the engine allow it to reach normal temperature, then stop the engine.
Caution: Do not run the engine in an enclosed space such as a garage or workshop.
✔ Support the motorcycle in an upright position, using an auxiliary stand if required, whilst checking the level. Make sure the motorcycle is on level ground.

Bike care:
● Use only the specified coolant mixture. It is important that anti-freeze is used in the system all year round, and not just in the winter. Do not top the system up using only water, as the system will become too diluted.
● Do not overfill the reservoir tank. If the coolant is significantly above the UPPER level line at any time, the surplus should be siphoned or drained off to prevent the possibility of it being expelled out of the overflow hose.
● If the coolant level falls steadily, check the system for leaks (see Chapter 1). If no leaks are found and the level continues to fall, it is recommended that the machine is taken to a dealer for a pressure test.

1 The coolant reservoir is located behind the main frame member on the right-hand side. The coolant UPPER and LOWER level lines (arrowed) are on the back of the reservoir (L, N and R models shown).

2 If the coolant level is not in between the UPPER and LOWER markings, remove the reservoir filler cap (arrowed – L, N and R models shown).

3 Top the coolant level up with the recommended coolant mixture, using a funnel to avoid spillage. Fit the cap securely.

4 Suspension, steering and final drive checks

Suspension and steering:
● Check that the front and rear suspension operate smoothly without binding.
● Check that the suspension is adjusted as required.

● Check that the steering moves smoothly from lock-to-lock.

Final drive:
● Check that the drive chain slack isn't excessive, and adjust if necessary (see Chapter 1).
● If the chain looks dry, lubricate it (see Chapter 1).

0•14 Daily (pre-ride) checks

5 Tyre checks

The correct pressures:
- The tyres must be checked when **cold**, not immediately after riding. Note that low tyre pressures may cause the tyre to slip on the rim or come off. High tyre pressures will cause abnormal tread wear and unsafe handling.
- Use an accurate pressure gauge.
- Proper air pressure will increase tyre life and provide maximum stability and ride comfort.

Tyre care:
- Check the tyres carefully for cuts, tears, embedded nails or other sharp objects and excessive wear. Operation of the motorcycle with excessively worn tyres is extremely hazardous, as traction and handling are directly affected.
- Check the condition of the tyre valve and ensure the dust cap is in place.
- Pick out any stones or nails which may have become embedded in the tyre tread. If left, they will eventually penetrate through the casing and cause a puncture.
- If tyre damage is apparent, or unexplained loss of pressure is experienced, seek the advice of a tyre fitting specialist without delay.

Tyre tread depth:
- At the time of writing UK law requires that tread depth must be at least 1 mm over 3/4 of the tread breadth all the way around the tyre, with no bald patches. Many riders, however, consider 2 mm tread depth minimum to be a safer limit. Honda recommend a minimum of 2 mm on both tyres.
- Many tyres now incorporate wear indicators in the tread. Identify the triangular pointer or 'TWI' mark on the tyre sidewall to locate the indicator bar and renew the tyre if the tread has worn down to the bar.

Loading	Front	Rear
Rider only	33 psi (2.25 Bar)	33 psi (2.25 Bar)
Rider and passenger	33 psi (2.25 Bar)	36 psi (2.50 Bar)

1 Check the tyre pressures when the tyres are **cold** and keep them properly inflated.

2 Measure tread depth at the centre of the tyre using a tread depth gauge.

3 Tyre tread wear indicator bar and its location marking (usually either an arrow, a triangle or the letters TWI) on the sidewall (arrowed).

6 Legal and safety checks

Lighting and signalling:
- Take a minute to check that the headlight, tail light, brake light, instrument lights and turn signals all work correctly.
- Check that the horn sounds when the switch is operated.
- A working speedometer graduated in mph is a statutory requirement in the UK.

Safety:
- Check that the throttle grip rotates smoothly and snaps shut when released, in all steering positions. Also check for the correct amount of freeplay (see Chapter 1).
- Check that the engine shuts off when the kill switch is operated.
- Check that sidestand return spring holds the stand securely up when retracted.
- Check that the clutch lever operates smoothly and with the correct amount of freeplay (see Chapter 1).

Fuel:
- This may seem obvious, but check that you have enough fuel to complete your journey. If you notice signs of fuel leakage – rectify the cause immediately.
- Ensure you use the correct grade unleaded fuel – see Chapter 4 Specifications.

Chapter 1
Routine maintenance and servicing

Contents

Air filter – cleaning	8	Engine/transmission – oil change and filter renewal	12
Air filter – renewal	26	Front forks – oil change	35
Battery – charging	see Chapter 9	Fuel hoses – renewal	30
Battery – check	14	Fuel system – check	13
Battery – removal, installation and inspection	see Chapter 9	Headlight aim – check and adjustment	18
Brake caliper seals and master cylinder seal renewal	28	Idle speed – check and adjustment	2
Brakes – fluid change	24	Nuts and bolts – tightness check	23
Brake hoses – renewal	29	Sidestand – check	19
Brake pads – wear check	3	Spark plugs – gap check and adjustment	5
Brake system – check	11	Spark plugs – renewal	16
Carburettors – synchronisation	17	Stand, lever pivots and cables – lubrication	6
Clutch – check and adjustment	4	Steering head bearings – check and adjustment	21
Cooling system – check	10	Steering head bearings – lubrication	33
Cooling system – draining, flushing and refilling	27	Suspension – check	20
Cylinder compression – check	31	Swingarm and suspension linkage bearings – lubrication	34
Drive chain and sprockets – check, adjustment and lubrication	1	Throttle and choke cables – check	15
Engine oil pressure – check	32	Valve clearances – check and adjustment	25
Engine/transmission – oil change	7	Wheels and tyres – general check	9
		Wheel bearings – check	22

Degrees of difficulty

Easy, suitable for novice with little experience	Fairly easy, suitable for beginner with some experience	Fairly difficult, suitable for competent DIY mechanic	Difficult, suitable for experienced DIY mechanic	Very difficult, suitable for expert DIY or professional 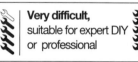

Specifications

Note: *Models are identified by their production code letter – refer to 'Identification numbers' at the front of this manual for details.*

Engine

Cylinder numbering (from left-hand to right-hand side of the bike) 1-2-3-4
Spark plugs
  Type
    Standard .... NGK CR8EH-9 or Nippondenso U24FER-9
    For extended high speed riding .... NGK CR9EH-9 or Nippondenso U27FER-9
  Electrode gap .... 0.8 to 0.9 mm
Engine idle speed .... 1300 ± 100 rpm
Carburettor synchronisation – max difference between carburettors .. 30 mmHg
Valve clearances (COLD engine)
  Inlet valves .... 0.12 to 0.18 mm
  Exhaust valves
    J, K, L and N models .... 0.17 to 0.23 mm
    R models .... 0.18 to 0.24 mm
Cylinder compression .... 156 to 213 psi (10.8 to 14.7 bar)
Oil pressure (with engine warm) .... 71 psi (5.0 Bar) @ 7000 rpm, oil @ 80°C

1•2 Specifications

Miscellaneous
Drive chain slack
    J models .... 10 to 20 mm
    K, L, N and R models .... 15 to 25 mm
Clutch cable freeplay .... 10 to 20 mm
Throttle cable freeplay .... 2 to 6 mm
Tyre pressures and tyre tread depth .... see Daily (pre-ride) checks

Recommended lubricants and fluids
Engine/transmission oil type .... API grade SE, SF or SG motor oil
Engine/transmission oil viscosity .... SAE 10W40
Engine/transmission oil capacity
  J and K models
    Oil change .... 2.9 litres
    Oil and filter change .... 3.1 litres
    Following engine overhaul – dry engine, new filter .... 3.5 litres
  L and N models
    Oil change .... 3.2 litres
    Oil and filter change .... 3.4 litres
    Following engine overhaul – dry engine, new filter .... 3.8 litres
  R models
    Oil change .... 3.0 litres
    Oil and filter change .... 3.2 litres
    Following engine overhaul – dry engine, new filter .... 3.8 litres
Coolant type .... 50% distilled water, 50% corrosion inhibited ethylene glycol anti-freeze
Coolant capacity
  Radiator and engine .... 2.0 litres
  Reservoir .... 0.3 litre
Front fork oil .... see Chapter 6
Brake fluid .... DOT 4
Drive chain .... SAE 80 or 90 gear oil or aerosol chain lubricant for O-ring chains

Miscellaneous
Steering head bearings .... Lithium-based multi-purpose grease
Wheel bearings (unsealed) .... Lithium-based multi-purpose grease
Swingarm pivot bearings .... Molybdenum disulphide grease
Suspension linkage bearings .... Molybdenum disulphide grease
Bearing seal lips .... Lithium-based multi-purpose grease
Gearchange lever/clutch lever/rear brake pedal pivots .... Molybdenum disulphide grease or dry film lubricant
Front brake lever pivot and piston tip .... Molybdenum disulphide grease or dry film lubricant
Cables .... Cable lubricant or 10W40 motor oil
Sidestand pivot .... Molybdenum disulphide grease
Throttle grip .... Multi-purpose grease or dry film lubricant

Torque settings
Note: *Where a specified setting is not given for a particular bolt, the general settings listed at the beginning apply. The dimension given applies to the diameter of the thread, not the head.*
5 mm bolt/nut .... 5 Nm
6 mm bolt/nut .... 10 Nm
8 mm bolt/nut .... 22 Nm
10 mm bolt/nut .... 35 Nm
12 mm bolt/nut .... 55 Nm
6 mm flange bolt with 8 mm head .... 9 Nm
6 mm flange bolt/nut with 10 mm head .... 12 Nm
8 mm flange bolt/nut .... 27 Nm
10 mm flange bolt/nut .... 40 Nm
Rear axle nut
  J and K models .... 90 Nm
  L, N and R models .... 95 Nm
Steering head bearing adjuster nut .... 22 Nm
Steering stem nut .... 105 Nm
Top yoke fork clamp bolts
  J and K models .... 11 Nm
  L, N and R models .... 23 Nm
Front brake master cylinder clamp bolts .... 12 Nm

Maintenance schedule

Note: *The daily (pre-ride) checks outlined in the owner's manual covers those items which should be inspected on a daily basis. Always perform the pre-ride inspection at every maintenance interval (in addition to the procedures listed). The intervals listed below are the intervals recommended by the manufacturer for each particular operation during the model years covered in this manual. Your owner's manual may have different intervals for your model.*

Daily (pre-ride)
- [ ] See *'Daily (pre-ride) checks'* at the beginning of this manual.

After the initial 600 miles (1000 km)
Note: *This check is usually performed by a dealer after the first 600 miles (1000 km) from new. Thereafter, maintenance is carried out according to the following intervals of the schedule.*

Every 600 miles (1000 km)
- [ ] Check, adjust and lubricate the drive chain (Section 1)

Every 4000 miles (6000 km) or 6 months (whichever comes sooner)
- [ ] Check and adjust the idle speed (Section 2)
- [ ] Check the brake pads (Section 3)
- [ ] Check and adjust the clutch (Section 4)
- [ ] Check the spark plug gaps (Section 5)
- [ ] Lubricate the clutch/gearchange/brake lever/brake pedal/sidestand pivots and the throttle/choke/clutch cables (Section 6)
- [ ] Change the engine oil (Section 7)
- [ ] Clean the air filter element (Section 8)
- [ ] Check the condition of the wheels and tyres (Section 9)
- [ ] Check the cooling system (Section 10)
- [ ] Check the brake system and brake light switch operation (Section 11)

Every 8000 miles (12,000 km) or 12 months (whichever comes sooner)
Carry out all the items under the 4000 mile (6000 km) check, plus the following
- [ ] Change the engine oil and filter (Section 12)
- [ ] Check the fuel system and hoses (Section 13)
- [ ] Check the battery terminals (Section 14)
- [ ] Check and adjust the throttle and choke cables (Section 15)

Every 8000 miles (12,000 km) or 12 months (whichever comes sooner) (continued)
- [ ] Renew the spark plugs (Section 16)
- [ ] Check/adjust the carburettor synchronisation (Section 17)
- [ ] Check and adjust the headlight aim (Section 18)
- [ ] Check the sidestand (Section 19)
- [ ] Check the suspension (Section 20)
- [ ] Check and adjust the steering head bearings (Section 21)
- [ ] Check the wheel bearings (Section 22)
- [ ] Check the tightness of all nuts, bolts and fasteners (Section 23)
- [ ] Change the brake fluid (Section 24)
- [ ] Check and adjust the valve clearances (Section 25)
- [ ] Check the chain and sprocket condition (Section 1)

Every 12,000 miles (18,000 km) or 18 months (whichever comes sooner)
Carry out all the items under the 4000 mile (6000 km) check, plus the following
- [ ] Renew the air filter element (Section 26)

Every 24,000 miles (36,000 km) or two years (whichever comes sooner)
Carry out all the items under the 12,000 mile (18,000 km) and 8000 mile (12,000 km) checks, plus the following
- [ ] Change the coolant (Section 27)
- [ ] Renew the brake master cylinder and caliper seals (Section 28)

Every four years
- [ ] Renew the brake hoses (Section 29)
- [ ] Renew the fuel hoses (Section 30)

Non-scheduled maintenance
- [ ] Check the cylinder compression (Section 31)
- [ ] Check the engine oil pressure (Section 32)
- [ ] Re-grease the steering head bearings (Section 33)
- [ ] Re-grease the swingarm and suspension linkage bearings (Section 34)
- [ ] Change the front fork oil (Section 35)

1•4 Component locations

J and K model component locations on right side

1. Rear brake fluid reservoir
2. Coolant reservoir
3. Clutch cable lower adjuster
4. Front brake fluid reservoir
5. Throttle cable upper adjuster
6. Radiator pressure cap
7. Timing mark inspection plug
8. Alternator bolt access plug
9. Engine/transmission oil dipstick
10. Rear brake light switch

J and K model component locations on left side

1. Clutch cable upper adjuster
2. Steering head bearings
3. Air filter
4. Idle speed adjuster
5. Fuel filter
6. Battery
7. Drive chain adjuster
8. Coolant drain bolt
9. Engine/transmission oil drain bolt
10. Engine/transmission oil filter
11. Front fork oil drain screw

Component locations

L, N and R component locations on right side

1. Rear brake fluid reservoir
2. Coolant reservoir
3. Clutch cable lower adjuster
4. Radiator pressure cap
5. Front brake fluid reservoir
6. Throttle cable upper adjuster
7. Front fork oil drain screw
8. Timing mark inspection plug
9. Alternator bolt access plug
10. Engine/transmission oil dipstick
11. Rear brake light switch

L, N and R component locations on left side

1. Clutch cable upper adjuster
2. Steering head bearings
3. Air filter
4. Idle speed adjuster
5. Fuel filter
6. Battery
7. Drive chain adjuster
8. Coolant drain bolt
9. Engine/transmission oil drain bolt
10. Engine/transmission oil filter

1•6 Introduction

1 This Chapter is designed to help the home mechanic maintain his/her motorcycle for safety, economy, long life and peak performance.
2 Deciding where to start or plug into the routine maintenance schedule depends on several factors. If the warranty period on your motorcycle has just expired, and if it has been maintained according to the warranty standards, you may want to pick up routine maintenance as it coincides with the next mileage or calendar interval. If you have owned the machine for some time but have never performed any maintenance on it, then you may want to start at the nearest interval and include some additional procedures to ensure that nothing important is overlooked. If you have just had a major engine overhaul, then you may want to start the maintenance routine from the beginning. If you have a used machine and have no knowledge of its history or maintenance record, you may desire to combine all the checks into one large service initially and then settle into the maintenance schedule prescribed.
3 Before beginning any maintenance or repair, the machine should be cleaned thoroughly, especially around the oil filter, spark plugs, valve cover, seat cowling, carburettors, etc. Cleaning will help ensure that dirt does not contaminate the engine and will allow you to detect wear and damage that could otherwise easily go unnoticed.
4 Certain maintenance information is sometimes printed on decals attached to the motorcycle. If the information on the decals differs from that included here, use the information on the decal.

> **HAYNES HiNT** *Models are identified by their production code letter – refer to 'Identification numbers' at the front of this manual for details.*

Every 600 miles (1000 km)

1 Drive chain and sprockets – check, adjustment and lubrication

Check – every 600 miles (1000 km)

1 A neglected drive chain won't last long and can quickly damage the sprockets. Routine chain adjustment and lubrication isn't difficult and will ensure maximum chain and sprocket life.
2 To check the chain, place the bike on its sidestand and shift the transmission into neutral. Make sure the ignition switch is OFF.
3 Push up on the bottom run of the chain and measure the slack midway between the two sprockets, then compare your measurement to that listed in this Chapter's Specifications **(see illustration)**. As the chain stretches with wear, adjustment will periodically be necessary (see below). Since the chain will rarely wear evenly, roll the bike forwards so that another section of chain can be checked; do this several times to check the entire length of chain.
4 In some cases where lubrication has been neglected, corrosion and galling may cause the links to bind and kink, which effectively shortens the chain's length. Such links should be thoroughly cleaned and worked free. If the chain is tight between the sprockets, rusty or kinked, it's time to renew it. If you find a tight area, mark it with felt pen or paint, and repeat the measurement after the bike has been ridden. If the chain's still tight in the same area, it may be damaged or worn. Because a tight or kinked chain can damage the transmission ouput shaft bearing, it's a good idea to renew it.
Caution: If the machine is ridden with excessive slack in the drive chain, the chain could contact the frame and swingarm, causing severe damage.

Check – every 8000 miles (12,000 km) or 12 months

5 Check the entire length of the chain for damaged rollers, loose links and pins, and missing O-rings and renew it if damage is found. **Note:** *Never install a new chain on old sprockets, and never use the old chain if you install new sprockets – renew the chain and sprockets as a set.*
6 Remove the front sprocket cover (see Chapter 6). Check the teeth on the front and rear sprockets for wear **(see illustration)**.
7 Inspect the drive chain slider on the swingarm for excessive wear and renew it if worn (see Chapter 6).

Adjustment

8 Rotate the rear wheel until the chain is positioned with the tightest point at the centre of its bottom run. If available, raise the rear wheel off the ground using an auxiliary stand or support.
9 Slacken the rear axle nut, and on L, N and R models the locknut on each chain adjuster **(see illustrations)**.
10 Turn the axle adjusters on both sides of the swingarm until the specified chain tension is obtained (get the adjuster on the chain side close, then set the adjuster on the opposite side) **(see illustrations 1.9a and b)**. Be sure to turn the adjusters evenly to keep the rear wheel in alignment. If the adjusters reach the end of their travel, the chain is excessively worn and should be renewed (see Chapter 6).

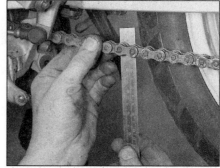
1.3 Push up on the chain and measure the slack

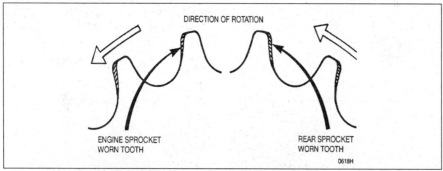
1.6 Check the sprockets in the areas indicated to see if they are worn excessively

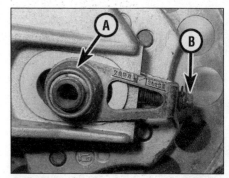
1.9a Rear axle nut (A), chain adjuster (B) – J and K models

Every 600 miles (1000 km)

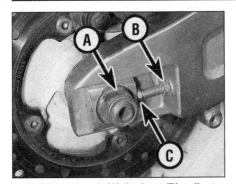

1.9b Rear axle nut (A), locknut (B), adjuster (C) – L, N and R models

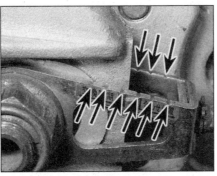

1.11a Alignment marks (arrowed) – J and K models

1.11b Alignment marks (A), notches (B) – L, N and R models

The chain wear decals will also indicate the need for chain renewal.

11 When the chain has the correct amount of slack, check that the wheel is correctly aligned by making sure the marks on each adjustment marker are in the same position relative to the back of the swingarm on J and K models, or the notches in the swingarm on L, N and R models **(see illustrations)**. If there is any discrepancy in the chain adjuster positions, adjust one of them so that its position is exactly the same as the other, then recheck the chain freeplay as described above. It is important each adjuster is identically aligned otherwise the rear wheel will be out of alignment with the front.

12 Tighten the axle nut to the torque setting specified at the beginning of the Chapter **(see illustration)**. On L, N and R models, tighten the chain adjuster locknuts securely.

Lubrication

13 If required, wash the chain in paraffin (kerosene), then wipe it off and allow it to dry, using compressed air if available. If the chain is excessively dirty it should be removed from the machine and allowed to soak in the paraffin (see Chapter 6).
Caution: Don't use petrol, solvent or other

1.12 Tighten the axle nut to the specified torque

cleaning fluids which might damage the internal sealing properties of the chain. Don't use high-pressure water. The entire process shouldn't take longer than ten minutes – if it does, the O-rings in the chain rollers could be damaged.

14 For routine lubrication, the best time to lubricate the chain is after the motorcycle has been ridden. When the chain is warm, the lubricant will penetrate the joints between the side plates better than when cold. *Note: Honda specifies SAE 80 to SAE 90 gear oil; if you do use aerosol chain lube ensure that it is*

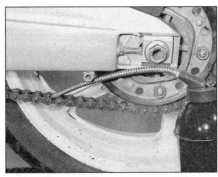

1.14 Apply the lubricant to the chain as described

suitable for O-ring chains. Apply the oil to the area where the side plates overlap – not the middle of the rollers **(see illustration)**.

> **HAYNES HiNT** *Apply the oil to the top of the lower chain run, so centrifugal force will work the oil into the chain when the bike is moving. After applying the lubricant, let it soak in a few minutes before wiping off any excess.*

Every 4000 miles (6000 km) or 6 months

2 Idle speed – check and adjustment

1 The idle speed should be checked and adjusted before and after the carburettors are synchronised (balanced) and when it is obviously too high or too low. Before adjusting the idle speed, make sure the valve clearances and spark plug gaps are correct. Also, turn the handlebars back-and-forth and see if the idle speed changes as this is done. If it does, the throttle cable may not be adjusted or routed correctly, or may be worn out. This is a dangerous condition that can cause loss of control of the bike. Be sure to correct this problem before proceeding.

2 The engine should be at normal operating temperature, which is usually reached after 10 to 15 minutes of stop-and-go riding. Place the motorcycle on its sidestand, and make sure the transmission is in neutral.

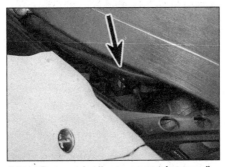

2.3a Idle speed adjuster screw (arrowed) – J and K models

3 The idle speed adjuster is located under the fuel tank on the left-hand side **(see illustrations)**. With the engine idling, adjust the idle speed by turning the adjuster screw

2.3b Idle speed adjuster screw (arrowed) – L, N and R models

1•8 Every 4000 miles (6000 km) or 6 months

3.1 Each pad has a cutout in its friction material which is visible by looking at the back of the caliper

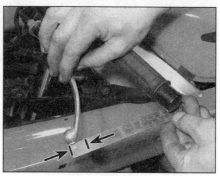

4.3 Measuring clutch cable freeplay

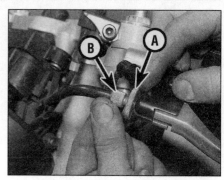

4.4 Slacken the lockring (A) and turn the adjuster (B) in or out as required

until the idle speed listed in this Chapter's Specifications is obtained. Turn the screw clockwise to increase idle speed, and anti-clockwise to decrease it.

4 Snap the throttle open and shut a few times, then recheck the idle speed. If necessary, repeat the adjustment procedure.

5 If a smooth, steady idle can't be achieved, the fuel/air mixture may be incorrect (see Chapter 4) or the carburettors may need synchronising (see Section 17).

3 Brake pads – wear check

1 A quick check of the brake pads can be made without removing them from the caliper. The amount of pad wear can be judged by looking at the pads from the rear of the caliper (both front and rear) **(see illustrations)**. A cutout in the friction material indicates the wear limit **(see illustration)**.

2 If either pad has worn down to, or beyond the cutout in the friction material, both pads must be renewed as a set. If the pads are dirty or if you are in doubt as to the amount of friction material remaining, remove them for inspection (see Chapter 7). **Note:** *Some aftermarket pads may use different indicators to those on the original equipment as shown.*

3 Refer to Chapter 7 for details of pad renewal.

4 Clutch – check and adjustment

1 Check that the clutch cable operates smoothly and easily.

2 If the clutch lever operation is heavy or stiff, remove the cable (see Chapter 2) and lubricate it (see Section 6). If the cable is still stiff, renew it. Install the lubricated or new cable (see Chapter 2).

3 With the cable operating smoothly, check that the clutch lever is correctly adjusted. Periodic adjustment is necessary to compensate for wear in the clutch plates and stretch of the cable. Check that the amount of freeplay at the clutch lever end is within the specifications listed at the beginning of the Chapter **(see illustration)**.

4 If adjustment is required, loosen the adjuster lockring at the top of the cable and turn the adjuster in or out until the required amount of freeplay is obtained **(see illustration)**. To increase freeplay, turn the adjuster clockwise. To reduce freeplay, turn the adjuster anti-clockwise. Tighten the locking ring securely.

5 If all the adjustment has been taken up at the lever, reset the adjuster to give the maximum amount of freeplay, then set the correct amount of freeplay using the adjuster nuts on each end of the threaded section in the cable bracket on the clutch cover on the right-hand side of the engine **(see illustration)**. Remove the right-hand fairing side panel and the lower fairing as required to access the adjuster nuts (see Chapter 8). To increase freeplay, slacken the front nut and tighten the rear nut until the freeplay is as specified, then tighten the front nut against the bracket. To reduce freeplay, slacken the rear nut and tighten the front nut until the freeplay is as specified, then tighten the rear nut against the bracket. Subsequent adjustments can now be made using the lever adjuster only.

5 Spark plug gaps – check and adjustment

1 Make sure your spark plug socket is the correct size before attempting to remove the plugs – a suitable one is supplied in the motorcycle's tool kit which is stored under the seat.

2 To access the spark plugs, remove the fairing side panels (see Chapter 8). Also remove the radiator lower mounting bolt(s), then release the radiator lower hose from its clip and swing the radiator forward **(see illustrations)**.

3 Clean the area around the plug caps to prevent any dirt falling into the spark plug channels.

4 Check that the cylinder location is marked

4.5 Clutch cable lower adjuster nuts (arrowed)

5.2a Remove the lower mounting bolt(s) . . .

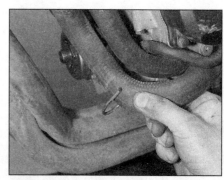

5.2b . . . and pull the hose out of the clip

Every 4000 miles (6000 km) or 6 months

5.4a Remove the spark plug cap . . .

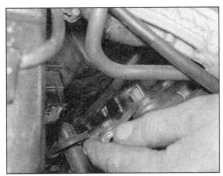

5.4b . . . then unscrew the spark plug

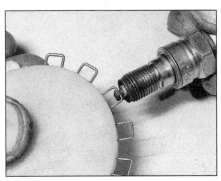

5.8a Using a wire type gauge to measure the spark plug electrode gap

on each plug lead and mark them accordingly if not. Pull the spark plug cap off each spark plug (see illustration). Clean the area around the base of the plugs to prevent any dirt falling into the engine. Using either the plug removing tool supplied in the bike's toolkit or a deep socket type wrench, unscrew the plugs from the cylinder head (see illustration). Lay each plug out in relation to its cylinder; if any plug shows up a problem it will then be easy to identify the troublesome cylinder.

5 Inspect the electrodes for wear. Both the centre and side electrodes should have square edges and the side electrodes should be of uniform thickness. Look for excessive deposits and evidence of a cracked or chipped insulator around the centre electrode. Compare your spark plugs to the colour spark plug reading chart at the end of this manual. Check the threads, the washer and the ceramic insulator body for cracks and other damage.

6 If the electrodes are not excessively worn, and if the deposits can be easily removed with a wire brush, the plugs can be re-gapped and re-used (if no cracks or chips are visible in the insulator). If in doubt concerning the condition of the plugs, renew them, as the expense is minimal.

7 Cleaning spark plugs by sandblasting is permitted, provided you clean the plugs with a high flash-point solvent afterwards.

8 Before installing the plugs, make sure they are the correct type and heat range and check the gap between the electrodes (see illustrations). Compare the gap to that specified and adjust as necessary. If the gap must be adjusted, bend the side electrodes only and be very careful not to chip or crack the insulator nose (see illustration). Make sure the washer is in place before installing each plug.

9 Since the cylinder head is made of aluminium, which is soft and easily damaged, thread the plugs into the heads turning the tool by hand (see illustration). Once the plugs are finger-tight, the job can be finished with a spanner on the tool supplied or a socket drive (see illustration 5.4b). Tighten the plugs an additional 1/4 to 1/2 turn or as directed on the manufacturer's packaging. Do not over-tighten them.

HAYNES HiNT *As the plugs are quite recessed, slip a short length of hose over the end of the plug to use as a tool to thread it into place. The hose will grip the plug well enough to turn it, but will start to slip if the plug begins to cross-thread in the hole – this will prevent damaged threads.*

10 Reconnect the spark plug caps, making sure they are securely connected to the correct cylinder (see illustration 5.4a). Install all other components previously removed.

HAYNES HiNT *Stripped plug threads in the cylinder head can be repaired with a thread insert – see 'Tools and Workshop Tips' in the Reference section.*

6 Stand, lever pivots and cables – lubrication

Pivot points

1 Since the controls, cables and various other components of a motorcycle are exposed to the elements, they should be lubricated periodically to ensure safe and trouble-free operation.

2 The footrests, clutch and brake levers, brake pedal, gearchange lever linkage and sidestand pivots should be lubricated frequently. In order for the lubricant to be applied where it will do the most good, the component should be disassembled. However, if chain and cable lubricant is being used, it can be applied to the pivot joint gaps and will usually work its way into the areas where friction occurs. If motor oil or light grease is being used, apply it sparingly as it may attract dirt (which could cause the controls to bind or wear at an accelerated rate). **Note:** *One of the best lubricants for the*

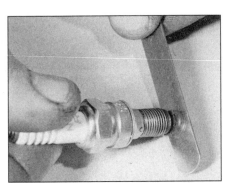

5.8b Using a feeler gauge to measure the spark plug electrode gap

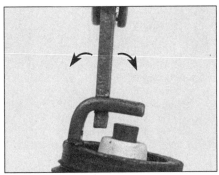

5.8c Adjust the electrode gap by bending the side electrode only

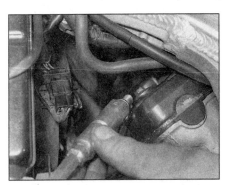

5.9 Thread the plug in as far as possible turning the tool by hand

1•10 Every 4000 miles (6000 km) or 6 months

6.3a Lubricating a cable with a cable oiler clamp. Make sure the tool seals around the inner cable

control lever pivots is a dry-film lubricant (available from many sources by different names).

Cables

3 To lubricate the cables, disconnect the relevant cable at its upper end, then lubricate the cable with a cable oiler clamp, or if one is not available, using the set-up shown **(see illustrations)**. See Chapter 4 for the choke and throttle cable removal procedures, and Chapter 2 for the clutch cable.

4 The speedometer cable should be removed (see Chapter 9) and the inner cable withdrawn from the outer cable and lubricated with motor oil or cable lubricant. Do not lubricate the upper few inches of the cable as the lubricant may travel up into the instrument head.

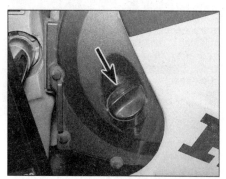

7.3 Unscrew the oil filler cap . . .

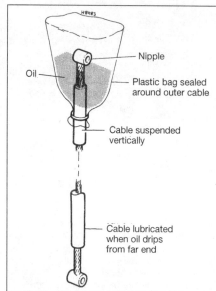

6.3b Lubricating a cable with a makeshift funnel and motor oil

 7 Engine/transmission – oil change

⚠️ **Warning: Be careful when draining the oil, as the exhaust pipes, the engine, and the oil itself can cause severe burns.**

1 Consistent routine oil and filter changes are the single most important maintenance procedure you can perform on a motorcycle. The oil not only lubricates the internal parts of the engine, transmission and clutch, but it also acts as a coolant, a cleaner, a sealant, and a protectant. Because of these demands, the oil takes a terrific amount of abuse and should be changed often with new oil of the recommended grade and type. Saving a little money on the difference in cost between a good oil and a cheap oil won't pay off if the engine is damaged.

2 Before changing the oil, warm up the engine so the oil will drain easily. Remove the lower fairing (see Chapter 8).

3 Put the motorcycle on its sidestand, and position a clean drain tray below the engine. Unscrew the oil filler cap from the alternator/clutch cover to vent the crankcase and to act as a reminder that there is no oil in the engine **(see illustration)**.

4 Next, unscrew the oil drain plug from the sump on the bottom of the engine and allow the oil to flow into the drain tray **(see illustrations)**. Check the condition of the sealing washer on the drain plug and obtain a new one if it is damaged or worn.

> **HAYNES HINT** *To help determine whether any abnormal or excessive engine wear is occurring, place a strainer between the engine and the drain tray so that any debris in the oil is filtered out and can be examined.*

5 When the oil has completely drained, fit the plug to the sump, using a new sealing washer if necessary, and tighten it securely **(see illustration)**. Avoid overtightening, as damage to the sump will result.

6 Refill the engine to the proper level using the recommended type and amount of oil (see *Daily (pre-ride) checks*). With the motorcycle vertical, the oil level should lie between the upper and lower level lines on the dipstick (see *Daily (pre-ride) checks*). Install the filler cap **(see illustration 7.3)**. Start the engine and let it run for two or three minutes (make sure that the oil pressure light extinguishes after a few seconds). Shut it off, wait a few minutes, then check the oil level. If necessary, add more oil to bring the level up to the upper level line on the dipstick. Check around the drain plug for leaks.

> **HAYNES HINT** *Saving a little money on the difference between good and cheap oils won't pay off if the engine is damaged as a result.*

7 The old oil drained from the engine cannot be re-used and should be disposed of properly. Check with your local refuse disposal company, disposal facility or

7.4a . . . and the oil drain plug (arrowed) . .

7.4b . . . and allow the oil to completely drain

7.5 Tighten the drain plug securely

Every 4000 miles (6000 km) or 6 months

environmental agency to see whether they will accept the used oil for recycling. Don't pour used oil into drains or onto the ground.

 Haynes Hint *Check the old oil carefully – if it is very metallic coloured, then the engine is experiencing wear from break-in (new engine) or from insufficient lubrication. If there are flakes or chips of metal in the oil, then something is drastically wrong internally and the engine will have to be disassembled for inspection and repair. If there are pieces of fibre-like material in the oil, the clutch is experiencing excessive wear and should be checked.*

Note: It is antisocial and illegal to dump oil down the drain. To find the location of your local oil recycling bank, call this number free.

8 Air filter – cleaning

Caution: *If the machine is continually ridden in continuously wet or dusty conditions, the filter should be renewed more frequently.*

1 Remove the fuel tank (see Chapter 4).
2 On L, N and R models, slacken the clamps securing the inlets on the air filter cover to the air ducts **(see illustration)**.
3 Remove the screws securing the air filter cover to the filter housing and remove the cover **(see illustrations)**. Remove the filter element from the housing **(see illustrations)**.
4 To clean the filter, tap the element on a hard surface to dislodge any dirt and use compressed air to clear the element, directing the air from the inside **(see illustration)**.
5 Check the element for signs of damage. If the element is torn or cannot be cleaned, renew it.
6 Remove the sub-air cleaner element from its housing (mounted on the back of the air filter housing), and clean or renew it as required.
7 Install the new filter by reversing the removal procedure **(see illustrations 8.3c and d)**. Make sure the filter is properly seated, then install the cover. On L, N and R models, make sure the inlets locate correctly onto the air ducts and tighten the clamps **(see illustration)**.
8 Install the fuel tank (see Chapter 4).

9 Wheels and tyres – general check

Tyres

1 Check the tyre condition and tread depth thoroughly – see *Daily (pre-ride) checks*.

Wheels

2 Cast wheels are virtually maintenance free, but they should be kept clean and checked periodically for cracks and other damage. Also check the wheel runout and alignment (see Chapter 7). Never attempt to repair damaged cast wheels; they must be renewed if damaged. Check the valve rubber for signs of damage or deterioration and have it renewed if necessary. Also, make sure the valve stem cap is in place and tight.

8.2 On L, N and R models, slacken the clamp screws (arrowed)

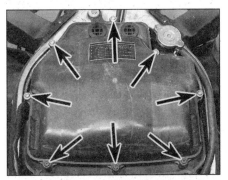
8.3a Air filter cover screws (arrowed) – J and K models

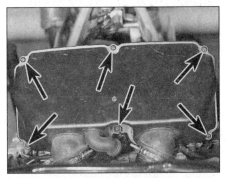
8.3b Air filter cover screws (arrowed) – L, N and R models

8.3c Removing the element – J and K models

8.3d Removing the element – L, N and R models

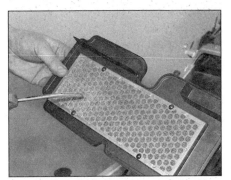
8.4 Direct the air from the inside out

8.7 On L, N and R models, locate the inlet tabs on the air duct lugs (arrowed)

1•12 Every 4000 miles (6000 km) or 6 months

10.6a Pressure cap (arrowed) – J and K models

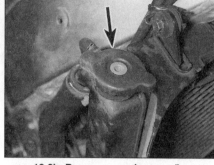

10.6b Pressure cap (arrowed) – L, N and R models

10 Cooling system – check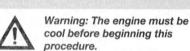

Warning: The engine must be cool before beginning this procedure.

1 Check the coolant level (see *Daily (pre-ride) checks*).
2 Remove the lower fairing (see Chapter 8). The entire cooling system should be checked for evidence of leakage. Examine each rubber coolant hose along its entire length. Look for cracks, abrasions and other damage. Squeeze each hose at various points. They should feel firm, yet pliable, and return to their original shape when released. If they are dried out or hard, renew them.
3 Check for evidence of leaks at each cooling system joint. Tighten the hose clips carefully to prevent future leaks.
4 Check the radiator for leaks and other damage. Leaks in the radiator leave tell-tale scale deposits or coolant stains on the outside of the core below the leak. If leaks are noted, remove the radiator (see Chapter 3) and have it repaired by a specialist.

Caution: Do not use a liquid leak stopping compound to try to repair leaks.

5 Check the radiator fins for mud, dirt and insects, which may impede the flow of air through the radiator. If the fins are dirty, remove the radiator (see Chapter 3) and clean it using water or low pressure compressed air directed through the fins from the backside. If the fins are bent or distorted, straighten them carefully with a screwdriver. If the air flow is restricted by bent or damaged fins over more than 30% of the radiator's surface area, renew the radiator.
6 On J and K models, remove the fuel tank (see Chapter 4). On L, N, and R models, remove the right-hand fairing side panel (if not already done), and, if required for improved access, the trim panel covering the pressure cap (see Chapter 8). Remove the pressure cap from the radiator filler neck by turning it anti-clockwise until it reaches a stop **(see illustrations)**. If you hear a hissing sound (indicating there is still pressure in the system), wait until it stops. Now press down on the cap and continue turning the cap until it can be removed. Check the condition of the coolant in the system. If it is rust-coloured or if accumulations of scale are visible, drain, flush and refill the system with new coolant (See Section 27). Check the cap seal for cracks and other damage. If in doubt about the pressure cap's condition, have it tested by a dealer or renew it. Install the cap by turning it clockwise until it reaches the first stop then push down on the cap and continue turning until it can turn no further.
7 Check the antifreeze content of the coolant with an antifreeze hydrometer. Sometimes coolant looks like it's in good condition, but might be too weak to offer adequate protection. If the hydrometer indicates a weak mixture, drain, flush and refill the system (see Section 27).
8 Start the engine and let it reach normal operating temperature, then check for leaks again. As the coolant temperature increases, the fan should come on automatically and the temperature should begin to drop. If it does not, refer to Chapter 3 and check the fan and fan circuit carefully.
9 If the coolant level is consistently low, and no evidence of leaks can be found, have the entire system pressure checked by a dealer.

11 Brake system – check

1 A routine general check of the brake system will ensure that any problems are discovered and remedied before the rider's safety is jeopardised.
2 Check the brake lever and pedal for loose connections, improper or rough action, excessive play, bends, and other damage. Renew any damaged parts (see Chapter 7).
3 Make sure all brake fasteners are tight. Check the brake pads for wear (see Section 3) and make sure the fluid level in the reservoirs is correct (see *Daily (pre-ride) checks*). Look for leaks at the hose connections and check for cracks in the hoses **(see illustration)**. If the lever or pedal is spongy, bleed the brakes (see Chapter 7).
4 Make sure the brake light operates when the front brake lever is pulled in. The front brake light switch, mounted on the underside of the master cylinder, is not adjustable. If it fails to operate properly, check it (see Chapter 9).
5 Make sure the brake light is activated just before the rear brake takes effect. If adjustment is necessary, hold the switch and turn the adjuster ring on the switch body until the brake light is activated when required **(see illustration)**. The switch is mounted on the inside of the rider's right-hand footrest bracket, just ahead of the master cylinder. If the brake light comes on too late, turn the ring clockwise. If the brake light comes on too soon or is permanently on, turn the ring anti-clockwise. If the switch doesn't operate the brake light, check it (see Chapter 9).
6 The front brake lever has a span adjuster which alters the distance of the lever from the handlebar **(see illustration)**. Each setting is identified by a notch in the adjuster which aligns with the arrow on the lever. Pull the lever away from the handlebar and turn the adjuster ring until the setting which best suits the rider is obtained. There are two settings.

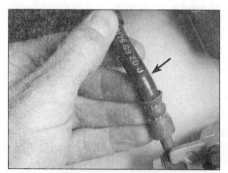

11.3 Flex the brake hoses and check for cracks, bulges and leaking fluid

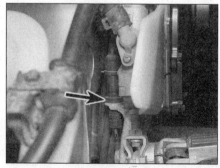

11.5 Rear brake light switch adjuster ring (arrowed)

11.6 Front brake lever span adjuster (arrowed)

Routine maintenance and servicing

Every 8000 miles (12,000 km) or 12 months

Carry out all the items under the 4000 mile (6000 km) check, plus the following:

12 Engine/transmission – oil and oil filter change

 Warning: Be careful when draining the oil, as the exhaust pipes, the engine, and the oil itself can cause severe burns.

1 Consistent routine oil and filter changes are the single most important maintenance procedure you can perform on a motorcycle. The oil not only lubricates the internal parts of the engine, transmission and clutch, but it also acts as a coolant, a cleaner, a sealant, and a protectant. Because of these demands, the oil takes a terrific amount of abuse and should be changed often with new oil of the recommended grade and type. Saving a little money on the difference in cost between a good oil and a cheap oil won't pay off if the engine is damaged.

2 Before changing the oil, warm up the engine so the oil will drain easily. Remove the lower fairing (see Chapter 8).

3 Put the motorcycle on its sidestand, and position a clean drain tray below the engine. Unscrew the oil filler cap from the clutch cover to vent the crankcase and to act as a reminder that there is no oil in the engine **(see illustration 7.3)**.

4 Next, unscrew the oil drain plug from the sump on the bottom of the engine and allow the oil to flow into the drain tray **(see illustrations 7.4a and b)**. Check the condition of the sealing washer on the drain plug and obtain a new one if it is damaged or worn.

 HAYNES HiNT To help determine whether any abnormal or excessive engine wear is occurring, place a strainer between the engine and the drain tray so that any debris in the oil is filtered out and can be examined.

5 When the oil has completely drained, fit the plug to the sump, using a new sealing washer if necessary, and tighten it securely **(see illustration 7.5)**. Avoid overtightening, as damage to the sump will result.

6 Now place the drain tray below the oil filter. Unscrew the oil filter using a filter adapter or a strap wrench and tip any residue oil into the drain tray **(see illustrations)**.

7 Smear clean engine oil onto the rubber seal on the new filter, then manoeuvre it into position and screw it onto the engine **(see illustrations)**. Tighten it securely using a filter wrench, or if one is not available, tighten the filter as tight as possible by hand **(see illustration)**.

8 Refill the engine to the proper level using the recommended type and amount of oil (see Daily (pre-ride) checks). With the motorcycle vertical, the oil level should lie between the upper and lower level lines on the dipstick (see Daily (pre-ride) checks). Install the filler cap **(see illustration 7.3)**. Start the engine and let it run for two or three minutes (make sure that the oil pressure light extinguishes after a few seconds). Shut it off, wait a few minutes, then check the oil level. If necessary, add more oil to bring the level up to the upper level line on the dipstick. Check around the drain plug and the oil filter for leaks.

 HAYNES HiNT Saving a little money on the difference between good and cheap oils won't pay off if the engine is damaged as a result.

9 The old oil drained from the engine cannot be re-used and should be disposed of properly. Check with your local refuse disposal company, disposal facility or environmental agency to see whether they will accept the used oil for recycling. Don't pour used oil into drains or onto the ground.

 HAYNES HiNT Check the old oil carefully – if it is very metallic coloured, then the engine is experiencing wear from break-in (new engine) or from insufficient lubrication. If there are flakes or chips of metal in the oil, then something is drastically wrong internally and the engine will have to be disassembled for inspection and repair. If there are pieces of fibre-like material in the oil, the clutch is experiencing excessive wear and should be checked.

Note: It is antisocial and illegal to dump oil down the drain. To find the location of your local oil recycling bank, call this number free.

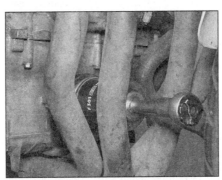

12.6a Unscrew the filter . . .

12.6b . . . and drain it

12.7a Smear the seal with oil and tighten it as described

12.7b Thread the new filter onto the engine

1•14 Every 8000 miles (12,000 km) or 12 months

13 Fuel system – check

Warning: *Petrol (gasoline) is extremely flammable, so take extra precautions when you work on any part of the fuel system. Don't smoke or allow open flames or bare light bulbs near the work area, and don't work in a garage where a natural gas-type appliance is present. If you spill any fuel on your skin, rinse it off immediately with soap and water. When you perform any kind of work on the fuel system, wear safety glasses and have a fire extinguisher suitable for a Class B type fire (flammable liquids) on hand.*

Check

1 Remove the fuel tank (see Chapter 4) and check the tank, the fuel tap, the fuel pump, the in-line fuel filter and the fuel hoses for signs of leakage, deterioration or damage; in particular check that there is no leakage from the fuel hoses. Renew any hoses which are cracked or deteriorated.

2 If the fuel tap is leaking, tightening the retaining nut and the assembly screws may help (see Chapter 4). If leakage persists, remove the tap and renew the O-ring. If the tap appears blocked, check the filter (see below). If a leakage or blockage cannot be cured, fit a new tap.

3 If the carburettor gaskets are leaking, the carburettors should be disassembled and rebuilt using new gaskets and seals (see Chapter 4).

Filter cleaning

4 Cleaning or renewal of the fuel filters is advised after a particularly high mileage has been covered. It is also necessary if fuel starvation is suspected.

5 A fuel filter is mounted in the tank and is integral with the fuel tap. Remove the fuel tank and the fuel tap (see Chapter 4). Clean the gauze filter to remove all traces of dirt and fuel sediment. Check the gauze for holes. If any are found, a new filter should be fitted (it is available separately). Check the condition of the O-ring and renew it if it is in any way damaged or deteriorated. It is advisable to renew it as a matter of course.

6 An in-line fuel filter is fitted in the hose from the fuel tap to the fuel pump **(see illustrations)**. If the filter is dirty or clogged or otherwise needs renewing, remove the fuel tank (see Chapter 4). Have a rag handy to soak up any residual fuel and disconnect the pipes from the filter. Slip the filter out of its bracket and install the new filter so that its arrow points in the direction of fuel flow (ie towards the pump). Secure the pipes to the filter with the retaining clips. Install the fuel tank (see Chapter 4), turn the tap ON and check that there are no leaks.

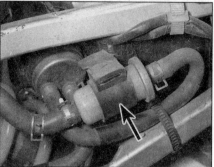

13.6a In-line fuel filter (arrowed) – J and K models

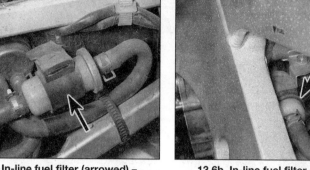

13.6b In-line fuel filter (arrowed) – L, N and R models

14 Battery – check

1 All models covered in this manual are fitted with a sealed maintenance-free battery which requires no maintenance. **Note:** *Do not attempt to remove the battery filler caps to check the electrolyte level or battery specific gravity. Removal will damage the caps, resulting in electrolyte leakage and battery damage.*

2 All that should be done is to check that its terminals are clean and tight and that the casing is not damaged or leaking. See Chapter 9 for further details.

3 If the machine is not in regular use, disconnect the battery and give it a refresher charge every month to six weeks, as described in Chapter 9.

15 Throttle and choke cables – check

Throttle cables

1 Make sure the throttle grip rotates easily from fully closed to fully open with the front wheel turned at various angles. The grip should return automatically from fully open to fully closed when released.

2 If the throttle sticks, this is probably due to a cable fault. Remove the cables (see Chapter 4) and lubricate them (see Section 6). Install the cables, making sure they are correctly routed. If this fails to improve the operation of the throttle, the cables must be renewed. Note that in very rare cases the fault could lie in the carburettors rather than the cables, necessitating the removal of the carburettors and inspection of the throttle linkage (see Chapter 4).

3 With the throttle operating smoothly, check for a small amount of freeplay in the cables, measured in terms of the amount of twistgrip rotation before the throttle opens, and compare the amount to that listed in this Chapter's Specifications **(see illustration)**. If it's incorrect, adjust the cables to correct it.

4 Freeplay adjustments can be made at the throttle end of the cable. Loosen the locknut on the accelerator cable where it leaves the handlebar **(see illustration)**. Turn the adjuster until the specified amount of freeplay is obtained (see this Chapter's Specifications), then retighten the locknut.

5 If the adjuster has reached its limit of adjustment, reset it so that the freeplay is at a maximum, then remove the fuel tank and air filter housing (see Chapter 4) and adjust the cable at the carburettor end. Slacken the adjuster locknut, then turn the adjuster out, making sure the lower nut remains captive in the bracket, thereby threading itself down the

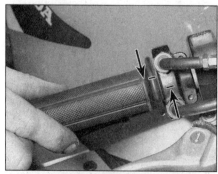

15.3 Throttle cable freeplay is measured in terms of twistgrip rotation

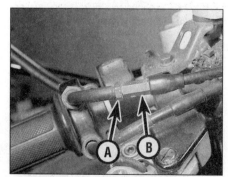

15.4 Throttle cable adjuster locknut (A) and adjuster (B) – throttle end

Every 8000 miles (12,000 km) or 12 months

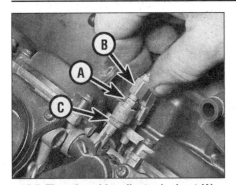

15.5 Throttle cable adjuster locknut (A), adjuster (B) and lower nut (C) – carburettor end

15.9 Slacken the screw and slide the outer cable end (arrowed) further into the bracket to create some freeplay

adjuster as you turn it **(see illustration)**. Turn the adjuster until the specified amount of freeplay is obtained, then tighten the locknut. Further adjustments can now be made at the throttle grip end. If the cable cannot be adjusted as specified, renew the cable (see Chapter 4).

⚠ **Warning: Turn the handlebars all the way through their travel with the engine idling. Idle speed should not change. If it does, the cable may be routed incorrectly. Correct this condition before riding the bike.**

6 Check that the throttle twistgrip operates smoothly and snaps shut quickly when released.

Choke cable

7 If the choke does not operate smoothly this is probably due to a cable fault. Remove the cable (see Chapter 4) and lubricate it (see Section 6). Install the cable, routing it so it takes the smoothest route possible.
8 If this fails to improve the operation of the choke, the cable must be renewed. Note that in very rare cases the fault could lie in the carburettors rather than the cable, necessitating the removal of the carburettors and inspection of the choke plungers (see Chapter 4).
9 Make sure there is a small amount of freeplay in the cable before the plungers move. If there isn't, check that the cable is correctly installed at both ends – remove the fuel tank and air filter housing to access the carburettor end of the cable (see Chapter 4). If it is, then slacken the choke outer cable bracket screw on the carburettor and slide the cable further into the bracket, creating some freeplay **(see illustration)**. Otherwise, renew the cable.

16 Spark plugs – renewal

1 Remove the old spark plugs as described in Section 5 and install new ones.

17 Carburettors – synchronisation

⚠ **Warning: Petrol (gasoline) is extremely flammable, so take extra precautions when you work on any part of the fuel system. Don't smoke or allow open flames or bare light bulbs near the work area, and don't work in a garage where a natural gas-type appliance is present. If you spill any fuel on your skin, rinse it off immediately with soap and water. When you perform any kind of work on the fuel system, wear safety glasses and have a fire extinguisher suitable for a Class B type fire (flammable liquids) on hand.**

⚠ **Warning: Take great care not to burn your hand on the hot engine unit when accessing the gauge take-off points on the inlet manifolds. Do not allow exhaust gases to build up in the work area; either perform the check outside or use an exhaust gas extraction system.**

1 Carburettor synchronisation is simply the process of adjusting the carburettors so they pass the same amount of fuel/air mixture to each cylinder. This is done by measuring the vacuum produced in each cylinder. Carburettors that are out of synchronisation will result in decreased fuel mileage, increased engine temperature, less than ideal throttle response and higher vibration levels. Before synchronising the carburettors, make sure the valve clearances are properly set.
2 To properly synchronise the carburettors, you will need a set of vacuum gauges or calibrated tubes (manometer) to indicate engine vacuum. The equipment used should be suitable for a four cylinder engine and come complete with the necessary adapters and hoses to fit the take-off points. **Note:** *Because of the nature of the synchronisation procedure and the need for special instruments, most owners leave the task to a dealer.*
3 Start the engine and let it run until it reaches normal operating temperature, then shut it off. Remove the fuel tank (see Chapter 4).
4 On J and K models, remove the blanking screws from the vacuum take-off points on the inlet manifolds **(see illustration)**. Install the take-off adapters provided with the vacuum gauges **(see illustration)**. Connect the vacuum gauge hoses to the adapters **(see illustration)**. Make sure they are a good fit because any air leaks will result in false readings.
5 On L, N and R models, remove the air filter housing and displace the carburettors to access the vacuum take-off points (see Chapter 4) – there is no need to disconnect the cables, though it may be necessary to either detach the fuel hose or displace the fuel pump to provide enough slack. Remove the

17.4a Remove the blanking screws (arrowed) (cylinders 1 and 2 shown) . . .

17.4b . . . then fit the adapters . . .

17.4c . . . and attach the hoses

1•16 Every 8000 miles (12,000 km) or 12 months

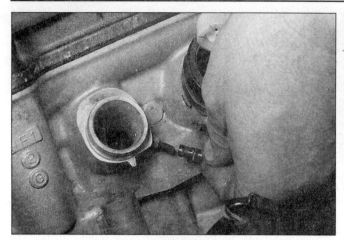
17.5a Remove the blanking caps . . .

17.5b . . . and attach the hoses

17.8 Carburettor synchronisation set-up

17.9a Synchronisation screws (arrowed) – J and K models (air filter housing removed for clarity)

blanking caps from the vacuum take-off points on the inlet manifolds **(see illustration)**. Connect the vacuum gauge hoses to the adapters or take-off points **(see illustration)**. Make sure they are a good fit because any air leaks will result in false readings. Install the carburettors and air filter housing (see Chapter 4).

6 Arrange a temporary fuel supply, using a small temporary tank mounted above the level of the carburettors **(see illustration 17.8)**. Connect the fuel supply hose from the tank to the inlet union on the in-line filter – do not connect it directly to the carburettors or you will by-pass the fuel pump.

7 Start the engine and make sure the idle speed is correct. If it isn't, adjust it (see Section 2). If the gauges are fitted with damping adjustment, set this so that the needle flutter is just eliminated but so that they can still respond to small changes in pressure.

8 The vacuum readings for all of the cylinders should be the same, or at least within the tolerance listed in this Chapter's Specifications **(see illustration)**. If the vacuum readings vary, adjust as necessary.

9 The carburettors are adjusted by turning the synchronising screws situated in-between the carburettors, in the throttle linkage **(see illustrations)**. Note: *Do not press down on the screws whilst adjusting them, otherwise a false reading will be obtained.* First synchronise the outer left carburettor (no. 1) to the inner left carburettor (no. 2) using the left-hand synchronising screw until the readings are the same. Then synchronise the outer right carburettor (no. 4) to the inner right carburettor (no. 3) using the right-hand synchronising screw. Finally synchronise the left-hand carburettors (nos. 1 and 2) to the right-hand carburettors (nos. 3 and 4) using the centre synchronising screw.

10 When all the carburettors are synchronised, open and close the throttle quickly to settle the linkage, and recheck the gauge readings, readjusting if necessary.

11 When the adjustment is complete, recheck the vacuum readings, then adjust the idle speed by turning the throttle stop screw (see Section 2) until the idle speed listed in this Chapter's Specifications is obtained. Stop the engine.

12 Remove the vacuum gauges and, where used, the adapters. Fit the blanking screws or caps onto the inlet manifolds.

13 Install the fuel tank (see Chapter 4).

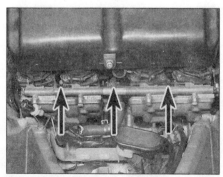
17.9b Synchronisation screws (arrowed) – L, N and R models

Every 8000 miles (12,000 km) or 12 months

18.2 Headlight beam adjusters (arrowed) – J and K models

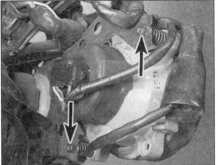

18.3 Headlight beam adjusters (arrowed) – L, N and R models

18 Headlight aim – check and adjustment

Note: *An improperly adjusted headlight may cause problems for oncoming traffic or provide poor, unsafe illumination of the road ahead. Before adjusting the headlight aim, be sure to consult with local traffic laws and regulations – refer to MOT Test Checks in the Reference section.*

1 The headlight beam can adjusted both horizontally and vertically. Before making any adjustment, check that the tyre pressures are correct and the suspension is adjusted as required. Make any adjustments to the headlight aim with the machine on level ground, with the fuel tank half full and with an assistant sitting on the seat. If the bike is usually ridden with a passenger on the back, have a second assistant to do this.

2 On J and K models, vertical adjustment is made by turning the adjuster screw on the top outer corner of each headlight unit **(see illustration)**. Turn it clockwise to move the beam up, and anti-clockwise to move it down. Horizontal adjustment is made by turning the adjuster screw on the bottom inner corner of each headlight unit. Turn it clockwise to move the beam in (towards the centre of the bike), and anti-clockwise to move it out.

3 On L, N and R models, vertical adjustment is made by turning the adjuster screw on the bottom inner corner of each headlight unit **(see illustration)**. Turn it clockwise to move the beam down, and anti-clockwise to move it up. Horizontal adjustment is made by turning the adjuster screw on the top outer corner of each headlight unit. Turn it clockwise to move the beam out (away from the centre of the bike), and anti-clockwise to move it in.

19 Sidestand – check

1 The sidestand return spring must be capable of retracting the stand fully and holding the stand retracted when the motorcycle is in use. If the spring is sagged or broken it must be renewed.

2 Lubricate the sidestand pivot regularly (see Section 6).

3 On J and K models, check the condition of the sidestand rubber. If it is worn down to or beyond the wear limit line, renew it **(see illustration)**.

4 On L, N and R models, the sidestand switch prevents the motorcycle being started if the stand is extended. Check its operation by shifting the transmission into neutral, retracting the stand and starting the engine. Pull in the clutch lever and select a gear. Extend the sidestand. The engine should stop as the sidestand is extended. If the sidestand switch does not operate as described, check its circuit (see Chapter 9).

20 Suspension – check

1 The suspension components must be maintained in top operating condition to ensure rider safety. Loose, worn or damaged suspension parts decrease the motorcycle's stability and control.

Front suspension

2 While standing alongside the motorcycle, apply the front brake and push on the handlebars to compress the forks several times. See if they move up-and-down smoothly without binding. If binding is felt, the forks should be disassembled and inspected (see Chapter 6).

3 Inspect the area around the dust seal for signs of oil leakage, then carefully lever up the dust seal using a flat-bladed screwdriver and inspect the area around the fork seal **(see illustration)**. If leakage is evident, the seals must be renewed (see Chapter 6).

4 Check the tightness of all suspension nuts and bolts to be sure none have worked loose.

Rear suspension

5 Inspect the rear shock for fluid leakage and tightness of its mountings. If leakage is found, the shock should be renewed (see Chapter 6).

6 With the aid of an assistant to support the bike, compress the rear suspension several times. It should move up and down freely without binding. If any binding is felt, the worn or faulty component must be identified and renewed. The problem could be due to either the shock absorber, the suspension linkage components or the swingarm components.

7 Support the motorcycle using an auxiliary stand so that the rear wheel is off the ground. Grab the swingarm and attempt to rock it from side to side – there should be no discernible movement at the rear **(see illustration)**. If there's a little movement or a slight clicking can be heard, inspect the tightness of all the rear suspension mounting bolts and nuts, referring to the torque settings specified at the beginning of Chapter 6, and re-check for movement. Next, grasp the top

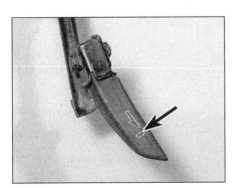

19.3 Renew the rubber if it is worn to or beyond the line (arrowed)

20.3 Check above and below the dust seal for signs of fluid leakage

20.7a Checking for play in the swingarm bearings

1•18 Every 8000 miles (12,000 km) or 12 months

20.7b Checking for play in the rear shock mountings and suspension linkage bearings

of the rear wheel and pull it upwards – there should be no discernible freeplay before the shock absorber begins to compress **(see illustration)**. Any freeplay felt in either check indicates worn bearings in the suspension linkage or swingarm, or worn shock absorber mountings. The worn components must be renewed (see Chapter 6).

8 To make an accurate assessment of the swingarm bearings, remove the rear wheel (see Chapter 7) and the bolt securing the suspension linkage assembly to the swingarm (see Chapter 6). Grasp the rear of the swingarm with one hand and place your other hand at the junction of the swingarm and the frame. Try to move the rear of the swingarm from side-to-side. Any wear (play) in the bearings should be felt as movement between the swingarm and the frame at the front. If there is any play the swingarm will be felt to move forward and backward at the front (not from side-to-side). Next, move the swingarm up and down through its full travel. It should move freely, without any binding or rough spots. If any play in the swingarm is noted or if the swingarm does not move freely, the bearings must be removed for inspection or renewal (see Chapter 6).

21 Steering head bearings – check and adjustment

1 This motorcycle is equipped with caged ball steering head bearings which can become dented, rough or loose during normal use of the machine. In extreme cases, worn or loose steering head bearings can cause steering wobble – a condition that is potentially dangerous.

Check

2 Support the motorcycle in an upright position using an auxiliary stand. Raise the front wheel off the ground by positioning the bike on an auxiliary stand and having an assistant push down on the rear or by placing a support under the engine.

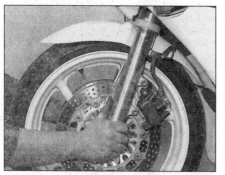

21.4 Checking for play in the steering head bearings

21.5 Unscrew the fork clamp bolts (arrowed)

3 Point the front wheel straight-ahead and slowly move the handlebars from side-to-side. Any dents or roughness in the bearing races will be felt, and if the bearings are too tight the bars will not move smoothly and freely. If the bearings are damaged or the action is rough, they should be renewed (see Chapter 6). If the bearings are too tight they should be adjusted as described below.

4 Next, grasp the fork sliders and try to move them forward and backward **(see illustration)**. Any looseness in the steering head bearings will be felt as front-to-rear movement of the forks. If play is felt in the bearings, adjust the steering head as follows.

HAYNES HiNT *Freeplay in the fork due to worn fork bushes can be misinterpreted for steering head bearing play – do not confuse the two.*

Adjustment

5 Unscrew and remove the fork clamp bolts in the top yoke, noting how the choke knob bracket locates **(see illustration)**. Depending on the tools available, access to the right-hand bolt may be restricted by the front brake master cylinder. If this is the case, unscrew the two master cylinder assembly clamp bolts and position the assembly clear of the handlebar, making sure no strain is placed on the hydraulic hose. Keep the master cylinder reservoir upright to prevent possible fluid leakage. If the master cylinder is displaced, the right-hand fork clamp bolt needs only to be slackened, not removed.

6 On L, N and R models, prise the plug out of the steering stem nut **(see illustration)**. On all models, unscrew and remove the nut using a 30 mm socket or spanner, and on J and K models remove the washer **(see illustration)**. Free the clutch cable from its guide on the top yoke.

7 Gently ease the top yoke upwards off the fork tubes and position it clear of the head bearings, using a rag to protect other components **(see illustration)**.

8 Prise the lockwasher tabs out of the

21.6a On L, N and R models, remove the plug

21.6b Unscrew and remove the steering stem nut

21.7 Ease the yoke up and off the forks

Every 8000 miles (12,000 km) or 12 months

21.8a Bend down the tabs securing the locknut . . .

21.8b . . . then unscrew the locknut . . .

21.8c . . . and remove the lockwasher

notches in the locknut **(see illustration)**. Unscrew the locknut using either a C-spanner or a suitable drift located in one of the notches **(see illustration)**. Remove the lockwasher, bending up the remaining tabs to release it from the adjuster nut if necessary **(see illustration)**. Inspect the tabs for cracks or signs of fatigue. If there are any, discard the lockwasher and use a new one; otherwise the old one can be reused.

9 Using either the C-spanner or drift, slacken the adjuster nut slightly until pressure is just released, then tighten it a little at a time until all freeplay is removed, yet the steering is able to move freely as described above **(see illustration)**. If the Honda adapter is available you can apply the torque setting specified at the beginning of the Chapter. Now turn the steering from lock to lock five times to settle the bearings, then recheck the adjustment or the torque setting. The object is to set the adjuster nut so that the bearings are under a very light loading, just enough to remove any freeplay. **Caution: Take great care not to apply excessive pressure because this will cause premature failure of the bearings. If the torque setting is applied and the bearings are too loose or tight, set them up according to feel.**

10 When the bearings are correctly adjusted, install the lockwasher, using a new one if the

21.9 Adjust the bearings using either a C-spanner or a drift as shown

tabs are weakened or cracked, onto the adjuster nut and fit two tabs into the slots in the adjuster nut **(see illustration 21.8c)**.

11 Install the locknut and tighten it finger-tight, then tighten it further (to a maximum of 90°) to align the remaining tabs on the lockwasher with the slots in the locknut **(see illustration 21.8b)**. Hold the adjuster nut to prevent it from moving if necessary. Bend up the lockwasher tabs to secure the locknut **(see illustration)**.

12 Fit the top yoke onto the steering stem, making sure the lug on each handlebar locates into its hole in the underside of the yoke **(see illustration 21.7)**. Install the nut,

21.11 Bend the tabs up into the notches in the locknut

not forgetting the washer on J and K models, and tighten it and both the fork clamp bolts to the torque settings specified at the beginning of the Chapter **(see illustration)**. If displaced, make sure the front brake master cylinder clamp is installed with the UP mark facing up and tighten the clamp bolts to the torque setting specified at the beginning of the Chapter. Do not forget to secure the choke knob with the left-hand fork clamp bolt **(see illustration)**. On L, N and R models, fit the plug into the stem nut **(see illustration 21.6a)**. Feed the clutch cable into its guide.

13 Check the bearing adjustment as described above and re-adjust if necessary.

21.12a Tighten the steering stem nut to the specified torque

21.12b Secure the choke knob with the left-hand fork clamp bolt

1•20 Every 8000 miles (12,000 km) or 12 months

22.2 Checking for play in the wheel bearings

25.5a Remove the timing inspection plug (A) and the centre plug (B)

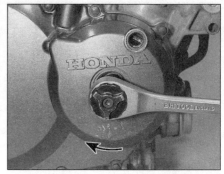

25.5b Turn the engine clockwise using a 14 mm socket on the alternator bolt

22 Wheel bearings – check

1 Wheel bearings will wear over a period of time and result in handling problems.
2 Support the motorcycle upright using an auxiliary stand. Check for any play in the bearings by pushing and pulling the wheel against the hub **(see illustration)**. Also rotate the wheel and check that it rotates smoothly.
3 If any play is detected in the hub, or if the wheel does not rotate smoothly (and this is not due to brake or transmission drag), the wheel bearings must be removed and inspected for wear or damage (see Chapter 7).

23 Nuts and bolts – tightness check

1 Since vibration of the machine tends to loosen fasteners, all nuts, bolts, screws, etc. should be periodically checked for proper tightness.
2 Pay particular attention to the following:
   Spark plugs
   Engine oil drain plug
   Gearchange pedal bolt
   Footrest and stand bolts
   Engine mounting bolts
   Shock absorber and suspension linkage bolts and swingarm pivot bolts
   Handlebar clamp bolts
   Front axle bolt and axle clamp bolts
   Front fork clamp bolts (top and bottom yoke)
   Rear axle nut
   Brake caliper mounting bolts
   Brake hose banjo bolts and caliper bleed valves
   Brake disc bolts
   Exhaust system bolts/nuts
3 If a torque wrench is available, use it along with the torque specifications at the beginning of this and other Chapters.

24 Brakes – fluid change

1 The brake fluid should be renewed at the prescribed interval or whenever a master cylinder or caliper overhaul is carried out. Refer to the brake bleeding section in Chapter 7, noting that all old fluid must be pumped from the fluid reservoir and hydraulic line before filling with new fluid.

> **HAYNES HiNT** *Old brake fluid is invariably much darker in colour than new fluid, making it easy to see when all old fluid has been expelled from the system.*

25 Valve clearances – check and adjustment

1 The engine must be completely cool for this maintenance procedure, so let the machine sit overnight before beginning.
2 Remove the spark plugs (see Section 5).
3 Remove the valve cover (see Chapter 2). The cylinders are identified by a number, and are numbered 1 to 4 from left to right.
4 Make a chart or sketch of all valve positions so that a note of each clearance can be made against the relevant valve.
5 Unscrew the timing inspection plug and the centre plug from the alternator cover on the right-hand side of the engine **(see illustration)**. Discard the plug O-rings as new ones should be used. The engine can be turned using a 14 mm socket on the alternator rotor bolt and turning it in a clockwise direction only **(see illustration)**. Alternatively, place the motorcycle on an auxiliary stand so that the rear wheel is off the ground, select a high gear and rotate the rear wheel by hand in its normal direction of rotation.

Check

J, K, L, and N models

6 Turn the engine until the line next to the 'T' mark on the alternator rotor aligns with the notches in the timing inspection hole **(see illustration)**. **Note:** *Turn the engine in the normal direction of rotation (clockwise), viewed from the right-hand end of the engine.* Now, check the position of the camshaft gears. If the 'IN' mark on the inlet camshaft gear and the 'EX' mark on the exhaust camshaft gear are aligned with the cylinder head surface with each mark the correct way up and on the outside of its respective gear, then no. 4 cylinder is at TDC on the compression stroke **(see illustration)**. If the 'IN' mark on the inlet camshaft gear and the 'EX' mark on the exhaust camshaft gear are aligned with the cylinder head surface with

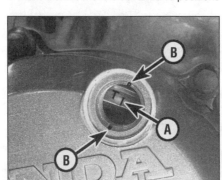

25.6a Turn the engine until the TI mark (A) aligns with the static marks (B)

25.6b No. 4 cylinder positioned at TDC on the compression stroke

Every 8000 miles (12,000 km) or 12 months

25.6c No. 1 cylinder positioned at TDC on the compression stroke

25.9 Measure the valve clearance using a feeler gauge as shown

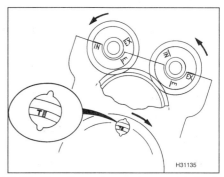

25.11 Align the marks as shown and check the inlet valves for cylinders 2 and 4

each mark upside down and on the inside of its respective gear, then no. 1 cylinder is at TDC on the compression stroke **(see illustration)**. If the marks are not positioned as required for the cylinders being checked, turn the crankshaft through 360° (one complete turn).

7 With no. 4 cylinder at TDC on the compression stroke, the following valves can be checked:
a) No. 2, inlet
b) No. 3, exhaust
c) No. 4, inlet and exhaust

8 With no. 1 cylinder at TDC on the compression stroke, the following valves can be checked:
a) No. 1, inlet and exhaust
b) No. 2, exhaust
c) No. 3, inlet

9 Insert a feeler gauge of the same thickness as the correct valve clearance (see Specifications) between the cam base and follower of each valve and check that it is a firm sliding fit – you should feel a slight drag when the you pull the gauge out **(see illustration)**. If not, use the feeler gauges to obtain the exact clearance. **Note:** *The inlet and exhaust valve clearances are different.* Record the measured clearance on the chart.

10 Rotate the crankshaft through 360° and measure the valve clearance of the remaining valves using the method described in Step 9.

R models

11 Turn the engine until the line next to the 'T' mark on the alternator rotor aligns with the static timing mark, which is a notch in the inspection plug hole in the cover **(see illustration 25.6a)**. **Note:** *Turn the engine in the normal direction of rotation (clockwise), viewed from the right-hand end of the engine.* Now, check the position of the camshaft gears – the 'IN' mark on the inlet camshaft gear and the 'EX' mark on the exhaust camshaft gear should be aligned with the cylinder head surface with each mark the correct way up and on the outside of its respective gear **(see illustration)**. If the marks are not correctly positioned, turn the crankshaft through 360° (one complete turn).

12 With the engine in this position, the inlet valves for Nos. 2 and 4 cylinders can be checked.

13 Insert a feeler gauge of the same thickness as the correct valve clearance (see Specifications) between the cam base and follower of each valve and check that it is a firm sliding fit – you should feel a slight drag when the you pull the gauge out **(see illustration 25.9)**. If not, use the feeler gauges to obtain the exact clearance. **Note:** *The inlet and exhaust valve clearances are different.* Record the measured clearance on the chart.

14 With the specified valves measured, rotate the crankshaft clockwise through 180° (half a turn) until the line next to the 'O' mark on the alternator rotor aligns with the static timing mark, and so that the 'E' mark on the exhaust camshaft gear aligns with the cylinder head surface, with the mark the correct way up and on the outside of its gear **(see illustration)**. The exhaust valves for Nos. 1 and 3 cylinders can be checked.

15 Rotate the crankshaft clockwise through another 180° (half a turn) until the line next to the 'T' mark on the rotor aligns with the static timing mark, and so that the 'IN' mark on the inlet camshaft gear and the 'EX' mark on the exhaust camshaft gear align with the cylinder head surface with each mark the wrong way up and on the inside of its respective gear **(see illustration)**. The inlet valves for Nos. 1 and 3 cylinders can be checked.

16 Rotate the crankshaft clockwise through 180° (half a turn) until the line next to the 'O' mark on the alternator rotor aligns with the static timing mark, and so that the 'E' mark on the exhaust camshaft gear aligns with the cylinder head surface, with the mark the wrong way up and on the inside of its gear **(see illustration)**. The exhaust valves for Nos. 2 and 4 cylinders can be checked.

Adjustment

17 When all clearances have been measured and charted, identify whether the clearance on any valve falls outside that specified. If it does, the shim between the follower and the valve must be replaced with one of a thickness which will restore the correct clearance.

18 Shim replacement requires removal of the camshafts (see Chapter 2). There is no need

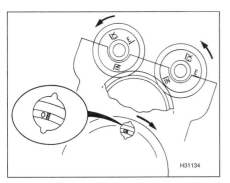

25.14 Align the marks as shown and check the exhaust valves for cylinders 1 and 3

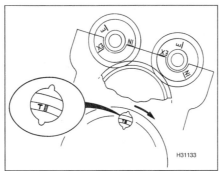

25.15 Align the marks as shown and check the inlet valves for cylinders 1 and 3

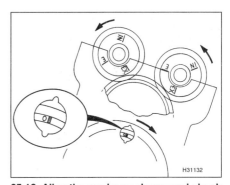

25.16 Align the marks as shown and check the exhaust valves for cylinders 2 and 4

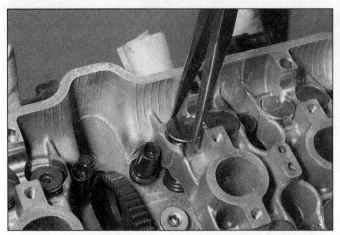

25.19a Remove the follower . . .

25.19b . . . and either retrieve the shim from inside the follower . . .

25.19c . . . or from the top of the valve, using either a magnet . . .

25.19d . . . or a screwdriver with grease on it

to remove both camshafts if shims from only one side of the engine need replacing.

19 With the camshaft removed, remove the cam follower of the valve in question using either a magnet or a pair of pliers (see illustration). Retrieve the shim from either the inside of the follower or pick it out of the top of the valve, using either a magnet, a small screwdriver with a dab of grease on it (the shim will stick to the grease), or a screwdriver and a pair of pliers (see illustrations). Do not allow the shim to fall into the engine.

20 The shim size should be stamped on its face, however, it is recommended that the shim is measured to check that it has not worn. The size marking is in the form of a three figure number, eg 180 indicating that the shim is 1.800 mm thick (see illustrations). Where the number does not equal a shim thickness, it should be rounded up or down, eg 182 or 183 both indicate that the shim is 1.825 mm thick. Shims are available in 0.025 mm increments from 1.200 to 2.800 mm. The new shim thickness required can then be calculated as follows. Note: *Always aim to get the clearance at the mid-point of the specified range.*

21 If the valve clearance was less than specified, subtract the measured clearance from the specified clearance then deduct the result from the original shim thickness. For example:

Sample calculation – inlet valve clearance too small

Clearance measured (A) – 0.08 mm
Specified clearance (B) – 0.16 mm (0.13 to 0.19 mm)
Difference (B – A) – 0.08 mm
Shim thickness fitted – 2.475 mm
Correct shim thickness required is 2.475 – 0.08 = 2.395 mm

Note: *If the required replacement shim is greater than 2.800 mm (the largest available), the valve is probably not seating correctly due to a build-up of carbon deposits and should be checked and cleaned or resurfaced as required (see Chapter 2).*

22 If the valve clearance was greater then specified, subtract the specified clearance from the measured clearance, and add the result to the thickness of the original shim. For example:

Sample calculation – exhaust valve clearance too large

Clearance measured (A) – 0.35 mm
Specified clearance (B) – 0.22 mm (0.19 to 0.25 mm)
Difference (A – B) – 0.13 mm
Shim thickness fitted – 1.975 mm
Correct shim thickness required is 1.975 + 0.13 = 2.105 mm

23 Obtain the correct thickness shims from your dealer. Where the required thickness is not equal to the available shim thickness, round off the measurement to the nearest available size. Shims are available in 0.025 mm increments from 1.200 mm to 2.800 mm.

24 Obtain the replacement shim, then lubricate it with molybdenum disulphide oil (a 50/50 mixture of molybdenum disulphide grease and engine oil) and fit it into its recess in the top of the valve, with the size marking

25.20a The shim size should be marked on its face . . .

25.20b . . . but measure it anyway

Every 8000 miles (12,000 km) or 12 months

25.24a Fit the shim with its size marking facing up . . .

25.24b . . . then install the follower

25.26 Fit new O-rings onto the plugs

on each shim facing up **(see illustration)**. Check that the shim is correctly seated, then lubricate the follower with molybdenum disulphide oil and install it onto the valve **(see illustration)**. Repeat the process for any other valves until the clearances are correct, then install the camshafts (see Chapter 2).
25 Rotate the crankshaft several turns to seat the new shim(s), then check the clearances again.
26 Install all disturbed components in a reverse of the removal sequence. Use new O-rings on the centre plug and the timing inspection plug and smear them and the plug threads with molybdenum disulphide oil (a 50/50 mixture of molybdenum disulphide grease and engine oil) **(see illustration)**.

Every 12,000 miles (18,000 km) or 18 months

Carry out all the items under the 4000 mile (6000 km) check:

26 Air filter – renewal

Caution: If the machine is continually ridden in continuously wet or dusty conditions, the filter should be renewed more frequently.

1 Remove the old air filter as described in Section 8 and install a new one.

Every 24,000 miles (36,000 km) or two years

Carry out all the items under the 8000 mile (12,000 km) and 12,000 mile (18,000 km) checks, plus the following:

27 Cooling system – draining, flushing and refilling

Warning: Allow the engine to cool completely before performing this maintenance operation. Also, don't allow antifreeze to come into contact with your skin or the painted surfaces of the motorcycle. Rinse off spills immediately with plenty of water. Antifreeze is highly toxic if ingested. Never leave antifreeze lying around in an open container or in puddles on the floor; children and pets are attracted by its sweet smell and may drink it. Check with local authorities (councils) about disposing of antifreeze. Many communities have collection centres which will see that antifreeze is disposed of safely. Antifreeze is also combustible, so don't store it near open flames.

Draining

1 Remove the lower fairing panel (see Chapter 8). On J and K models, remove the fuel tank (see Chapter 4). On L, N, and R models, if required for improved access, remove the trim panel covering the pressure cap (see Chapter 8). Remove the pressure cap by turning it anti-clockwise until it reaches a stop **(see illustrations 10.6a and b)**. If you hear a hissing sound (indicating there is still pressure in the system), wait until it stops. Now press down on the cap and continue turning the cap until it can be removed.
2 Position a suitable container beneath the water pump on the left-hand side of the engine. Remove the coolant drain plug and its sealing washer and allow the coolant to completely drain from the system **(see illustrations)**. Retain the old sealing washer for use during flushing.

Flushing

3 Flush the system with clean tap water by inserting a garden hose in the radiator filler neck. Allow the water to run through the system until it is clear and flows cleanly out of

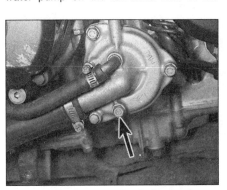

27.2a Unscrew the drain plug (arrowed) . . .

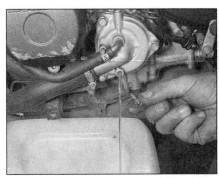

27.2b . . . and allow the coolant to drain

Every 24,000 miles (36,000 km) or two years

27.11 Use a new sealing washer on the drain plug

27.12 Fill the system using the specified mixture

the drain holes. If the radiator is extremely corroded, remove it (see Chapter 3) and have it cleaned at a radiator shop.

4 Clean the drain hole then install the drain plug using the old sealing washer.

5 Fill the cooling system with clean water mixed with a flushing compound. Make sure the flushing compound is compatible with aluminium components, and follow the manufacturer's instructions carefully.

6 Start the engine and allow it to reach normal operating temperature. Let it run for about ten minutes.

7 Stop the engine. Let it cool for a while, then cover the pressure cap with a heavy rag and turn it anti-clockwise to the first stop, releasing any pressure that may be present in the system. Once the hissing stops, push down on the cap and remove it completely.

9 Drain the system once again.

10 Fill the system with clean water and repeat the procedure in Steps 6 to 9.

Refilling

11 Fit a new sealing washer onto the drain plug and tighten it securely **(see illustration)**.

12 Fill the system with the proper coolant mixture (see this Chapter's Specifications) **(see illustration)**. *Note: Pour the coolant in slowly to minimise the amount of air entering the system.*

13 When the system is full (all the way up to the top of the radiator filler neck), install the pressure cap. Also top up the coolant reservoir to the UPPER level mark (see *Daily (pre-ride) checks)*. On J and K models, install the fuel tank (see Chapter 4).

14 Start the engine and allow it to idle for 2 to 3 minutes. Flick the throttle twistgrip part open 3 or 4 times, so that the engine speed rises to approximately 4000 – 5000 rpm, then stop the engine.

15 Let the engine cool then remove the pressure cap as described in Step 1. Check that the coolant level is still up to the radiator filler neck. If it's low, add the specified mixture until it reaches the top of the filler neck. Refit the cap.

16 Check the coolant level in the reservoir and top up if necessary.

17 Check the system for leaks.

18 Do not dispose of the old coolant by pouring it down the drain. Instead pour it into a heavy plastic container, cap it tightly and take it into an authorised disposal site or service station – see **Warning** at the beginning of this Section.

28 Brake caliper seals and master cylinder seals – renewal

1 Brake seals will deteriorate over a period of time and lose their effectiveness, leading to sticking operation or fluid loss, or allowing the ingress of air and dirt. Refer to Chapter 7 and dismantle the components for seal renewal.

Every four years

29 Brake hoses – renewal

1 The hoses will in time deteriorate with age and should be renewed regardless of their apparent condition.

2 Refer to Chapter 7 and disconnect the brake hoses from the master cylinders and calipers. Always renew the banjo union sealing washers.

30 Fuel hoses – renewal

Warning: *Petrol (gasoline) is extremely flammable, so take extra precautions when you work on any part of the fuel system. Don't smoke or allow open flames or bare light bulbs near the work area, and don't work in a garage where a natural gas-type appliance is present. If you spill any fuel on your skin, rinse it off immediately with soap and water. When you perform any kind of work on the fuel system, wear safety glasses and have a fire extinguisher suitable for a Class B type fire (flammable liquids) on hand.*

1 The fuel delivery and vacuum hoses should be renewed regardless of their condition.

2 Remove the fuel tank (see Chapter 4). Disconnect the fuel hoses from the fuel tap and from the carburettors, noting the routing of each hose and where it connects (see Chapter 4 if required). It is advisable to make a sketch of the various hoses before removing them to ensure they are correctly installed.

3 Secure each new hose to its unions using new clamps. Run the engine and check for leaks before taking the machine out on the road.

Non-scheduled maintenance

31 Cylinder compression – check

1 Among other things, poor engine performance may be caused by leaking valves, incorrect valve clearances, a leaking head gasket, or worn pistons, rings and/or cylinder walls. A cylinder compression check will help pinpoint these conditions and can also indicate the presence of excessive carbon deposits in the cylinder heads.

2 The only tools required are a compression gauge and a spark plug wrench. A compression gauge with a 10 mm threaded adapter for the spark plug hole is preferable to the type which requires hand pressure to maintain a tight seal. Depending on the outcome of the initial test, a squirt-type oil can may also be needed.

3 Make sure the valve clearances are correctly set (see Section 25) and that the cylinder head nuts are tightened to the correct torque setting (see Chapter 2).

4 Refer to *Fault Finding Equipment* in the Reference section for details of the compression test.

Non-scheduled maintenance

32 Engine – oil pressure check

1 The oil pressure warning light should come on when the ignition (main) switch is turned ON and extinguish a few seconds after the engine is started – this serves as a check that the warning light bulb is sound. If the oil pressure light comes on whilst the engine is running, low oil pressure is indicated – stop the engine immediately and carry out an oil level check *(see Daily (pre-ride) checks)*.

2 An oil pressure check must be carried out if the warning light comes on when the engine is running yet the oil level is good (Step 1). It can also provide useful information about the condition of the engine's lubrication system.

3 To check the oil pressure, a suitable gauge (which screws into the crankcase) will be needed. Honda provide a tool (part no. 07501-4220100) for this purpose.

4 Warm the engine up to normal operating temperature then stop it.

5 Remove the oil pressure switch (see Chapter 9) and screw the adapter into the crankcase threads. Connect the gauge to the adapter.

6 Start the engine and increase the engine speed briefly to 7000 rpm whilst watching the gauge reading. The oil pressure should be similar to that given in the Specifications at the start of this Chapter.

7 If the pressure is significantly lower than the standard, either the pressure regulator is stuck open, the oil pump is faulty, the oil strainer or filter is blocked, or there is other engine damage. Begin diagnosis by checking the oil filter, strainer and regulator, then the oil pump (see Chapter 2). If those items check out okay, chances are the bearing oil clearances are excessive and the engine needs to be overhauled.

8 If the pressure is too high, either an oil passage is clogged, the regulator is stuck closed or the wrong grade of oil is being used.

9 Stop the engine and unscrew the gauge and adapter from the crankcase.

10 Install the oil pressure switch (see Chapter 9). Check the oil level (see *Daily (pre-ride) checks*).

33 Steering head bearings – lubrication

1 Over a period of time the grease will harden or may be washed out of the bearings by incorrect use of jet washes.

2 Disassemble the steering head for re-greasing of the bearings. Refer to Chapter 6 for details.

34 Swingarm and suspension linkage bearings – lubrication

1 Over a period of time the grease will harden or dirt will penetrate the bearings due to failed dust seals.

2 The swingarm is not equipped with grease nipples. Remove the swingarm and suspension linkage as described in Chapter 6 for greasing of the bearings.

35 Front forks – oil change

1 Fork oil degrades over a period of time and loses its damping qualities.

Fork oil drain screw (arrowed)

2 These models are equipped with drain screws in the fork sliders **(see illustration)** and therefore changing the fork oil is a relatively straightforward task.

3 Refer to the appropriate part of Chapter 6, Section 7 and remove the fork top bolts, spacers and springs. There is no need to remove the fork legs from the yokes.

4 Hold a piece of thick card to act as a chute beneath the fork drain screw on one fork slider, then remove the drain screw and allow the oil to drain. Gently pump the fork to expel all of the oil. Now do the same on the other fork leg. Refit the drain screws with their sealing washers.

5 Slowly pour in the correct quantity of the specified grade of fork oil and pump the fork at least ten times to distribute it evenly; the oil level should also be measured and adjustment made by adding or subtracting oil. Fully compress the fork tube into the slider and measure the fork oil level from the top of the tube. Add or subtract fork oil until it is at the level specified at the beginning of Chapter 6.

6 Refer to the appropriate part of Chapter 6, Section 7 and install the springs, spacers and top bolts.

Routine maintenance and servicing

Notes

Chapter 2
Engine, clutch and transmission

Contents

Alternator – removal and installation . . . . see Chapter 9	Neutral switch – check, removal and installation . . . . see Chapter 9
Camshafts – removal, inspection and installation 9	Oil and filter – change . . . . see Chapter 1
Camshaft drive gear assembly – removal, inspection and installation 13	Oil cooler – removal and installation 7
Clutch cable – replacement 17	Oil level – check . . . . .see Daily (pre-ride) checks
Clutch – check . . . . .see Chapter 1	Oil pressure – check . . . . .see Chapter 1
Clutch – removal, inspection and installation 16	Oil pressure switch – check, removal and installation . . .see Chapter 9
Connecting rods – removal, inspection and installation 24	Oil pump – removal, inspection and installation 19
Crankcase halves and cylinder bores – inspection and servicing . . 22	Oil sump, oil strainer and pressure relief valve – removal, inspection and installation 18
Crankcase halves – separation and reassembly 21	Operations possible with the engine in the frame 2
Crankshaft and main bearings – removal, inspection and installation 27	Operations requiring engine removal 3
Cylinder block (J and K models) – removal, inspection and installation 14	Pistons – removal, inspection and installation 25
Cylinder head – removal and installation 10	Piston rings – inspection and installation 26
Cylinder head and valves – disassembly, inspection and reassembly 12	Pulse generator coil assembly – removal and installation . . . . .see Chapter 5
Engine – compression check . . . . .see Chapter 1	Recommended running-in procedure 32
Engine – removal and installation 5	Selector drum and forks – removal, inspection and installation 30
Engine disassembly and reassembly – general information 6	Spark plug gap – check and adjustment . . . . .see Chapter 1
Gearchange mechanism – removal, inspection and installation 20	Starter clutch and idle/reduction gear – removal, nspection and installation 15
General information 1	Starter motor – removal and installation . . . . .see Chapter 9
Idle speed – check and adjustment . . . . .see Chapter 1	Transmission shafts and bearings – removal and installation 28
Initial start-up after overhaul 31	Transmission shafts – disassembly, inspection and reassembly . . . 29
Main and connecting rod bearings – general note 23	Valve clearances – check and adjustment . . . . .see Chapter 1
Major engine repair – general note 4	Valve cover – removal and installation 8
	Valves/valve seats/valve guides – servicing 11

Degrees of difficulty

| Easy, suitable for novice with little experience | Fairly easy, suitable for beginner with some experience | Fairly difficult, suitable for competent DIY mechanic | Difficult, suitable for experienced DIY mechanic | Very difficult, suitable for expert DIY or professional |

Specifications

Note: *Models are identified by their production code letter – refer to 'Identification numbers' at the front of this manual for details.*

General

Type	Four-stroke, dohc, in-line four cylinder
Capacity	399 cc
Bore	55 mm
Stroke	42 mm
Compression ratio	11.3 to 1
Cylinder numbering	1 to 4 from left to right
Cooling system	Liquid cooled
Clutch	Wet multi-plate
Transmission	Six-speed constant mesh
Final drive	Chain

Camshafts and followers

Inlet lobe height
  J and K models
    Standard .... 33.570 to 33.730 mm
    Service limit (min) .... 33.520 mm
  L and N models
    Standard .... 33.530 to 33.770 mm
    Service limit (min) .... 33.480 mm
  R models
    Standard .... 33.610 to 33.690 mm
    Service limit (min) .... 33.560 mm

Exhaust lobe height
  J and K models
    Standard .... 33.120 to 33.280 mm
    Service limit (min) .... 33.070 mm
  L and N models
    Standard .... 33.080 to 33.320 mm
    Service limit (min) .... 33.030 mm
  R models
    Standard .... 33.660 to 33.740 mm
    Service limit (min) .... 33.610 mm

Journal diameter
  J models
    Standard .... Not available
    Service limit (min) .... 22.945 mm
  K models
    Standard .... 22.959 to 22.980 mm
    Service limit (min) .... 22.910 mm
  L, N and R models
    Standard .... 22.959 to 22.980 mm
    Service limit (min) .... 22.950 mm

Journal oil clearance
  Standard .... 0.020 to 0.062 mm
  Service limit (min) .... 0.100 mm
Runout (max) .... 0.05 mm

Cylinder head

Warpage (max) .... 0.10 mm

Valves, guides and springs

Valve clearances .... see Chapter 1
Inlet valve
  Stem diameter
    Standard .... 3.775 to 3.790 mm
    Service limit (min) .... 3.70 mm
  Guide bore diameter
    Standard .... 3.800 to 3.812 mm
    Service limit (max) .... 3.89 mm
  Stem-to-guide clearance
    Standard .... 0.010 to 0.037 mm
    Service limit (max) .... 0.04 mm
  Seat width
    Standard .... 0.9 to 1.1 mm
    Service limit .... 1.5 mm
  Valve guide height above cylinder head
    J and K models .... 11.52 to 11.92 mm
    L, N and R models .... 15.47 to 15.67 mm
Exhaust valve
  Stem diameter
    J and K models
      Standard .... 3.755 to 3.770 mm
      Service limit (min) .... 3.69 mm
    L, N and R models
      Standard .... 3.765 to 3.780 mm
      Service limit (min) .... 3.70 mm
  Guide bore diameter
    Standard .... 3.800 to 3.812 mm
    Service limit (max) .... 3.89 mm

Valves, guides and springs (continued)

Exhaust valve (continued)
  Stem-to-guide clearance
    Standard .... 0.020 to 0.047 mm
    Service limit (max) .... 0.05 mm
  Seat width
    Standard .... 0.9 to 1.1 mm
    Service limit .... 1.5 mm
  Valve guide height above cylinder head
    J and K models .... 11.52 to 11.92 mm
    L, N and R models .... 13.12 to 13.14 mm
Valve springs free length (inlet and exhaust)
  J and K models
    Standard .... 39.4 mm
    Service limit (min) .... 38.6 mm
  L, N and R models
    Standard .... 39.5 mm
    Service limit (min) .... 38.5 mm

Clutch

Friction plates
  J and K models 6
  L, N and R models 7
Plain plates
  J and K models 5
  L, N and R models 6
Friction plate thickness
  J and K models
    Outermost friction plate
      Standard .... 3.42 to 3.58 mm
      Service limit (min) .... 3.1 mm
    All other friction plates
      Standard .... 3.22 to 3.38 mm
      Service limit (min) .... 2.9 mm
  L, N and R models
    Standard .... 2.92 to 3.08 mm
    Service limit (min) .... 2.6 mm
Plain plate warpage (max) .... 0.3 mm
Spring free height
  J and K models
    Standard .... 37.5 mm
    Service limit (min) .... 36.0 mm
  L, N and R models
    Standard .... 43.2 mm
    Service limit (min) .... 41.7 mm
Clutch housing bush OD
  Standard .... 29.993 to 30.007 mm
  Service limit (min) .... 29.96 mm
Clutch housing bush ID
  Standard .... 21.980 to 21.993 mm
  Service limit (max) .... 22.03 mm
Transmission input shaft OD at bush point
  Standard .... 21.967 to 21.980 mm
  Service limit (max) .... 21.95 mm

Lubrication system

Oil pressure .... see Chapter 1
Oil pump
  Inner rotor tip-to-outer rotor clearance
    Standard .... 0.10 mm
    Service limit (max) .... 0.15 mm
  Outer rotor-to-body clearance
    Standard .... 0.15 to 0.22 mm
    Service limit (max) .... 0.35 mm
  Rotor end-float
    Standard .... 0.02 to 0.07 mm
    Service limit (max) .... 0.10 mm

Cylinder bores
Bore
  Standard .... 55.000 to 55.015 mm
  Service limit (max) .... 55.10 mm
Warpage (max) .... 0.10 mm
Ovality (out-of-round) (max) .... 0.10 mm
Taper (max) .... 0.10 mm
Cylinder compression .... see Chapter 1

Connecting rods
Small-end internal diameter
  Standard .... 14.016 to 14.034 mm
  Service limit (max) .... 14.07 mm
Small-end-to-piston pin clearance
  Standard .... 0.016 to 0.040 mm
  Service limit (max) .... 0.06 mm
Big-end side clearance
  J and K models
    Standard .... 0.05 to 0.20 mm
    Service limit (max) .... 0.30 mm
  L, N and R models
    Standard .... 0.1 to 0.25 mm
    Service limit (max) .... 0.35 mm
Big-end oil clearance
  Standard .... 0.020 to 0.052 mm
  Service limit (max) .... 0.06 mm

Pistons
Piston diameter (measured 10.0 mm (J and K models) or 8.5 mm (L, N and R models) up from skirt, at 90° to piston pin axis)
  J and K models
    Standard .... 54.960 to 54.990 mm
    Service limit (min) .... 54.90 mm
  L, N and R models
    Standard .... 54.970 to 54.990 mm
    Service limit (min) .... 54.90 mm
Piston-to-bore clearance
  Standard .... 0.010 to 0.045 mm
  Service limit (max) .... 0.10 mm
Piston pin diameter
  Standard .... 13.994 to 14.000 mm
  Service limit (min) .... 13.98 mm
Piston pin bore diameter in piston
  Standard .... 14.002 to 14.008 mm
  Service limit (max) .... 14.05 mm
Piston pin-to-piston pin bore clearance
  Standard .... 0.002 to 0.014 mm
  Service limit (max) .... 0.04 mm

Piston rings
Ring end gap (installed)
  Top ring
    Standard .... 0.18 to 0.28 mm
    Service limit (max) .... 0.50 mm
  2nd ring
    Standard .... 0.18 to 0.33 mm
    Service limit (max) .... 0.50 mm
  Oil ring side-rail
    Standard .... 0.20 to 0.70 mm
    Service limit (max) .... 1.10 mm
Ring-to-groove clearance
  Top ring
    Standard .... 0.015 to 0.050 mm
    Service limit (max)
      J and K models .... 0.08 mm
      L, N and R models .... 0.10 mm
  2nd ring
    Standard .... 0.015 to 0.050 mm
    Service limit (max)
      J and K models .... 0.08 mm
      L, N and R models .... 0.10 mm

Crankshaft and bearings

Main bearing oil clearance
    Standard .... 0.022 to 0.046 mm
    Service limit (max) .... 0.05 mm
Runout (max) .... 0.05 mm

Transmission

Gear ratios (no. of teeth)
  J models
    Primary reduction .... 2.181 to 1 (96/44T)
    1st gear .... 3.370 to 1 (43/13T)
    2nd gear .... 2.352 to 1 (40/17T)
    3rd gear .... 1.875 to 1 (30/16T)
    4th gear .... 1.591 to 1 (35/22T)
    5th gear .... 1.435 to 1 (33/23T)
    6th gear .... 1.280 to 1 (32/25T)
    Final reduction .... 2.800 to 1 (42/15T)
  K models
    Primary reduction .... 2.181 to 1 (96/44T)
    1st gear .... 3.370 to 1 (43/13T)
    2nd gear .... 2.352 to 1 (40/17T)
    3rd gear .... 1.875 to 1 (30/16T)
    4th gear .... 1.591 to 1 (35/22T)
    5th gear .... 1.435 to 1 (33/23T)
    6th gear .... 1.333 to 1 (32/24T)
    Final reduction .... 2.666 to 1 (40/15T)
  L and N models
    Primary reduction .... 2.117 to 1 (72/34T)
    1st gear .... 3.307 to 1 (43/13T)
    2nd gear .... 2.352 to 1 (40/17T)
    3rd gear .... 1.875 to 1 (30/16T)
    4th gear .... 1.591 to 1 (35/22T)
    5th gear .... 1.435 to 1 (33/23T)
    6th gear .... 1.318 to 1 (29/22T)
    Final reduction .... 2.600 to 1 (39/15T)
  R models
    Primary reduction .... 2.117 to 1 (72/34T)
    1st gear .... 3.307 to 1 (43/13T)
    2nd gear .... 2.352 to 1 (40/17T)
    3rd gear .... 1.875 to 1 (30/16T)
    4th gear .... 1.578 to 1 (30/19T)
    5th gear .... 1.435 to 1 (33/23T)
    6th gear .... 1.318 to 1 (29/22T)
    Final reduction .... 2.600 to 1 (39/15T)
Input shaft 5th and 6th gears ID
    Standard .... 25.000 to 25.021 mm
    Service limit (max) .... 25.05 mm
Input shaft 5th and 6th gears bush OD
    Standard .... 24.959 to 24.980 mm
    Service limit (min) .... 24.92 mm
Input shaft 5th and 6th gears gear-to-bush clearance
    Standard .... 0.020 to 0.062 mm
    Service limit (max) .... 0.10 mm
Input shaft 5th gear bush ID
    Standard .... 21.985 to 22.006 mm
    Service limit (max) .... 22.07 mm
Input shaft OD at 5th gear bush point
    Standard .... 21.959 to 21.980 mm
    Service limit (min) .... 21.93 mm
Input shaft-to-bush clearance at 5th gear bush point
    Standard .... 0.005 to 0.047 mm
    Service limit (max) .... 0.14 mm
Output shaft 2nd, 3rd and 4th gears ID
    Standard .... 28.000 to 28.021 mm
    Service limit (max) .... 28.05 mm
Output shaft 2nd, 3rd and 4th gears bush OD
    Standard .... 27.959 to 27.980 mm
    Service limit (min) .... 27.92 mm

Transmission (continued)

Output shaft 2nd, 3rd and 4th gears gear-to-bush clearance
    Standard .... 0.020 to 0.062 mm
    Service limit (max) .... 0.10 mm
Output shaft 2nd gear bush ID
    Standard .... 24.985 to 25.006 mm
    Service limit (max) .... 25.07 mm
Output shaft OD at 2nd gear bush point
    Standard .... 24.967 to 24.980 mm
    Service limit (min) .... 24.93 mm
Output shaft-to-bush clearance at 2nd gear bush point
    Standard .... 0.005 to 0.039 mm
    Service limit (max) .... 0.06 mm

Selector drum and forks

Selector fork end thickness
    Standard .... 5.93 to 6.00 mm
    Service limit (min) .... 5.60 mm
Selector fork bore ID
    Standard .... 12.000 to 12.021 mm
    Service limit (max) .... 12.06 mm
Selector fork shaft OD
    Standard .... 11.969 to 11.980 mm
    Service limit (min) .... 11.90 mm

Torque settings

Note: *Where a specified setting is not given for a particular bolt, the general settings listed at the beginning apply. The dimension given applies to the diameter of the thread, not the head.*

5 mm bolt/nut .... 5 Nm
6 mm bolt/nut .... 10 Nm
8 mm bolt/nut .... 22 Nm
10 mm bolt/nut .... 35 Nm
12 mm bolt/nut .... 55 Nm
6 mm flange bolt with 8 mm head .... 9 Nm
6 mm flange bolt/nut with 10 mm head .... 12 Nm
8 mm flange bolt/nut .... 27 Nm
10 mm flange bolt/nut .... 40 Nm
Engine mountings
    Mounting bracket bolts .... 27 Nm
    Mounting bolts and nuts .... 40 Nm
    Adjusting bolt (L, N and R models) .... 11 Nm
    Adjusting bolt locknut (L, N and R models) .... 55 Nm
Oil cooler bolt
    J and K models .... 55 Nm
    L, N and R models .... 65 Nm
Valve cover bolts .... 10 Nm
Camshaft holder bolts .... 12 Nm
Timing inspection plug .... 7 Nm
Centre plug (in alternator cover) .... 7 Nm
Cylinder head nuts
    J and K models .... 30 Nm
    L, N and R models .... 34 Nm
Cylinder head bolts .... 12 Nm
Cylinder block bolt (J and K models) .... 10 Nm
Timing rotor/starter clutch bolt .... 85 Nm
Starter clutch cover bolts .... 12 Nm
Clutch nut .... 85 Nm
Alternator/clutch cover bolts .... 12 Nm
Oil pump driven sprocket bolt .... 15 Nm
Gearchange mechanism centralising spring locating pin .... 23 Nm
Crankcase 8 mm bolts .... 24 Nm
Crankcase 6 mm bolts .... 12 Nm
Transmission input shaft bearing retainer plate bolts .... 12 Nm
Connecting rod big-end cap nuts .... 24 Nm
Selector drum bearing retainer plate bolts (L, N and R models) ...... 12 Nm
Selector fork shaft retaining bolt (L, N and R models) .... 12 Nm

Engine, clutch and transmission 2•7

1 General information

The engine/transmission unit is a liquid-cooled in-line four cylinder design, fitted transversely across the frame. The sixteen valves are operated by double overhead camshafts which are gear driven off the crankshaft. The engine/transmission assembly is constructed from aluminium alloy. The crankcase is divided horizontally.

The crankcase incorporates a wet sump, pressure-fed lubrication system which uses a chain-driven oil pump, an oil filter and by-pass valve assembly, a relief valve and an oil pressure switch. The lubrication system includes an oil cooler mounted between the crankcase and oil filter, which is fed off the main cooling system by the water pump.

The alternator is on the right-hand end of the crankshaft and the starter clutch is on the left-hand end. The oil pump is chain driven off a sprocket mounted behind the primary driven gear on the clutch housing. The water pump is driven by the oil pump.

Power from the crankshaft is routed to the transmission via the clutch. The clutch is of the wet, multi-plate type and is gear-driven off the crankshaft. The transmission is a six-speed constant-mesh unit. Final drive to the rear wheel is by chain and sprockets.

 Models are identified by their production code letter – refer to 'Identification numbers' at the front of this manual for details.

2 Operations possible with the engine in the frame

The components and assemblies listed below can be removed without having to remove the engine/transmission assembly from the frame. If however, a number of areas require attention at the same time, removal of the engine is recommended.

Valve cover
Camshafts
Ignition rotor and pulse generator coil assembly
Clutch
Gearchange mechanism (external components)
Alternator
Oil filter and oil cooler
Oil sump, oil pump, oil strainer and oil pressure relief valve
Starter motor
Starter clutch
Water pump

3 Operations requiring engine removal

It is necessary to remove the engine/transmission assembly from the frame to gain access to the following components:
Cylinder head
Camshaft drive gears
Cylinder block (J and K models only)
Pistons
Connecting rods and bearings
Transmission shafts
Selector drum and forks
Crankshaft and bearings

4 Major engine repair – general note

1 It is not always easy to determine when or if an engine should be completely overhauled, as a number of factors must be considered.
2 High mileage is not necessarily an indication that an overhaul is needed, while low mileage, on the other hand, does not preclude the need for an overhaul. Frequency of servicing is probably the single most important consideration. An engine that has regular and frequent oil and filter changes, as well as other required maintenance, will most likely give many miles of reliable service. Conversely, a neglected engine, or one which has not been run in properly, may require an overhaul very early in its life.
3 Exhaust smoke and excessive oil consumption are both indications that piston rings and/or valve guides are in need of attention, although make sure that the fault is not due to oil leakage.
4 If the engine is making obvious knocking or rumbling noises, the connecting rod and/or main bearings are probably at fault.
5 Loss of power, rough running, excessive valve train noise and high fuel consumption rates may also point to the need for an overhaul, especially if they are all present at the same time. If a complete tune-up does not remedy the situation, major mechanical work is the only solution.
6 An engine overhaul generally involves restoring the internal parts to the specifications of a new engine. The piston rings and main and connecting rod bearings are usually renewed and the cylinder walls honed during a major overhaul. Generally the valve seats are re-ground, since they are usually in less than perfect condition at this point. The end result should be a like new engine that will give as many trouble-free miles as the original.
7 Before beginning the engine overhaul, read through the related procedures to familiarise yourself with the scope and requirements of the job. Overhauling an engine is not all that difficult, but it is time consuming. Plan on the motorcycle being tied up for a minimum of two weeks. Check on the availability of parts and make sure that any necessary special tools, equipment and supplies are obtained in advance.
8 Most work can be done with typical workshop hand tools, although a number of precision measuring tools are required for inspecting parts to determine if they must be renewed. Often a dealer will handle the inspection of parts and offer advice concerning reconditioning and renewal. As a general rule, time is the primary cost of an overhaul so it does not pay to install worn or substandard parts.
9 As a final note, to ensure maximum life and minimum trouble from a rebuilt engine, everything must be assembled with care in a spotlessly clean environment.

5 Engine – removal and installation

Caution: The engine is very heavy. Engine removal and installation should be carried out with the aid of at least one assistant; personal injury or damage could occur if the engine falls or is dropped. An hydraulic or mechanical floor jack should be used to support and lower or raise the engine if possible.
Note: *On L, N and R models, a peg spanner is required to slacken and tighten the adjuster bolt locknut on the upper rear engine mounting bolts. If the Honda service tool (ask your dealer) or a suitable peg spanner is not available, one will have to be obtained, or fabricated out of a piece of 30 mm (OD) steel tubing or an old 22 mm socket (see **Tool Tip**).*

Removal

1 Support the bike securely in an upright position using an auxiliary stand. Work can be made easier by raising the machine to a suitable working height on an hydraulic ramp or a suitable platform. Make sure the motorcycle is secure and will not topple over (see *Tools and Workshop Tips* in the Reference section).
2 If the engine is dirty, particularly around its mountings, wash it thoroughly before starting any major dismantling work. This will make work much easier and rule out the possibility of caked on lumps of dirt falling into some vital component.
3 Remove the rider's seat (see Chapter 8). Disconnect the battery negative (-ve) lead from the battery, then disconnect the positive (+ve) lead from the battery (see Chapter 9 if required).
4 Remove the lower fairing (see Chapter 8). On J and K models, also remove the lower fairing brackets on each side.
5 Drain the engine oil (see Chapter 1). Also drain the cooling system (see Chapter 1).

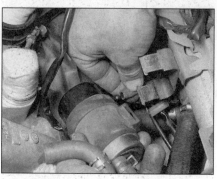

5.9 Displace the pump and detach the drain hose

5.10a On J and K models, slacken the clamps (arrowed) and detach the hoses

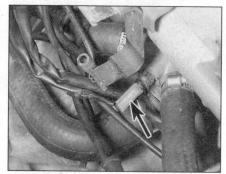

5.10b On L, N and R models, detach the temperature sender wiring connector

6 Remove the fuel tank and the air filter housing (see Chapter 4).
7 Remove the radiator (see Chapter 3).
8 Remove the exhaust system (see Chapter 4).
9 Remove the carburettors (see Chapter 4). Plug the engine inlet manifolds with clean rag. On L, N and R models, displace the fuel pump and detach the drain hose from its underside to prevent it becoming entangled in the coolant hoses **(see illustration)**.
10 On J and K models, release the clamps and detach the hoses from the thermostat housing to the engine **(see illustration)**. On L, N and R models, detach the wiring connector from the temperature sender mounted in the thermostat housing on the top of the crankcase and secure it clear of the engine **(see illustration)**.
11 Trace the alternator wiring from the alternator/clutch cover on the right-hand side of the engine and disconnect it at the white (J and K models) or red (L, N and R models) 3-pin connector **(see illustrations)**. Release the wiring from any clips or ties, noting its routing, and coil it so that it does not impede engine removal.
12 Trace the ignition pulse generator wiring from the cover on the left-hand side of the engine and disconnect it at the red 4-pin connector (J and K models) or brown 2-pin connector (L, N and R models) **(see illustrations)**. Release the wiring from any clips or ties, noting its routing, and coil it on top of the crankcase so that it does not impede engine removal.
13 Pull back the rubber cover on the oil pressure switch, then remove the screw securing the wiring connector and detach it from the switch **(see illustration)**. Secure it clear of the engine.
14 Disconnect all the HT leads from the spark plugs and secure them clear of the engine.
15 Pull back the rubber cover on the starter motor terminal, then unscrew the nut and disconnect the lead **(see illustration)**. Secure it clear of the engine. Also remove the front mounting bolt and detach the earth lead.
16 Unscrew the bolts securing the clutch cable bracket to the clutch cover, then detach

5.11a Alternator wiring connector – J and K models

5.11b Alternator wiring connector – L, N and R models

5.12a Pulse generator wiring connector (arrowed) – J and K models

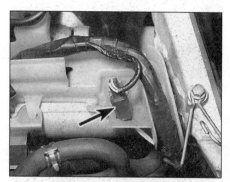

5.12b Pulse generator wiring connector (arrowed) – L, N and R models

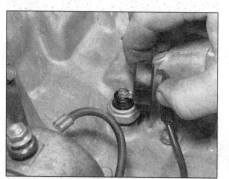

5.13 Pull back the rubber cover and remove the wiring connector screw

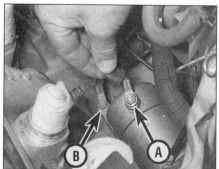

5.15 Detach the starter motor lead (A) and the earth lead (B)

Engine, clutch and transmission 2•9

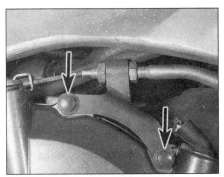

5.16a Unscrew the bolts (arrowed) and detach the bracket . . .

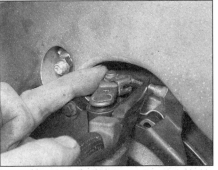

5.16b . . . and the clutch cable

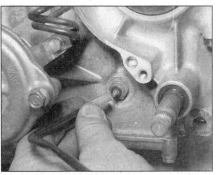

5.18 Detach the neutral switch wiring connector

the bracket and slip the cable end out of the release lever **(see illustrations)**.

17 Remove the front sprocket (see Chapter 6).

18 Disconnect the neutral switch wiring connector from the switch and secure it clear of the engine **(see illustration)**.

19 At this point, position an hydraulic or mechanical jack under the engine with a block of wood between the jack head and sump **(see illustration)**. Make sure the jack is centrally positioned so the engine will not topple when the last mounting bolt is removed. Take the weight of the engine on the jack.

J and K models

20 Remove the screw and detach the trim panel on each side of the frame **(see illustration)**.

21 Unscrew and remove the engine upper front mounting bolts, and remove the oval spacer/washer fitted between the frame and engine on the right-hand bolt **(see illustrations)**.

22 Unscrew the nuts on the engine lower front mounting bolts and withdraw the bolts, and remove the oval spacer/washer fitted between the frame and engine on the right-hand bolt and the hose guide fitted similarly on the left-hand bolt **(see illustrations 5.21a, b and c and 5.22)**.

23 Make sure the engine is properly supported on the jack, and have an assistant support it as well.

24 Unscrew the nut on the engine lower rear mounting bolt and withdraw the bolt, and

remove the small spacer fitted between the engine and the frame on the left-hand side **(see illustrations 5.21a and b and 5.24)**.

5.19 Position a jack under the engine

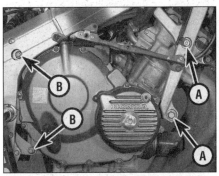

5.21a Engine mounting bolts (A) and nuts (B) – right-hand side

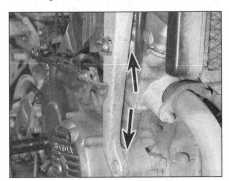

5.21c Oval spacers/washers (arrowed) with front right-hand bolts

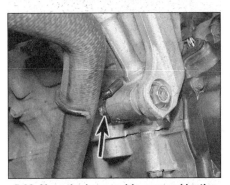

5.22 Note the hose guide secured by the nut on the lower front left-hand bolt

Unscrew the bolts securing the mounting bolt bracket to the right-hand side of the frame and remove the bracket.

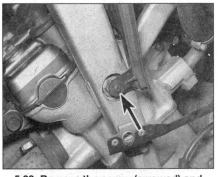

5.20 Remove the screw (arrowed) and detach the trim panel on each side

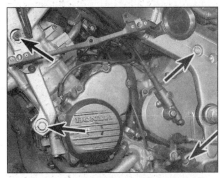

5.21b Engine mounting bolts (arrowed) – left-hand side

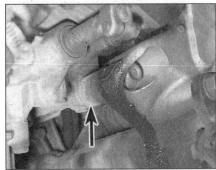

5.24 Note the spacer (arrowed) on the lower rear mounting bolt

2•10 Engine, clutch and transmission

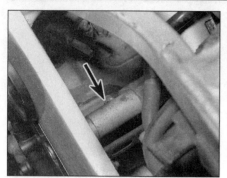

5.25 Note the spacer (arrowed) on the upper rear mounting bolt

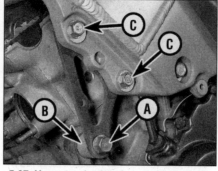

5.27 Unscrew the front mounting bolt (A) on each side and remove the spacer (B). Each bracket is secured by two bolts (C)

5.28 Unscrew the middle mounting bolts and remove the spacer with the right-hand bolt

25 Unscrew the nut on the engine upper rear mounting bolt and withdraw the bolt, and remove the large spacer fitted between the engine and the frame on the left-hand side **(see illustrations 5.21a and b and 5.25)**. Unscrew the bolts securing the mounting bolt bracket to the right-hand side of the frame and remove the bracket.
26 The engine can now be removed from the frame. Check that all relevant wiring, cables and hoses are disconnected and secured well clear, then carefully lower the engine and manoeuvre it out of the side of the frame (see *Caution* above).

L, N and R models

27 Unscrew and remove the engine front mounting bolts and remove the 19 mm spacers fitted between the brackets and engine **(see illustration)**. To improve clearance when removing the engine, also remove the brackets.
28 Unscrew and remove the engine middle mounting bolts and remove the 23 mm spacer fitted between the frame and engine on the right-hand side **(see illustration)**.
29 Make sure the engine is properly supported on the jack, and have an assistant support it as well.
30 Unscrew the nut on the engine lower rear mounting bolt and withdraw the bolt **(see illustration)**.
31 Unscrew the nut on the engine upper rear mounting bolt and remove the bolt and the

55 mm spacers fitted between the frame and engine **(see illustration 5.30)**. Using either the Honda service tool, a suitable peg spanner or a fabricated tool (see **Tool Tip**), unscrew the locknut on the adjusting bolt fitted in the right-hand side of the frame **(see illustration)**. Using a suitable Allen key, unscrew and remove the adjusting bolt **(see illustration)**.

A peg spanner can be made by cutting an old 22 mm socket as shown – measure the width and depth of the slots in the locknut to determine the size of the castellations on the socket. If an old socket is not available, a piece of steel tubing with an outside diameter of 30 mm can be used – its wall thickness should be the same as the depth of the slots in the locknut.

32 The engine can now be removed from the frame. Check that all relevant wiring, cables and hoses are disconnected and secured well clear, then carefully lower the engine and manoeuvre it out of the side of the frame (see *Caution* above).
33 With the aid of an assistant lift the engine unit off the jack and lower it carefully onto the work surface, taking care not to break the long fins cast onto the bottom of the oil pan. These fins are there specifically for the engine to stand on and keep it in an upright position.

Installation

34 With the aid of an assistant place the engine unit on top of the jack and block of wood and carefully raise the engine unit into position in the frame. Make sure no wires, cables or hoses become trapped between the engine and the frame.

J and K models

35 Fit the engine mounting brackets onto the frame and tighten the bolts to the torque setting specified at the beginning of the Chapter.
36 Position the large spacer for the upper rear mounting bolt between the frame and the engine on the left-hand side and slide the bolt through from the left **(see illustration 5.25)**. Fit the nut, tightening it hand-tight only at this stage.
37 Position the small spacer for the lower

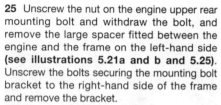

5.30 Lower rear mounting bolt (A), upper rear bolt (B). Note the spacer (C) on each side

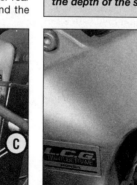

5.31a Remove the locknut . . .

5.31b . . . and the adjusting bolt

Engine, clutch and transmission 2•11

5.41a Install the upper rear mounting bolt with its left-hand . . .

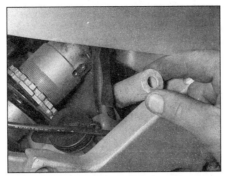

5.41b . . . and right-hand spacers

5.41c Tighten the adjusting bolt to the specified torque

5.41d Make reference marks between the adjusting bolt and the frame . . .

5.41e . . . and tighten the locknut to the specified torque

5.41f Fit the nut and tighten it hand-tight only at this stage

rear mounting bolt between the frame and the engine on the left-hand side and slide the bolt through from the left (see illustration 5.24). Fit the nut, tightening it hand-tight only at this stage.

38 Position the oval spacer/washer for the lower front right-hand bolt and the cable guide for the left-hand bolt between the frame and engine, then install the bolts and tighten the nuts hand-tight only (see illustrations 5.21c and 5.22).

39 Position the oval spacer/washer for the right-hand upper front mounting bolt, then install the bolt and tighten it finger-tight (see illustration 5.21c). Also install the left-hand upper front bolt and tighten it finger-tight.

40 Now tighten all the nuts and bolts to the torque settings specified at the beginning of the Chapter.

L, N and R models

41 Position the 55 mm spacers for the upper rear mounting bolt between the frame and the engine, and slide the bolt through from the left (see illustrations). Slide the adjusting bolt onto the right-hand end (see illustration 5.31b) and tighten it to the torque setting specified at the beginning of the Chapter – you will have to push the mounting bolt back out a bit to allow an Allen key into the adjusting bolt (see illustration). Make a reference mark between the adjusting bolt and the frame as a check against the bolt turning while the locknut is being tightened (see illustration), then fit the

locknut (see illustration 5.31a) and tighten it to the specified torque setting using the peg spanner or socket (see illustration). Check that the reference marks still align – if they don't, repeat the installation and tightening procedure. Now push the mounting bolt fully through from the left and fit the nut onto its right-hand end (see illustration). Tighten it hand-tight only at this stage.

42 Slide the lower rear mounting bolt through from the left-hand side (see illustration 5.30). Fit the nut, tightening it hand-tight only at this stage.

43 Position the 23 mm spacer for the right-hand middle bolt between the frame and engine, then install the middle bolts and tighten them hand-tight only (see illustration 5.28).

44 If removed, install the front mounting bolt brackets (see illustration 5.27). Position the 19 mm spacers for the front bolts between the brackets and engine, then install the bolts and tighten them hand-tight only.

45 Now tighten all the nuts and bolts to the torque settings specified at the beginning of the Chapter.

All models

46 The remainder of the installation procedure is the reverse of removal, noting the following points.
a) Make sure all wires, cables and hoses are correctly routed and connected, and secured by the relevant clips or ties.

b) Adjust the throttle and clutch cable freeplay (see Chapter 1).
c) Adjust the drive chain (see Chapter 1).
d) Refill the engine with oil and coolant (see Chapter 1).
e) Prior to installing the lower fairing panels start the engine and check that there are no signs of coolant/oil leakage.

6 Engine disassembly and reassembly – general information

Disassembly

1 Before disassembling the engine, the external surfaces of the unit should be thoroughly cleaned and degreased. This will prevent contamination of the engine internals, and will also make working a lot easier and cleaner. A high flash-point solvent, such as paraffin (kerosene) can be used, or better still, a proprietary engine degreaser. Use old paintbrushes and toothbrushes to work the solvent into the various recesses of the engine casings. Take care to exclude solvent or water from the electrical components and inlet and exhaust ports.

 Warning: The use of petrol (gasoline) as a cleaning agent should be avoided because of the risk of fire.

2•12 Engine, clutch and transmission

7.3 Slacken the hose clamps and detach the hoses (A), then unscrew the bolt (B) – L, N and R models shown

7.5a Fit a new O-ring . . .

7.5b . . . then install the cooler, locating the bracket over the lug (arrowed) . . .

2 When clean and dry, arrange the unit on the workbench, leaving suitable clear area for working. Gather a selection of small containers and plastic bags so that parts can be grouped together in an easily identifiable manner. Some paper and a pen should be on hand so that notes can be made and labels attached where necessary. A supply of clean rag is also required.

3 Before commencing work, read through the appropriate section so that some idea of the necessary procedure can be gained. When removing components it should be noted that great force is seldom required, unless specified. In many cases, a component's reluctance to be removed is indicative of an incorrect approach or removal method – if in any doubt, re-check with the text.

4 When disassembling the engine, keep 'mated' parts together (including gears, pistons, connecting rods, valves, etc. that have been in contact with each other during engine operation). These 'mated' parts must be reused or renewed as an assembly.

5 A complete engine/transmission disassembly should be done in the following general order with reference to the appropriate Sections.
   Remove the valve cover
   Remove the camshafts
   Remove the cylinder head
   Remove the thermostat housing (L, N and R models) (see Chapter 3)
   Remove the cylinder block and pistons (J and K models)
   Remove the starter clutch
   Remove the clutch
   Remove the alternator (see Chapter 9)
   Remove the starter motor (see Chapter 9)
   Remove the water pump (see Chapter 3)
   Remove the gearchange mechanism external components
   Remove the oil sump
   Remove the oil pump
   Separate the crankcase halves
   Remove the connecting rods (and pistons on L, N and R models)
   Remove the crankshaft
   Remove the transmission shafts
   Remove the selector drum and forks

Reassembly

6 Reassembly is accomplished by reversing the general disassembly sequence.

7 Oil cooler – removal and installation

Note: *The oil cooler can be removed with the engine in the frame. If the engine has been removed, ignore the steps which do not apply.*

Removal

1 Remove the lower fairing (see Chapter 8).
2 Drain the engine oil and remove the oil filter (see Chapter 1). Drain the cooling system (see Chapter 1), or have some means of blocking or clamping the hoses to avoid excessive loss of coolant (see 'Section 9 Hoses' in the Tools and Workshop Tips section at the end of this manual).
3 Slacken the clamp securing each hose to the cooler and detach the hoses **(see illustration)**.
4 Unscrew the cooler bolt using a 30 mm socket and remove the cooler, noting how the cutout in the bracket locates over the lug on the crankcase **(see illustration 7.5b)**. Discard the cooler O-ring as a new one must be used.

Installation

5 Installation is the reverse of removal, noting the following:

7.5c . . . and tighten the bolt to the specified torque

a) Always use a new O-ring when installing the cooler **(see illustration)**.
b) Locate the bracket on the cooler over the lug on the crankcase **(see illustration)**.
c) Tighten the cooler bolt to the torque setting specified at the beginning of the Chapter **(see illustration)**.
d) Make sure the coolant hoses are pressed fully onto their unions and are secured by the clamps **(see illustration 7.3)**.
e) Fit a new oil filter and fill the engine with oil (see Chapter 1).
f) Refill the cooling system if it was drained, or check the level in both the radiator and the reservoir and top up if necessary (see Chapter 1).

8 Valve cover – removal and installation

Note: *The valve cover can be removed with the engine in the frame. If the engine has been removed, ignore the steps which do not apply.*

Removal

1 Remove the lower fairing (see Chapter 8).
2 Release the lower radiator hose from its clip on the left-hand side of the engine, then unscrew the radiator lower mounting bolt(s) and swing the bottom of the radiator forward **(see illustrations)**. To provide better

8.2a Release the hose from its clip . . .

Engine, clutch and transmission 2•13

8.2b ... then unscrew the bolt(s) and swing the radiator forward

8.6 Unscrew the bolts (arrowed) and remove the cover

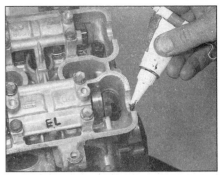

8.9a Apply the sealant as described ...

clearance, and to avoid scratching anything, it is better to remove the radiator (see Chapter 3).
3 Disconnect the spark plug caps from the plugs and secure them clear of the engine, noting which fits where.
4 For best access, remove the air filter housing (see Chapter 4). On L, N and R models, also remove the ignition HT coils (see Chapter 5), then remove the heat shield, noting how it fits.
5 If the air filter housing was not removed, release the clamp securing the breather hose to the breather on the top of the valve cover and detach the hose.
6 Unscrew the eight bolts securing the valve cover then lift the cover off the cylinder head (see illustration). If it is stuck, do not try to lever it off with a screwdriver. Tap it gently around the sides with a rubber hammer or block of wood to dislodge it. Also remove the gasket. Note the rubber washers fitted in the cover and remove them if they are loose.

Installation

7 Examine the valve cover gasket and the rubber washers for signs of damage or deterioration and renew them if necessary.
8 Clean the mating surfaces of the cylinder head and the valve cover with lacquer thinner, acetone or brake system cleaner.
9 Apply a smear of a suitable sealant into the grooves in the valve cover and into the cutouts in the cylinder head. Install the gasket onto the valve cover, making sure it fits correctly into the groove (see illustrations).
10 The valve cover must be installed with the arrow on its top pointing to the front. Position the valve cover on the cylinder head, making sure the gasket stays in place (see illustration). If removed, fit the rubber washers into the cover, using new ones if required, and making sure they are installed with the 'UP' mark facing up. Install the cover bolts and tighten them to the torque setting specified at the beginning of the Chapter.
11 Install the remaining components in the reverse order of removal.

9 Camshafts – removal, inspection and installation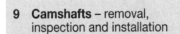

Note: *The camshafts can be removed with the engine in the frame. Place rags over the spark plug holes and the cam gear train hole to prevent any component from dropping into the engine on removal.*

Removal

1 Remove the valve cover (see Section 8).
2 Unscrew the timing inspection plug and the centre plug from the alternator cover on the

8.9b ... then fit the gasket onto the cover ...

right-hand side of the engine (see illustration). Discard the plug O-rings as new ones should be used. The engine can be turned using a 14 mm socket on the alternator rotor bolt and turning it in a clockwise direction only (see illustration). Alternatively, place the motorcycle on an auxiliary stand so that the rear wheel is off the ground, select a high gear and rotate the rear wheel by hand in its normal direction of rotation.
3 Turn the engine until the line next to the 'T' mark on the alternator rotor aligns with the notches in the timing inspection hole, and the 'IN' mark on the inlet camshaft gear and the 'EX' mark on the exhaust camshaft gear are aligned with the cylinder head surface with each mark the correct way up and on the outside of its respective gear (see

8.10 ... and install the cover

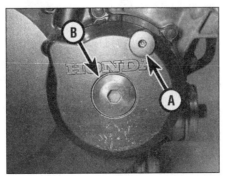

9.2a Remove the timing inspection plug (A) and the centre plug (B)

9.2b Turn the engine clockwise using a 14 mm socket on the alternator rotor bolt

2•14 Engine, clutch and transmission

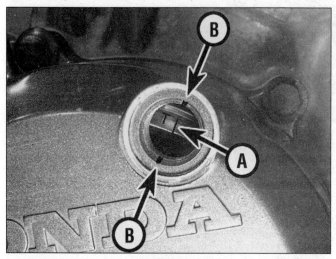

9.3a Turn the engine until the T1 mark (A) aligns with the static marks (B) . . .

9.3b . . . and the camshaft gear marks align as described and shown

9.5a Camshaft holder bolts (arrowed)

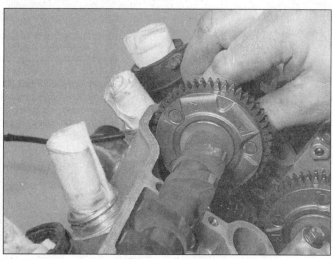

9.5b Remove the camshafts . . .

illustrations). If the 'IN' and 'EX' marks are upside down and on the inside of their respective gears, turn the engine one full turn (360°) until the line next to the 'T' mark again aligns with the notches. The 'IN' and 'EX' marks will now be correctly aligned.

4 Before disturbing the camshaft holders, check them for identification markings. The inlet camshaft holders are marked INR (right-hand side) and INL (left-hand side) and the exhaust camshaft holders marked EXR (right-hand side) and EXL (left-hand side). These markings ensure that the holders can be matched up to their original locations on installation. If no markings are visible, make your own using a felt pen. If necessary, make a sketch of the layout as a further aid for installation.

5 Unscrew the camshaft holder bolts for the camshaft being worked on, evenly and a little at a time in a criss-cross pattern, until they are all loose **(see illustration)**. While slackening the bolts make sure that the holder is lifting squarely away from the cylinder head and is not sticking on the locating dowels.

Caution: If the bolts are carelessly loosened and the holder does not come squarely away from the head, the holder is likely to break. If this happens the complete cylinder head assembly must be renewed; the holders are matched to the cylinder head and cannot be renewed separately. Also, a camshaft could break if the holder bolts are not slackened evenly and the pressure from a depressed valve causes the shaft to bend.

Remove the bolts, then lift off the camshaft holders, noting how they fit. Retrieve the dowels from either the holder or the cylinder head if they are loose. Lift each shaft out of the head **(see illustration)**. The camshafts are marked for identification. The inlet camshaft is marked 'IN' and the exhaust camshaft is marked 'EX' **(see illustration)**.

Inspection

6 Inspect the bearing surfaces of the camshaft holders and the corresponding

9.5c . . . noting the identification marks

Engine, clutch and transmission 2•15

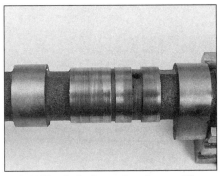

9.6 Check the journal surfaces of the camshaft for scratches or wear

9.7a Check the lobes of the camshaft for wear – here's an example of damage requiring camshaft repair or renewal

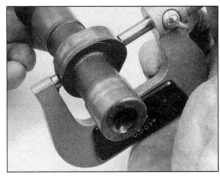

9.7b Measure the height of the camshaft lobes with a micrometer

journals on the camshaft. Look for score marks, deep scratches and evidence of spalling (a pitted appearance) **(see illustration)**.

7 Check the camshaft lobes for heat discoloration (blue appearance), score marks, chipped areas, flat spots and spalling **(see illustration)**. Measure the height of each lobe with a micrometer **(see illustration)** and compare the results to the minimum lobe height listed in this Chapter's Specifications. If damage is noted or wear is excessive, the camshaft must be renewed. Also check the condition of the cam followers in the cylinder head (see Section 12).

8 Check the amount of camshaft runout by supporting each end of the camshaft on V-blocks, and measuring any runout using a dial gauge. If the runout exceeds the specified limit the camshaft must be renewed.

 HAYNES HiNT *Refer to Tools and Workshop Tips in the Reference section for details of how to read a micrometer and dial gauge.*

9 The camshaft bearing oil clearance should now be checked. There are two possible ways of doing this, either by direct measurement (see Step 10) or by the use of a product known as Plastigauge (see Steps 11 to 14).

10 If the direct measurement method is to be used, make sure the camshaft holder dowels are in position then fit the holders, making sure they are in their correct location (see Step 4). Tighten the holder bolts evenly and a little at a time in a criss-cross pattern to the torque setting specified at the beginning of the Chapter. Using telescoping gauges and a micrometer (see *Tools and Workshop Tips*), measure the journal holder diameter. Now measure the diameter of the camshaft journal with a micrometer **(see illustration)**. To determine the journal oil clearance, subtract the holder diameter from the journal diameter and compare the result to the clearance specified. If the clearance is greater than specified, and the journal diameter is not worn below the service limit, the cylinder head/holder bearing surfaces are worn.

11 If the Plastigauge method is to be used, clean the camshafts, the bearing surfaces in the cylinder head and camshaft holder with a clean lint-free cloth, then lay the camshafts in place in the cylinder head.

12 Cut strips of Plastigauge and lay one piece on each bearing journal, parallel with the camshaft centreline **(see illustration)**. Make sure the camshaft holder dowels are in position then fit the holders, making sure they are in their correct location (see Step 4). Tighten the holder bolts evenly and a little at a time in a criss-cross pattern to the torque setting specified at the beginning of the Chapter. While doing this, don't let the camshafts rotate.

13 Now unscrew the bolts evenly and a little at a time in a criss-cross pattern, and carefully lift off the camshaft holders.

14 To determine the oil clearance, compare the crushed Plastigauge (at its widest point) on each journal to the scale printed on the Plastigauge container **(see illustration)**. Compare the results to this Chapter's Specifications. If the oil clearance is greater than specified, measure the diameter of the camshaft journal with a micrometer **(see illustration 9.10)**. If the journal diameter is less than the specified limit, renew the camshaft and recheck the clearance. If the clearance is still too great, renew the cylinder head and camshaft holders.

 HAYNES HiNT *Before renewing camshafts or holders because of damage, check with local machine shops specialising in motorcycle engine work. In the case of the camshafts, it may be possible for cam lobes to be welded, reground and hardened, at a cost far lower than that of a new camshaft. If the bearing surfaces in the holders are damaged, it may be possible for them to be bored out to accept bearing inserts. Due to the cost of new components it is recommended that all options be explored.*

9.10 Measure the cam bearing journals with a micrometer

9.12 Lay a strip of Plastigauge across each bearing journal, parallel with the camshaft centreline

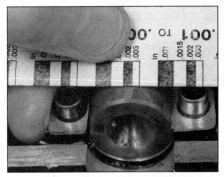

9.14 Compare the width of the crushed Plastigauge to the scale printed on the Plastigauge container

2•16 Engine, clutch and transmission

9.16 Apply molybdenum oil to the camshaft journals and holders . . .

9.18 . . . then install the camshafts . . .

9.19a . . . and the holders . . .

15 Check the camshaft drive gear on the drive gear assembly and the driven gear on each camshaft for wear, cracks and other damage, renewing them if necessary. The driven gears are integral with the camshafts, whilst the drive gear assembly can be removed from the engine (see Section 13). If wear this severe is apparent, the entire engine should be disassembled for inspection.

Installation

16 Make sure the bearing surfaces on the camshafts and in the holders are clean, then apply molybdenum disulphide oil (a 50/50 mixture of molybdenum disulphide grease and engine oil) to each of them **(see illustration)**. Also apply it to the camshaft lobes and the followers.
17 The camshafts and holders must be installed in their correct location according to their identification marks (see Steps 5 and 4).
18 Verify that the line next to the 'T' mark on the alternator rotor is still aligned with the notch **(see illustration 9.3a)**, then lay each camshaft into place **(see illustration)**, engaging them with the drive gear assembly, and making sure all the marks are in exact alignment (see Step 3) **(see illustration 9.3b)**. Take extra care at this stage as it is easy to be one tooth out on the timing without it appearing as a drastic misalignment of the timing marks.
19 Make sure the camshaft holder dowels are in position then fit the holders, making sure they are in their correct location (see Step 4) **(see illustration)**. Tighten the holder bolts evenly and a little at a time in a criss-cross pattern to the torque setting specified at the beginning of the Chapter, starting with the bolts that are closest to any valve which will be compressed by the camshaft lobes as the holders are tightened down **(see illustration)**. This is to avoid placing undue strain on the camshafts. Whilst tightening the bolts, make sure the holders are being pulled squarely down and are not binding on the dowels. Check that each camshaft is not pinched by turning the engine a few degrees in each direction.

Caution: The camshaft is likely to break if it is tightened down onto the closed valves before the open valves. The holders are likely to break if they are not tightened down evenly and squarely.

20 With all holders tightened down, check again that the valve timing marks still align (see Step 3).

Caution: If the marks are not aligned exactly as described, the valve timing will be incorrect and the valves may strike the pistons, causing extensive damage to the engine.

21 Check the valve clearances and adjust them if necessary (see Chapter 1). This is essential if new components have been installed or if the camshafts were removed to change the shims.
22 Use new O-rings on the timing inspection plug and centre plug and smear them and the plug threads with molybdenum disulphide oil (a 50/50 mixture of molybdenum disulphide grease and engine oil) **(see illustration)**. Tighten the plugs to the torque setting specified at the beginning of the Chapter.
23 Install the valve cover (see Section 8).

10 Cylinder head – removal and installation

Caution: The engine must be completely cool before beginning this procedure or the cylinder head may become warped.
Note: To remove the cylinder head the engine must be removed from the frame.

Removal

1 Remove the engine from the frame (see Section 5).
2 Remove the camshafts (see Section 9).
3 Obtain a container which is divided into sixteen compartments, and label each compartment with the location of its corresponding valve in the cylinder head and whether it belongs with an inlet or an exhaust valve. If a container is not available, use labelled plastic bags. Lift each cam follower out of the cylinder head using either a magnet or a pair of pliers and store it in its corresponding compartment in the container **(see illustration)**. Retrieve the shim from either the inside of the follower or pick it out of

9.19b . . . and tighten the holder bolts to the specified torque

9.22 Install the plugs using new O-rings

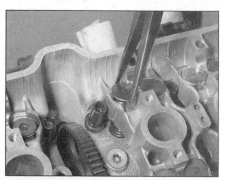

10.3a Remove the follower . . .

Engine, clutch and transmission 2•17

10.3b ... and either retrieve the shim from inside the follower ...

10.3c ... or from the top of the valve, using either a magnet ...

10.3d ... or a screwdriver with grease on it

10.4 Unscrew the bolts (arrowed) and detach the pipes from the head

10.5a Cylinder head bolts (A) and nuts (B)

the top of the valve, using either a magnet, a small screwdriver with a dab of grease on it (the shim will stick to the grease), or a screwdriver and a pair of pliers **(see illustrations)**. Do not allow the shim to fall into the engine.

4 Unscrew the bolt securing each coolant pipe in the rear of the cylinder head **(see illustration)**. Pull each pipe out of the head. Discard the O-rings as new ones must be used.

5 Each cylinder head is secured by two bolts and twelve nuts **(see illustration)**. First slacken and remove the two bolts located in the cam drive gear tunnel. Now slacken the nuts evenly and a little at a time in a criss-cross sequence, working from the outside towards the centre, until they are all slack, and remove the nuts by sliding them towards the centre as they are held captive by the casting **(see illustration)**.

6 Pull the cylinder head up off the block **(see illustration)**. If it is stuck, tap around the joint faces of the cylinder head with a soft-faced mallet to free the head. Do not attempt to free the head by inserting a screwdriver between the head and cylinder block – you'll damage the sealing surfaces. Remove the old cylinder head gasket and discard it as a new one must be used.

7 If they are loose, remove the two dowels from the cylinder block. If they appear to be missing they are probably stuck in the underside of the cylinder head.

8 Check the cylinder head gasket and the mating surfaces on the cylinder head and block for signs of leakage, which could indicate warpage. Refer to Section 12 and check the flatness of the cylinder head.

9 Clean all traces of old gasket material from the cylinder head and block. If a scraper is used, take care not to scratch or gouge the soft aluminium. Be careful not to let any of the gasket material fall into the crankcase, the cylinder bores or the oil passages.

Installation

10 If removed, install the dowels onto the outer rear studs and press them into the cylinder block **(see illustration 10.11)**. Lubricate the cylinder bores with engine oil.

11 Ensure both cylinder head and block mating surfaces are clean, then lay the new head gasket in place on the cylinder block, making sure all the holes are correctly aligned,

10.5b Slide the nuts towards the centre so the base rim clears the casting

10.6 Lift the head up off the block

2•18 Engine, clutch and transmission

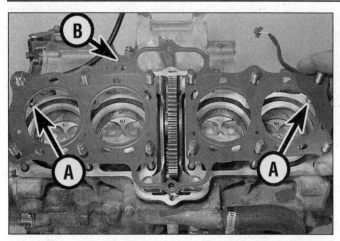

10.11 Make sure the gasket locates onto the dowels (A) and the UP mark (B) reads correctly

10.13 Tighten the cylinder head nuts as described to the specified torque

it locates correctly onto the dowels, and, where appropriate, the 'UP' mark stamped out of the gasket is positioned along the rear edge and is the correct way up **(see illustration)**. Never re-use the old gasket.

12 Carefully fit the cylinder head onto the block, making sure it locates correctly onto the dowels **(see illustration 10.6)**.

13 Install the twelve nuts and the two bolts and tighten them all finger-tight **(see illustration 10.5a)**. When installing the nuts, locate them slightly in from the stud, then slide them onto the stud so the base rim fits under the casting **(see illustration 10.5b)**. First tighten the nuts evenly and a little at a time in a criss-cross pattern, starting from the centre and working outwards, to the torque setting specified the beginning of the Chapter **(see illustration)**. Now tighten the bolts to the specified torque.

14 Fit a new O-ring onto each coolant pipe, then press the pipes into their bores and secure them with the bolts **(see illustration)**.

15 Lubricate each shim with molybdenum disulphide oil (a 50/50 mixture of molybdenum disulphide grease and engine oil) and fit it into its recess in the top of the valve, with the size marking on each shim facing up **(see illustration)**.Check that the shim is correctly seated, then lubricate the follower with molybdenum disulphide oil and install it onto the valve **(see illustration)**. **Note:** *It is most important that the shims and followers are returned to their original valves otherwise the valve clearances will be inaccurate.*

16 Install the camshafts (see Section 9).
17 Install the engine (see Section 5).

11 Valves/valve seats/valve guides – servicing

1 Because of the complex nature of this job and the special tools and equipment required, most owners leave servicing of the valves, valve seats and valve guides to a professional.
2 The home mechanic can, however, remove the valves from the cylinder head, clean and check the components for wear and assess the extent of the work needed, and, unless a valve service is required, grind in the valves (see Section 12).
3 The dealer or motorcycle engineer will remove the valves and springs, renew the valves and guides, recut the valve seats, check and renew the valve springs, spring retainers and collets (as necessary), renew the valve seals and reassemble the valve components.

4 After the valve service has been performed, the head will be in like-new condition. When the head is returned, be sure to clean it again very thoroughly before installation on the engine to remove any metal particles or abrasive grit that may still be present from the valve service operations. Use compressed air, if available, to blow out all the holes and passages.

12 Cylinder head and valves – disassembly, inspection and reassembly

1 As mentioned in the previous section, valve servicing, valve seat re-cutting and valve guide renewal should be left to a dealer or motorcycle engineer. However, disassembly, cleaning and inspection of the valves and related components can be done (if the necessary special tools are available) by the home mechanic. This way no expense is incurred if the inspection reveals that overhaul is not required at this time.
2 To disassemble the valve components without the risk of damaging them, a valve spring compressor is absolutely necessary.

10.14 Fit the pipes using new O-rings

10.15a Fit the shim with its size marking facing up . . .

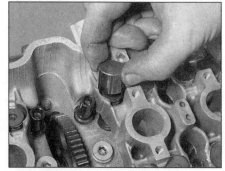

10.15b . . . then install the follower

Engine, clutch and transmission 2•19

Disassembly

3 Before proceeding, arrange to label and store the valves along with their related components in such a way that they can be returned to their original locations without getting mixed up **(see illustration)**. A good way to do this is to use the same container as the followers and shims are stored in (see Section 10), or to obtain a separate container which is divided into sixteen compartments, and to label each compartment with the identity of the valve which will be stored in it (ie number of cylinder, inlet or exhaust side, inner or outer valve). Alternatively, labelled plastic bags will do just as well.

4 If not already done, clean all traces of old gasket material from the cylinder head. If a scraper is used, take care not to scratch or gouge the soft aluminium.

 Haynes HiNT *Refer to Tools and Workshop Tips for details of gasket removal methods.*

5 Compress the valve spring on the first valve with a spring compressor, making sure it is correctly located onto each end of the valve assembly **(see illustration)**. Do not compress the springs any more than is absolutely necessary. Remove the collets, using either needle-nose pliers, tweezers, a magnet or a screwdriver with a dab of grease on it **(see illustration)**. Carefully release the valve spring compressor and remove the spring retainer, noting which way up it fits, the spring, the spring seat, and the valve from the head **(see illustration 12.3)**. If the valve binds in the guide (won't pull through), push it back into the head and deburr the area around the collet groove with a very fine file or whetstone **(see illustration)**.

6 Repeat the procedure for the remaining valves. Remember to keep the parts for each valve together and separate from the other valves so they can be reinstalled in the same location.

7 Once the valves have been removed and labelled, pull the valve stem seals off the top of the valve guides with pliers and discard them (the old seals should never be reused).

8 Next, clean the cylinder head with solvent and dry it thoroughly. Compressed air will speed the drying process and ensure that all holes and recessed areas are clean.

9 Clean all of the valve springs, collets, retainers and spring seats with solvent and dry them thoroughly. Do the parts from one valve at a time so that no mixing of parts between valves occurs.

10 Scrape off any deposits that may have formed on the valve, then use a motorised wire brush to remove deposits from the valve heads and stems. Again, make sure the valves do not get mixed up.

Inspection

11 Inspect the head very carefully for cracks and other damage. If cracks are found, a new head will be required. Check the cam bearing surfaces for wear and evidence of seizure. Check the camshafts for wear as well (see Section 9).

12 Inspect the outer surfaces of the cam followers for evidence of scoring or other damage. If a follower is in poor condition, it is probable that the bore in which it works is also damaged. Check for clearance between the followers and their bores. Whilst no specifications are given, if slack is excessive, renew the followers. If the bores are seriously out-of-round or tapered, the cylinder head and the followers must be renewed.

13 Using a precision straight-edge and a feeler gauge set to the warpage limit listed in the specifications at the beginning of the Chapter, check the head gasket mating surface for warpage. Refer to *Tools and Workshop Tips* in the Reference section for details of how to use the straight-edge.

14 Examine the valve seats in the combustion chamber. If they are pitted, cracked or burned, the head will require work

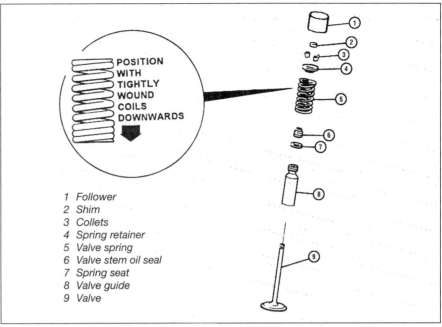

12.3 Valve components

1 Follower
2 Shim
3 Collets
4 Spring retainer
5 Valve spring
6 Valve stem oil seal
7 Spring seat
8 Valve guide
9 Valve

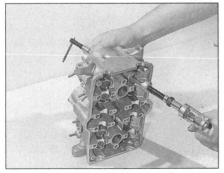

12.5a Compressing the valve spring using a valve spring compressor

12.5b Remove the collets with needle-nose pliers, tweezers, a magnet or a screwdriver with a dab of grease on it

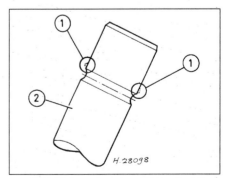

12.5c If the valve stem (2) won't pull through the guide, deburr the area (1) above the collet groove

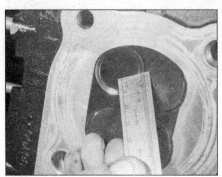

12.14 Measure the valve seat width with a ruler (or for greater precision use a vernier caliper)

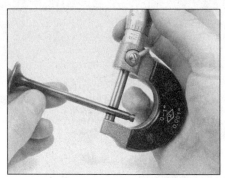

12.15a Measure the valve stem diameter with a micrometer

12.15b Insert a small-hole gauge into the valve guide and expand it so there's a slight drag when it's pulled out

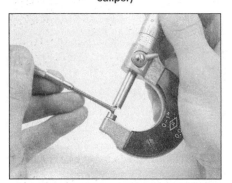

12.15c Measure the small-hole gauge with a micrometer

12.16 Check the valve face (A), stem (B) and collet groove (C) for signs of wear and damage

beyond the scope of the home mechanic. Measure the valve seat width and compare it to this Chapter's Specifications **(see illustration)**. If it exceeds the service limit, or if it varies around its circumference, valve overhaul is required. If available, use Prussian blue to determine the extent of valve seat wear. Uniformly coat the seat with the Prussian blue, then install the valve and rotate it back and forth using a lapping tool. Remove the valve and check whether the ring of blue on the valve is uniform and continuous around the valve, and of the correct width as specified.

15 Measure the valve stem diameter **(see illustration)**. Clean the valve guides to remove any carbon build-up, then measure the inside diameters of the guides (at both ends and the centre of the guide) with a small-hole gauge and micrometer **(see illustrations)**. The guides are measured at the ends and at the centre to determine if they are worn in a bell-mouth pattern (more wear at the ends). Subtract the stem diameter from the valve guide diameter to obtain the valve stem-to-guide clearance. If the stem-to-guide clearance is greater than listed in this Chapter's Specifications, the guides and valves will have to be renewed. If the valve stem or guide is worn beyond its limit, or if the guide is worn unevenly, it must be renewed.

16 Carefully inspect each valve face for cracks, pits and burned spots. Check the valve stem and the collet groove area for cracks **(see illustration)**. Rotate the valve and check for any obvious indication that it is bent. Check the end of the stem for pitting and excessive wear. The presence of any of the above conditions indicates the need for valve servicing. The stem end can be ground down, provided that the amount of stem above the collet groove after grinding is sufficient.

17 Check the end of each valve spring for wear and pitting. Measure the spring free length and compare it to that listed in the specifications **(see illustration)**. If any spring is shorter than specified it has sagged and must be renewed. Also place the spring upright on a flat surface and check it for bend by placing a ruler against it **(see illustration)**. If the bend in any spring is excessive, it must be renewed.

18 Check the spring retainers and collets for obvious wear and cracks. Any questionable parts should not be reused, as extensive damage will occur in the event of failure during engine operation.

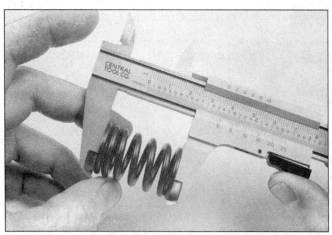

12.17a Measure the free length of the valve springs

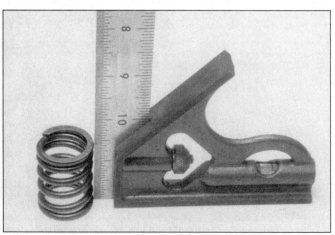

12.17b Check the valve springs for squareness

Engine, clutch and transmission 2•21

12.21 Apply the grinding compound very sparingly, in small dabs, to the valve face only

12.22a Rotate the valve grinding tool back and forth between the palms of your hands

12.22b The valve face and seat should show a uniform unbroken ring . . .

19 If the inspection indicates that no overhaul work is required, the valve components can be reinstalled in the head.

Reassembly

20 Unless a valve service has been performed, before installing the valves in the head they should be ground in (lapped) to ensure a positive seal between the valves and seats. This procedure requires coarse and fine valve grinding compound and a valve grinding tool. If a grinding tool is not available, a piece of rubber or plastic hose can be slipped over the valve stem (after the valve has been installed in the guide) and used to turn the valve.

21 Apply a small amount of coarse grinding compound to the valve face, then slip the valve into the guide **(see illustration)**. **Note:** *Make sure each valve is installed in its correct guide and be careful not to get any grinding compound on the valve stem.*

22 Attach the grinding tool (or hose) to the valve and rotate the tool between the palms of your hands. Use a back-and-forth motion (as though rubbing your hands together) rather than a circular motion (ie so that the valve rotates alternately clockwise and anti-clockwise rather than in one direction only) **(see illustration)**. Lift the valve off the seat and turn it at regular intervals to distribute the grinding compound properly. Continue the grinding procedure until the valve face and seat contact area is of uniform width and unbroken around the entire circumference of the valve face and seat **(see illustrations)**.

23 Carefully remove the valve from the guide and wipe off all traces of grinding compound. Use solvent to clean the valve and wipe the seat area thoroughly with a solvent soaked cloth.

24 Repeat the procedure with fine valve grinding compound, then repeat the entire procedure for the remaining valves.

25 Lay the spring seat for the first valve in place in the cylinder head, with its narrower shouldered side up so that it fits into the base of the spring, then install new valve stem seal onto the guide **(see illustration 12.3)**. Use an appropriate size deep socket to push the seal over the end of the valve guide until it is felt to clip into place **(see illustration)**. Don't twist or cock it, or it will not seal properly against the valve stem. Also, don't remove it again or it will be damaged.

26 Coat the valve stem with molybdenum disulphide grease, then install it into its guide, rotating it slowly to avoid damaging the seal **(see illustration)**. Check that the valve moves up and down freely in the guide. Next, fit the spring, with its closer-wound coils facing down into the cylinder head, onto the spring seat, then fit the spring retainer, with its shouldered side facing down so that it fits into the top of the spring **(see illustrations)**.

27 Apply a small amount of grease to the collets to help hold them in place **(see**

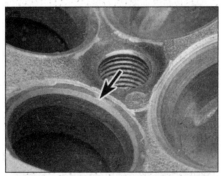

12.22c . . . and the seat (arrowed) should be the specified width all the way round

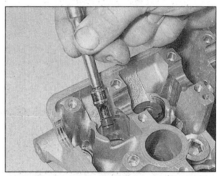

12.25 Press the valve stem seal into position using a suitable deep socket

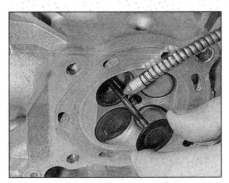

12.26a Lubricate the stem and slide the valve into its correct location

12.26b Fit the valve spring with its closer-wound coils facing down . . .

12.26c . . . then fit the spring retainer

2•22 Engine, clutch and transmission

illustration). Compress the spring with the valve spring compressor and install the collets **(see illustration).** When compressing the spring, depress it only as far as is absolutely necessary to slip the collets into place. Make certain that the collets are securely locked in the retaining groove.

28 Repeat the procedure for the remaining valves. Remember to keep the parts for each valve together and separate from the other valves so they can be reinstalled in the same location.

29 Support the cylinder head on blocks so the valves can't contact the workbench top, then very gently tap each of the valve stems with a soft-faced hammer. This will help seat the collets in their grooves.

> **HAYNES HiNT** *Check for proper sealing of the valves by pouring a small amount of solvent into each of the valve ports. If the solvent leaks past any valve into the combustion chamber area the valve grinding operation on that valve should be repeated.*

13 Camshaft drive gear assembly – removal, inspection and installation

Note: *To remove the camshaft drive gear, the engine must be removed from the frame.*

13.3a On J and K models, the bolt (arrowed) is on the front

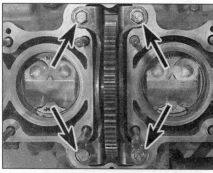

13.4a Remove the four bolts (arrowed) . . .

12.27a A small dab of grease will help to keep the collets in place on the valve while the spring is released

Removal

1 Remove the engine (see Section 5).
2 Remove the cylinder head (see Section 10).
3 Unscrew the single bolt securing the camshaft drive gear assembly to the outside of the cylinder block – on J and K models the bolt is situated on the front of the block, while on L, N and R models it is on the back **(see illustrations).** Check the condition of the sealing washer and renew it if necessary.
4 Unscrew and remove the four bolts securing the assembly to the cylinder block **(see illustration).** Carefully lift the assembly out of the cylinder head, noting how it fits **(see illustration).** Remove the two dowels if they are loose **(see illustration).**

13.3b On L, N and R models, the bolt is on the back

13.4b . . . then lift the assembly out . . .

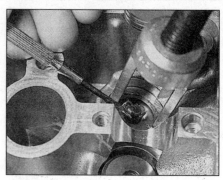

12.27b Compress the springs and install the collets, making sure they locate in the groove

Inspection

5 Wash the assembly in clean solvent and dry it off.
6 Check the teeth on both gears for cracks and other damage and make sure the gears turn smoothly and freely. If the teeth are damaged or worn, the assembly must be renewed. If this is the case, the corresponding teeth on both the camshafts and the crankshaft should be inspected.

Installation

7 If removed, fit the two dowels into the cylinder block, making sure they are fully pressed home **(see illustration 13.4c)**.
8 Install the gear assembly, making sure it is fitted with the bolt hole for the cylinder block bolt facing in the correct direction for your model (see Step 3). Press the assembly squarely down onto its dowels, making sure the lower gear teeth mesh with those of the drive gear on the crankshaft **(see illustration 13.4b)**. Due to the split sprung gear, the assembly will probably not sit flat onto the block as it cannot mesh fully with the drive gear on the crankshaft. Check that the gears have at least partially meshed either by turning the upper gear back and forth – the whole assembly should be felt to rise off the cylinder head slightly, or by turning the crankshaft – the gears should turn accordingly.
9 Install the four mounting bolts **(see illustration 13.4a)** and tighten them all finger-tight at first, then tighten them evenly and a

13.4c . . . and remove the dowels (arrowed) if loose (L, N and R models shown)

little at a time, making sure the assembly is being drawn squarely down onto the dowels as the gears mesh fully. Install and tighten the cylinder block bolt, using a new sealing washer if required **(see illustrations 13.3a or b)**.

10 Install the cylinder head (see Section 10) and the engine (see Section 5).

14 Cylinder block (J and K models) – removal, inspection and installation

Note: *To remove the cylinder block, the engine must be removed from the frame (see Section 5).*

Removal

1 Remove the engine (see Section 5).
2 Remove the cylinder head (see Section 10) and the camshaft drive gear assembly (see Section 13).
3 Slacken the clamp securing the coolant hose to the union on the back of the cylinder block and detach the hose **(see illustration)**.
4 Unscrew and remove the single bolt located in the back of the camshaft drive gear tunnel **(see illustration)**.
5 Lift the cylinder block up off the crankcase **(see illustration)**. If the block is stuck, tap around the joint faces of the block with a soft-faced mallet to free it from the crankcase. Don't attempt to free the block by inserting a screwdriver between it and the crankcase – you'll damage the sealing surfaces. When the block is removed, stuff clean rags around the pistons to prevent anything falling into the crankcase. Remove the dowels from the mating surface of the crankcase or the underside of the block if they are loose. Be careful not to let these drop into the engine.
6 Remove the gasket and clean all traces of old gasket material from the cylinder block and crankcase mating surfaces. If a scraper is used, take care not to scratch or gouge the soft aluminium. Be careful not to let any of the gasket material fall into the crankcase or the oil passages.

14.3 Slacken the clamp and detach the hose . . .

> **HAYNES HINT** *Rotate the crankshaft until the inner pistons are uppermost and feed them into the block first. This makes access to the lower pistons easier when compressing the rings and feeding them into the bores.*

Inspection

7 Refer to Section 22.

Installation

8 Check that the mating surfaces of the cylinder block and crankcase are free from oil or pieces of old gasket. If removed, fit the dowels into the crankcase.
9 Remove the rags from around the pistons, and lay the new base gasket in place on the crankcase making sure all the holes are correctly aligned. Never re-use the old gasket.
10 If required, install piston ring clamps onto the pistons to ease their entry into the bores as the block is lowered. This is not essential as each cylinder has a good lead-in enabling the piston rings to be hand-fed into the bore. If possible, have an assistant support the block while this is done.
11 Lubricate the cylinder bores, pistons and piston rings, and the connecting rod big- and

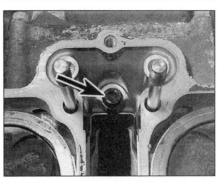

14.4 . . . then unscrew the bolt (arrowed) . . .

small-ends, with clean engine oil, then install the block down onto the studs until the uppermost piston crowns fit into the bores **(see illustration)**.
12 Gently push down on the cylinder block, making sure the pistons enter the bore squarely and do not get cocked sideways. If piston ring clamps are not being used, carefully compress and feed each ring into the bore as the block is lowered. If necessary, use a soft mallet to gently tap the block down, but do not use force if the block appears to be stuck as the pistons and/or rings will be damaged. If clamps are used, remove them once the pistons are in the bore.
13 When the pistons are correctly installed in the cylinders, press the block down onto the base gasket, making sure it locates correctly onto the dowels.
14 Install the single bolt into the rear of the camshaft drive gear tunnel and tighten it to the torque setting specified at the beginning of the Chapter **(see illustration 14.4)**.
15 Attach the coolant hose to the union on the back of the cylinder block and tighten the clamp **(see illustration 14.3)**.
16 Install the camshaft drive gear assembly (see Section 13) and the cylinder head (see Section 10).
17 Install the engine (see Section 5).

14.5 . . . and remove the block

14.11 Lower the block and feed the pistons and rings into the bores

2•24 Engine, clutch and transmission

15.4 Unscrew the bolts (arrowed) and remove the cover

15.6a Unscrew the bolt and remove the timing rotor (arrowed) . . .

15.6b . . . then slide the starter clutch off

15 Starter clutch and idle/reduction gear – removal, inspection and installation

Note: *The starter clutch can be removed with the engine in the frame. If the engine has been removed, ignore the steps which don't apply.*

1 Remove the lower fairing (see Chapter 8).
2 Drain the engine oil (see Chapter 1).
3 Remove the fuel tank (see Chapter 4). Trace the ignition pulse generator wiring from the left-hand side of the engine and disconnect it at the red 4-pin connector (J and K models) or brown 2-pin connector (L, N and R models) **(see illustrations 5.12a and b).** Release the wiring from any clips or ties, and feed it through to the cover, noting its routing.

15.7 Withdraw the shaft and remove the gear

4 Unscrew the bolts securing the starter clutch cover and remove the cover, being prepared to catch any residue oil **(see illustration).** Discard the gasket as a new one must be used. Remove the dowel from either the cover or the crankcase if it is loose.
5 To slacken the timing rotor/starter clutch bolt the engine must be prevented from turning. If the engine is still in the frame, select a gear and have an assistant apply the rear brake. If the engine has been removed, remove the clutch/alternator cover from the right-hand side of the engine (see Section 16) and hold the alternator rotor using a rotor strap.

Caution: Do not hold the rotor using a socket on the rotor bolt as it could become loose.

6 Unscrew the bolt and remove the washer and the timing rotor **(see illustration).** Draw the starter clutch off the end of the crankshaft, and remove the splined thrust washer from behind it **(see illustration).** Note how the wider grooves in the timing rotor, starter clutch and thrust washer fit over the wider spline on the shaft.
7 Withdraw the starter idle/reduction gear shaft from the centre of the gear, then remove the gear, noting how it fits **(see illustration).**

Inspection

8 With the starter clutch face-down on a workbench, check that the starter driven gear rotates freely in an anti-clockwise direction and locks against the rotor in a clockwise direction **(see illustration).** If it doesn't, renew the starter clutch.
9 Withdraw the needle roller bearing and starter driven gear from the starter clutch **(see illustration).** If the gear appears stuck, rotate it anti-clockwise as you withdraw it to free it from the starter clutch.
10 Check the condition of the rollers inside the clutch body **(see illustration).** If they are damaged, marked or flattened at any point, the starter clutch should be renewed – individual components are not available.
11 Check the bearing surface of the starter driven gear hub and the needle roller bearing. If the bearing surface shows signs of excessive wear or the or the bearing itself is worn or damaged, they should be renewed.
12 Check the teeth of the starter idle/reduction gear and the corresponding teeth of the starter driven gear and starter motor drive shaft. Renew the gears and/or starter motor if worn or chipped teeth are discovered on related gears. Also check the idle/reduction gear shaft for damage, and check that the gear is not a loose fit on the shaft. Renew the shaft if necessary.

Installation

13 Lubricate the hub of the starter driven gear with clean engine oil, then install the starter driven gear into the clutch, rotating it anti-clockwise as you do so to spread the rollers and allow the hub of the gear to enter

15.8 Check the starter clutch as described

15.9 Withdraw the gear and bearing . . .

15.10 . . . and check the rollers (two arrowed)

Engine, clutch and transmission 2•25

15.15 Slide the washer . . .

15.16 . . . the starter clutch . . .

15.17a . . . and the timing rotor onto the shaft, . . .

(see illustration 15.9). Fit the needle roller bearing into the driven gear.

14 Lubricate the idle/reduction gear shaft with clean engine oil. Position the gear, making sure the teeth of the larger pinion mesh correctly with the teeth of the starter motor shaft, then slide the shaft into the gear (see illustration 15.7).

15 Slide the splined thrust washer onto the end of the crankshaft, aligning the wider groove in the washer with the wider spline on the shaft (see illustration).

16 Slide the starter clutch assembly onto the end of the crankshaft, aligning the wider groove in the starter clutch with the wider spline on the shaft (see illustration).

17 Slide the timing rotor onto the end of the crankshaft, aligning the wider groove in the rotor with the wider spline on the shaft (see illustration). Install the timing rotor/starter clutch bolt and its washer and, using the method employed on removal to stop the engine turning, tighten the bolt to the torque setting specified at the beginning of the Chapter (see illustrations). If preferred, a socket or spanner on the alternator rotor bolt can now be used to hold the engine, as the counter-effect of tightening the starter clutch bolt will now be to tighten the alternator bolt,

15.17b . . . then install the bolt . . .

rather than loosen it. The torque setting for the alternator bolt is the same on J and K models and higher on L, N and R models, so there should be no danger of stripping it.

18 If removed, insert the dowel in the crankcase, then install the cover using a new gasket, making sure it locates correctly onto the dowel and the idle/reduction gear shaft (see illustration). Tighten the cover bolts evenly in a criss-cross sequence to the specified torque setting.

19 Feed the pulse generator coil wiring back to its connector, making sure it is correctly routed,

15.17c . . . and tighten it to the specified torque

and reconnect it (see illustrations 5.12a and b). Install the fuel tank (see Chapter 4).
20 Refill the engine with oil (see Chapter 1).
21 Install the lower fairing (see Chapter 8).

16 Clutch – removal, inspection and installation

Note: *The clutch can be removed with the engine in the frame. If the engine has been removed, ignore the steps which don't apply.*

15.18a Locate the gasket over the dowel (arrowed) . . .

15.18b . . . and install the cover

2•26 Engine, clutch and transmission

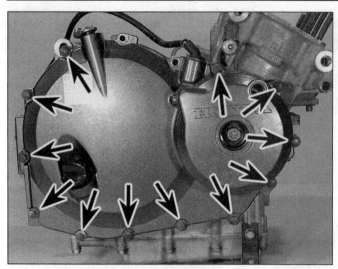

16.5 Clutch cover bolts (arrowed) – two cable bracket bolts already removed

16.6 Unscrew the bolts (arrowed) and remove the plate and springs

16.7a Unstake the nut (arrowed) . . .

16.7b . . . and slacken it as described

Removal

1 Remove the lower fairing (see Chapter 8).
2 Drain the engine oil (see Chapter 1).
3 Unscrew the two bolts securing the clutch cable bracket to the alternator/clutch cover on the right-hand side of the engine, then free the cable end from the release lever **(see illustrations 17.2a and b)**. Position the cable clear of the engine.
4 Remove the fuel tank (see Chapter 4). Trace the alternator wiring from the alternator/clutch cover and disconnect it at the white (J and K models) or red (L, N and R models) 3-pin connector **(see illustrations 5.11a and b)**. Release the wiring from any clips or ties, noting its routing, and feed it through to the cover.
5 Unscrew the cover bolts and remove the cover, being prepared to catch any residual oil **(see illustration)**. Discard the gasket as a new one must be used. Remove the two dowels from either the cover or the crankcase if they are loose. Note the pushrod fitted in the bottom of the shaft in the cover and remove it for safekeeping if required **(see illustration 16.34)**. It is possible that it is stuck to the release bearing in the centre of the clutch release plate.
6 Working evenly in a criss-cross pattern, gradually slacken the clutch release plate bolts until spring pressure is released, then remove the bolts, plate and springs **(see illustration)**.

7 The clutch nut is staked against the transmission input shaft **(see illustration)**. Unstake the nut using a screwdriver, a punch, or a small drill. To slacken the clutch nut the input shaft must be locked. This can be done in several ways:

a) *If the engine is in the frame, engage 1st gear and have an assistant hold the rear brake on hard with the rear tyre in firm contact with the ground.*
b) *Use the Honda service tool, available from a dealer.*
c) *A home-made tool made from two strips of steel bent at the ends and bolted together in the middle (see **Tool tip**), can be used to stop the clutch centre from turning whilst the nut is slackened **(see illustration)**.*
d) *If the engine is out of the frame, fit a close-fitting ring spanner or socket over the output shaft splines, select top gear and hold the spanner or socket while the nut is slackened.*

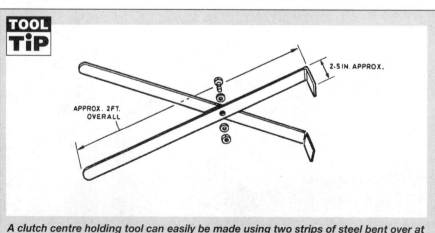

TOOL TIP

A clutch centre holding tool can easily be made using two strips of steel bent over at the ends and bolted together in the middle

Engine, clutch and transmission 2•27

16.9 Grasp the pressure plate and clutch centre and draw the assembly off

16.11 Draw the housing off the shaft

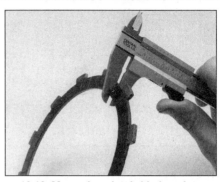

16.13 Measuring clutch friction plate thickness

8 Unscrew the clutch nut and remove the Belleville washer from the shaft, noting which way round it fits. On R models, also remove the thrust washer, again noting which way round it fits. Discard the nut as a new one must be used.

9 Grasp the clutch centre with the complete set of clutch plates and the pressure plate and remove them as a pack **(see illustration)**. Unless the plates are being renewed, keep them in their original order. Note that of the friction plates, there are two types, the outermost plate having a larger internal diameter to the other friction plates. This is to accommodate the anti-judder spring and spring seat that fit between the outer friction plate and the clutch centre. On J and K models, the outermost plain plate is also slightly different. Take care not to mix them up.

10 On L, N and R models, remove the thrust plate from the input shaft, noting which way round it fits **(see illustration 16.27)**.

11 Remove the clutch housing **(see illustration)**.

12 If required, withdraw the clutch housing bush from the centre of the oil pump drive sprocket to provide slack in the chain, then slip the chain off the driven sprocket and remove the drive sprocket and chain, and the thrust washer **(see illustrations 16.24 and 16.23b and a)**. Note the pins on the drive sprocket which locate in the holes in the back of the housing.

Inspection

13 After an extended period of service the clutch friction plates will wear and promote clutch slip. Measure the thickness of each plate using a vernier caliper **(see illustration)**. If the thickness is less than the service limit given in the Specifications at the beginning of the Chapter, the friction plates must be renewed as a set. Also, if any of the plates smell burnt or are glazed, they must be renewed.

14 The plain plates should not show any signs of excess heating (bluing). Check for warpage using a surface plate and feeler gauges **(see illustration)**. If any plate appears warped, or shows signs of bluing, all plain plates must be renewed as a set.

15 Measure the free length of each clutch spring **(see illustration)**. If any spring is below the service limit specified, renew all the springs as a set.

16 Inspect the clutch assembly for burrs and indentations on the edges of the protruding tangs of the friction plates and/or slots in the edge of the housing with which they engage. Similarly check for wear between the inner tongues of the plain plates and the slots in the clutch centre. Wear of this nature will cause clutch drag and slow disengagement during gear changes, since the plates will snag when the pressure plate is lifted. With care, a small amount of wear can be corrected by dressing with a fine file, but if this is excessive the worn components should be renewed.

17 Check the release plate, release bearing, and pushrod for signs of roughness, wear or damage, and renew any parts as necessary. Check that the bearing outer race is a tight fit in the centre of the plate, and that the inner race rotates freely without any rough spots.

18 Measure the clutch housing bush inner and outer diameters, and the diameter of the input shaft where the bush fits. Compare the measurements to the specifications at the beginning of the Chapter and renew the bush and/or the shaft if they are worn beyond their service limits. Also check all the above components, and the needle bearing the clutch housing, for signs of damage or scoring, and renew if necessary.

19 Check the pressure plate, and on L, N and R models the thrust plate for signs of roughness, wear or damage, and renew any parts as necessary.

20 Check the anti-judder spring and spring seat for distortion, wear or damage, and renew them if necessary.

21 Check that the release shaft rotates smoothly in the clutch clover. If the action is rough or stiff, free the spring end from the spring pin in the shaft and draw the shaft out of the cover **(see illustration)**. Check the condition of the two needle bearings – the action may be improved by cleaning them with solvent and applying some oil, but otherwise they should be renewed. Also renew the oil seal if it is damaged or deteriorated, or shows any signs of leakage.

16.14 Check the plain plates for warpage

16.15 Measuring clutch spring free length

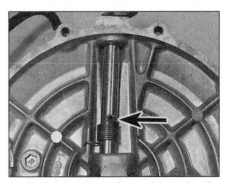

16.21 Release the spring end from the pin (arrowed) and withdraw the release shaft

2•28 Engine, clutch and transmission

16.23a Slide on the thrust washer...

16.23b ... and the drive sprocket and chain...

16.24 ... and fit the bush into the centre of the sprocket

16.26a The pins (A) must locate in the holes (B)

16.26b Align the teeth of the pinions as described and slide the housing home

Installation

22 Remove all traces of old gasket from the crankcase and alternator/clutch cover surfaces.

23 If removed, slide the thrust washer onto the shaft, followed by the oil pump drive sprocket, with its pins facing out, then loop the chain around both the drive and driven sprockets (see illustrations).

24 Smear the clutch housing bush with clean engine oil, then slide it onto the shaft and into the middle of the oil pump drive sprocket (see illustration).

25 Turn the engine using a socket or spanner on the alternator rotor bolt until the line next to the 'T' on the rotor is positioned roughly between the 1 and 2 o'clock positions – the idea is to position the crank web so that it does not interfere with the clutch housing as it is slid onto the shaft.

26 Slide the clutch housing onto the bush on the input shaft, making sure that the teeth of the primary driven gear on the back of the housing engage with those of the primary drive gear, and that the pins on the oil pump drive sprocket engage with the holes in the back of the housing (see illustration). As the primary driven gear is a split sprung gear, the teeth will be felt to engage with the drive gear and then go no further. At this point the teeth of the solid pinion and the sprung pinion have to be aligned by inserting a screwdriver blade between them and twisting it – the housing will be felt to slide further in when they align and engage fully with the drive gear (see illustration). Use a screwdriver to turn the oil pump driven sprocket to locate the drive sprocket pins in their holes.

27 On L, N and R models, slide the thrust plate onto the shaft (see illustration).

28 Slide the pressure plate onto the shaft (see illustration).

29 Coat each clutch plate with engine oil, then build up the plates in the housing, making sure the outermost friction plate, and on J and K models also the outermost plain plate, are correctly identified and fitted last (see Step 9). Start with a friction plate, then a plain plate and alternate friction and plain plates until all are installed (see illustrations).

16.27 On L, N and R models, slide on the thrust plate

16.28 Slide the pressure plate on the shaft

16.29a Fit a friction plate first...

Engine, clutch and transmission 2•29

16.29b ...then a plain plate

16.29c The outermost friction plate has a larger internal diameter

16.29d Align the plain plate tabs as shown

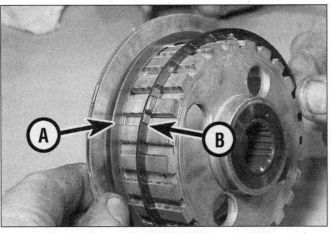
16.30a Fit the spring seat (A) and spring (B)...

16.30b ...so that the outer edge of the spring is off the seat

Align the tabs of the plain plates to ease fitting the clutch centre **(see illustration)**.

30 If removed, fit the spring seat onto the clutch centre, followed by the anti-judder spring, making sure its outer edge is raised off the spring seat – if it is touching the seat and the inner edge is raised, it is the wrong way round **(see illustrations)**.

31 Slide the clutch centre onto the shaft splines and into the clutch plates, wiggling it as required to align the tabs **(see illustration)**.

32 On R models, slide the thrust washer onto the shaft. On all models, slide on the Belleville washer with its OUTSIDE marking facing out, then fit a new clutch nut **(see illustrations)**.

16.31 Slide the centre into the clutch

16.32a Slide on the Belleville washer...

16.32b ...then fit a new clutch nut...

16.32c ...and tighten it to the specified torque

16.32d Stake the nut rim against the indent in the shaft

2•30 Engine, clutch and transmission

16.33a Fit the springs . . .

16.33b . . . and the release plate and tighten the bolts as described

16.34 Turn the shaft to align the seat and fit the pushrod

Using the method employed on removal to lock the input shaft (see Step 7), tighten the nut to the torque setting specified at the beginning of the Chapter **(see illustration)**. Stake the rim of the nut into the indent in the end of the shaft using a suitable punch **(see illustration and 16.7a)**. Note: *Check that the clutch centre rotates freely after tightening.*

33 Install the clutch springs, release plate and release plate bolts and tighten them evenly in a criss-cross sequence **(see illustrations)**.

34 Apply some grease to the release pushrod, then turn the shaft until the pushrod seat is aligned and insert the pushrod **(see illustration)**.

35 If removed, insert the dowels in the crankcase. Install the alternator/clutch cover using a new gasket, making sure it locates correctly onto the dowels, then install all the cover bolts with the exception of the two which secure the cable bracket **(see illustrations)**. Tighten the bolts evenly in a criss-cross sequence to the specified torque setting.

36 If the engine is in the frame, fit the clutch cable end into the release lever and position the bracket on the cover, then fit the two bolts and tighten them to the specified torque. If the engine has been removed, fit them after the engine has been installed.

37 Feed the alternator wiring back to its connector, making sure it is correctly routed, and reconnect it **(see illustrations 5.11a and b)**. Install the fuel tank (see Chapter 4).

38 Refill the engine with oil (see Chapter 1).

39 Install the lower fairing (see Chapter 8).

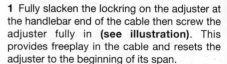

17 Clutch cable – replacement

1 Fully slacken the lockring on the adjuster at the handlebar end of the cable then screw the adjuster fully in **(see illustration)**. This provides freeplay in the cable and resets the adjuster to the beginning of its span.

2 Slacken the adjuster nuts on the threaded section in the cable bracket on the alternator/clutch cover on the right-hand side of the engine **(see illustration)**. Unscrew the two bolts securing the cable bracket to the cover, then free the cable end from the release lever, noting how it fits **(see illustration)**. Thread the rear nut off the

16.35a Locate the gasket onto the dowels (arrowed) . . .

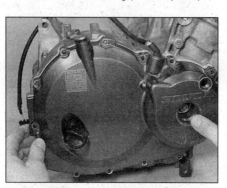

16.35b . . . and fit the cover

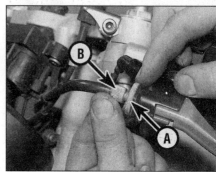

17.1 Fully slacken the lockring (A) and thread the adjuster (B) into the bracket

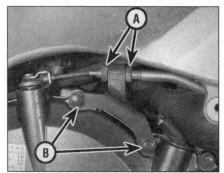

17.2a Slacken the locknuts (A), then remove the bracket bolts (B) . . .

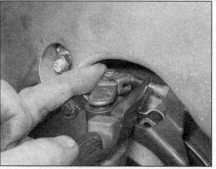

17.2b . . . and free the cable end from the release lever

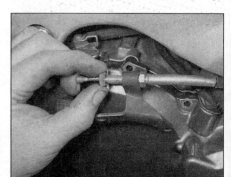

17.2c Unscrew the nut . . .

Engine, clutch and transmission 2•31

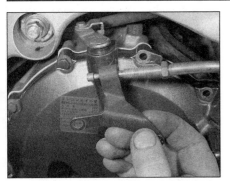

17.2d ... and slide the bracket off

17.3a Align the slots as shown ...

17.3b ... and detach the inner cable

cable and draw the bracket off **(see illustrations)**.

3 Align the slots in the adjuster and lockwheel at the handlebar end of the cable with that in the lever bracket, then pull the outer cable end from the socket in the adjuster and release the inner cable from the lever **(see illustrations)**. Remove the cable from the machine, noting its routing through the guide on the top yoke.

 Before removing the cable from the bike, tape the lower end of the new cable to the upper end of the old cable. Slowly pull the lower end of the old cable out, guiding the new cable down into position. Using this method will ensure the cable is routed correctly.

4 Installation is the reverse of removal. Apply grease to the cable ends. Make sure the cable is correctly routed. Adjust the amount of clutch lever freeplay (see Chapter 1).

18 Oil sump, oil strainer and pressure relief valve – removal, inspection and installation

Note: *The oil sump, strainer and pressure relief valve can be removed with the engine in the frame. If the engine has been removed, ignore the steps which don't apply.*

Removal

1 Remove the lower fairing (see Chapter 8).
2 Drain the engine oil (see Chapter 1).
3 Remove the exhaust system (see Chapter 4).
4 Unscrew the sump bolts, slackening them evenly in a criss-cross sequence to prevent distortion, and remove the sump **(see illustration)**. Discard the gasket as a new one must be used.
5 Unscrew the bolt(s) securing the oil strainer and pull it out of the oil pump **(see illustration)**. Remove the O-ring and discard it as a new one must be used.
6 The pressure relief valve is a push-fit into its socket in the crankcase. Pull it out, then discard the O-ring as a new one must be used **(see illustration)**.

Inspection

7 Remove all traces of gasket from the sump and crankcase mating surfaces, and clean the inside of the sump with solvent.
8 Clean the oil strainer in solvent and remove any debris caught in the mesh. Inspect the strainer for any signs of wear or damage and renew it if necessary.
9 Push the relief valve plunger into the valve body and check that it moves smoothly and freely against the spring pressure **(see illustration)**. If the valve operation is rough or sticky, remove the circlip, noting that it is under spring pressure, and withdraw the washer, spring and piston. Clean all the parts in solvent and inspect the plunger and bore for wear and damage. Apply oil to all parts and reassemble the valve. Check its operation again as above – if it is still rough or sticky it must be renewed – individual components are not available.

Installation

10 Fit a new O-ring onto the relief valve and smear it with clean oil, then push the valve into its socket in the crankcase **(see illustration 18.6)**.
11 Fit a new O-ring onto the oil strainer and smear it with clean oil, then align the bolt hole(s) and push the strainer into the oil pump

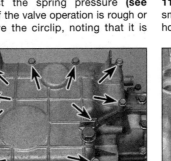

18.4 Unscrew the bolts (arrowed) and remove the sump

18.5 Unscrew the oil strainer bolt(s) (arrowed) and remove the strainer – L, N and R models shown

18.6 The pressure relief valve is a push-fit

18.9 Check the operation of the valve as described

18.11 Fit the strainer using a new O-ring

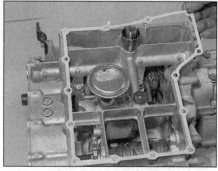

18.12 Fit a new gasket . . .

18.13 . . . then install the cover

(**see illustration**). Install the bolt(s) and tighten them.

12 Lay a new gasket onto the sump (if the engine is in the frame) or onto the crankcase (if the engine has been removed and is positioned upside down on the work surface) (**see illustration**). Make sure the holes in the gasket align correctly with the bolt holes.

13 Position the sump onto the crankcase and install the bolts finger-tight (**see illustration**). Tighten the bolts evenly in a criss-cross pattern.

14 Install the exhaust system (see Chapter 4).

15 Refill the engine with oil (see Chapter 1). Start the engine and check for leaks around the sump.

16 Install the lower fairing (see Chapter 8).

19.3 Lock the sprocket as shown and unscrew the bolt (arrowed)

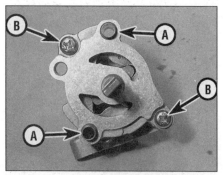

19.5 Remove the dowels (A) and the screws (B) and remove the cover

19 Oil pump – removal, inspection and installation

Note: *The oil pump can be removed with the engine in the frame. If the engine has been removed, ignore the steps which don't apply.*

Removal

1 Remove the sump and oil strainer (see Section 18).
2 Remove the clutch and the oil pump drive sprocket and chain (see Section 16).
3 Using a screwdriver or steel rod located as

19.4 Unscrew the two bolts (arrowed) and remove the pump

shown, unscrew the driven sprocket bolt and remove the sprocket from the pump (**see illustration**). Note the scribe marks on the back of the oil pump driven sprocket which must face inwards.

4 Unscrew the two bolts securing the pump to the crankcase, then remove the pump, noting how it fits (**see illustration**).

Inspection

5 Remove the dowels, then remove the screws securing the pump cover to the body and separate them (**see illustration**).
6 Withdraw the pump driveshaft along with the inner rotor, then remove the outer rotor from the pump body (**see illustration**). Retrieve the thrust washer if it isn't with the shaft. Note how the pin locates through the shaft and in the notches in the inner rotor, and how the punch mark on the outer rotor faces the pump cover. Clean all the components in solvent.
7 Inspect the pump body and rotors for scoring and wear (**see illustration**). If any damage, scoring or uneven or excessive wear is evident, renew the pump (individual components are not available). If the engine is being rebuilt, it is advisable to fit a new oil pump as a matter of course.
8 Measure the clearance between the inner rotor tip and the outer rotor with a feeler gauge and compare it to the maximum clearance listed in the specifications at the

19.6 Withdraw the driveshaft and inner rotor

19.7 Look for scoring and wear, such as on this outer rotor

Engine, clutch and transmission 2•33

19.8 Measuring inner rotor tip-to-outer rotor clearance

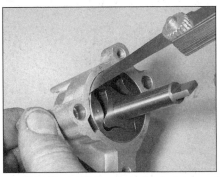
19.9 Measuring outer rotor-to-body clearance

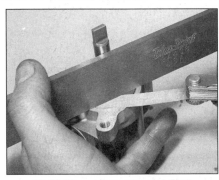
19.10 Measuring rotor end-float

beginning of the Chapter **(see illustration)**. If the clearance measured is greater than the maximum listed, renew the pump.

9 Measure the clearance between the outer rotor and the pump body with a feeler gauge and compare it to the maximum clearance listed in the specifications at the beginning of the Chapter **(see illustration)**. If the clearance measured is greater than the maximum listed, renew the pump.

10 Lay a straight-edge across the rotors and the pump body and, using a feeler gauge, measure the rotor end-float (the gap between the rotors and the straight-edge **(see illustration)**. If the clearance measured is greater than the maximum listed, renew the pump.

11 Check the pump drive chain and sprockets for wear or damage, and renew them as a set if necessary.

12 If the pump is good, make sure all the components are clean, then lubricate them with new engine oil.

13 Install the outer rotor into the pump body with the punch mark facing out towards the pump cover **(see illustration)**.

14 Fit the drive pin into the pump driveshaft and slide the inner rotor onto the thin-tabbed end of the shaft, making sure the pin ends fit into the notches in the bottom of the rotor. Slide the thrust washer onto the thick-tabbed end of the shaft and against the bottom of the inner rotor. Fit the shaft and inner rotor into the pump body **(see illustration)**.

15 Install the cover and tighten the screws securely, then fit the dowels into the pump body **(see illustrations)**.

Installation

16 Position the water pump and oil pump driveshaft slots vertically so that they engage easily when the pump is installed.

17 Manoeuvre the pump into position, making sure it engages correctly with the water pump and the dowels locate correctly **(see illustration)**. Fit the pump mounting bolts and tighten them securely **(see illustration)**.

18 Fit the driven sprocket onto the oil pump, making sure that the scribe lines face the

19.13 Fit the outer rotor with its punch mark (arrowed) facing out

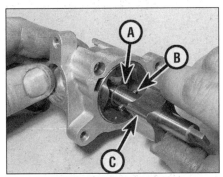
19.14 Slide the shaft with the thrust washer (A), drive pin (B) and inner rotor (C) into the pump

19.15a Fit the cover . . .

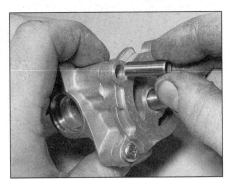
19.15b . . . then push in the dowels

19.17a Align the drive tab with the slot when fitting the pump . . .

19.17b . . . then install the bolts

2•34 Engine, clutch and transmission

19.18a Fit the sprocket ...

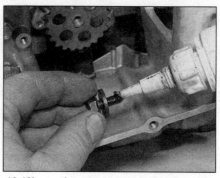
19.18b ... then apply a thread-lock to the bolt ...

19.18c ... and tighten it to the specified torque

engine **(see illustration)**. Clean the bolt threads, then apply a suitable non-permanent thread locking compound and tighten the bolt to the torque setting specified at the beginning of the chapter, using a screwdriver as before to prevent the sprocket turning **(see illustrations)**.
19 Install the oil pump chain and driven sprocket and the clutch (see Section 16).
20 Install the oil strainer and sump (see Section 18).

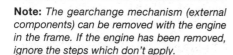

20 Gearchange mechanism – removal, inspection and installation

Note: *The gearchange mechanism (external components) can be removed with the engine in the frame. If the engine has been removed, ignore the steps which don't apply.*

J and K models

Removal

1 Make sure the transmission is in neutral. Remove the clutch and the oil pump drive sprocket and chain (see Section 16).
2 Unscrew the gearchange lever linkage arm pinch bolt and remove the arm from the shaft, noting the alignment punch marks **(see illustration 20.22)**. If no marks are visible, make your own before removing the arm so that it can be correctly aligned with the shaft on installation.

3 Wrap some insulating tape around the gearchange shaft splines to avoid damaging the oil seal as it is removed.
4 Note how the gearchange shaft centralising spring ends fit on each side of the locating pin in the casing, and how the eye of the selector arm locates over the collar on the drum shift assembly **(see illustration 20.18b)**. Withdraw the gearchange shaft and its thrust washer from the casing **(see illustration)**. Remove the collar from the pin on the drum selector assembly **(see illustration)**.
5 Unscrew the two bolts securing the drum selector guide plate, then remove the plate along with the drum selector assembly, noting that the pawls of the selector assembly are under spring pressure **(see illustration)**. Use

your fingers to prevent them from pinging out. Note the correct fitted position of all components. Remove the spacer from the guide plate upper mounting dowel.
6 Press the stopper arm down off the stopper plate on the selector drum and remove the arm with its return spring and collar, noting how they fit **(see illustrations 20.14e and d)**. Also remove the thrust washer and, if required, the dowel **(see illustrations 20.14b and a)**.
7 If required, unscrew the pin bolt securing the drum selector cam to the selector drum, then remove the cam **(see illustrations)**. Note the locating pin between the selector cam and the selector drum and remove it for safe keeping if required.

20.4a Withdraw the shaft ...

20.4b ... and remove the collar

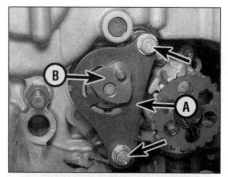
20.5 Unscrew the bolts (arrowed) and remove the plate (A) and drum selector assembly (B)

20.7a Unscrew the bolt (arrowed) ...

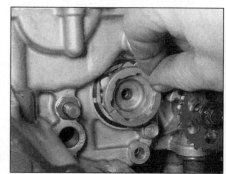
20.7b ... then remove the cam

Engine, clutch and transmission 2•35

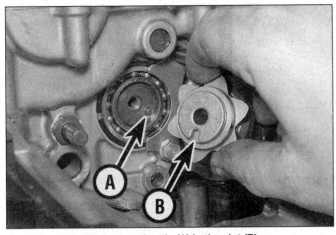

20.12a Locate the pin (A) in the slot (B) . . .

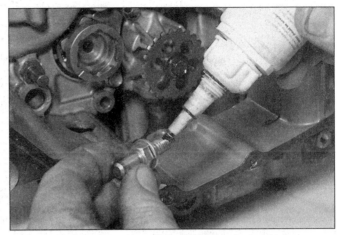

20.12b . . . and install the bolt using thread-lock

Inspection

8 Inspect the stopper arm return spring and the shaft centralising spring. If they are fatigued, worn or damaged they must be renewed. Also check that the centralising spring locating pin in the casing is securely tightened. If it is loose, remove it and apply a suitable non-permanent thread-locking compound, then tighten it to the specified torque.

9 Check the gearchange shaft for straightness and damage to the splines. If the shaft is bent you can attempt to straighten it, but if the splines are damaged the shaft must be renewed. Also check the condition of the shaft oil seal in the casing and renew it if damaged or deteriorated. Lever the old seal out using a screwdriver and drive the new seal in squarely **(see illustrations 20.30a and b)**.

10 Inspect the selector arm eye and the drum selector collar, and the stopper arm roller and the stopper plate. If they are worn or damaged they must be renewed.

11 Check the drum selector, pawls, pins and springs for wear and damage. Renew them if defects are found.

Installation

12 If removed, install the locating pin in the offset hole in the end of the selector drum. Install the drum selector cam, locating the slot in its back over the locating pin in the selector drum **(see illustration)**. Clean the threads of the pin bolt, then apply a suitable non-permanent thread-locking compound **(see illustration)**. Install the pin bolt and tighten it securely.

13 Fit the spacer onto the guide plate upper mounting dowel.

14 If removed, fit the stopper arm dowel into the lower bolt hole, and slide the thrust washer onto it **(see illustrations)**. Fit the collar into the stopper arm **(see illustration)**. Fit the stopper arm assembly onto the lower mounting dowel, positioning the stopper arm onto the neutral detent on the stopper plate (identified by the line on the face of the cam) then fit the spring, making sure the ends are located correctly over the stopper arm and against the casing **(see illustrations)**. Make sure the stopper arm is free to move and is returned by the pressure of the spring.

15 If the drum selector assembly was disassembled, install the springs, pins and pawls, making sure the rounded end of each pawl fits into the rounded cut-out in the drum

20.14a Install the dowel . . .

20.14b . . . and slide the washer (arrowed) onto it

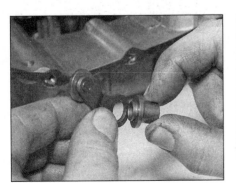

20.14c Fit the collar into the arm . . .

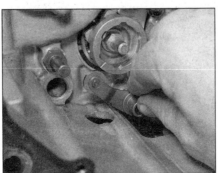

20.14d . . . and slide them onto the dowel . . .

20.14e . . . then position the arm and fit the spring

2•36 Engine, clutch and transmission

20.15a Drum selector pawl assembly

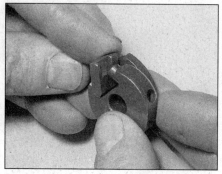

20.15b Note that the pawl cut-out is offset – ensure they are fitted correctly

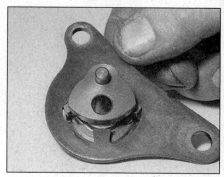

20.16a Fit the drum selector into the guide plate . . .

selector, and that the pins locate correctly in the cut-outs in the pawls **(see illustrations)**.

16 Depress the pawls and fit the drum selector assembly into the guide plate **(see illustration)**. Fit the assembly into the selector cam, locating the pin bolt through the hole in the drum selector **(see illustration)**. Make sure the pawls locate correctly in the cutouts in the cam. Clean the threads of the guide plate bolts, then apply a suitable non-permanent thread locking compound and tighten them securely **(see illustration)**.

17 Fit the collar onto the pin on the drum selector assembly **(see illustration 20.4b)**.

18 Check that the gearchange shaft centralising spring is correctly positioned, then fit the thrust washer onto the end of the shaft **(see illustration)**. Fit the shaft into its

hole in the casing **(see illustration 20.4a)**. Make sure the centralising spring ends are correctly located on each side of the tab on the arm and the pin in the casing, and that the eye of the arm locates around the collar **(see illustration)**. Remove the tape from around the splines.

19 Install the gearchange linkage arm onto the end of the shaft on the left-hand side of the engine, aligning the punch marks, and check that the mechanism works correctly **(see illustrations 20.22)**. Tighten the pinch bolt.

20 Install the oil pump drive sprocket and chain, and the clutch (see Section 16).

L, N and R models
Removal

21 Make sure the transmission is in neutral.

Remove the clutch, the oil pump drive sprocket and chain (see Section 16).

22 Unscrew the gearchange lever linkage arm pinch bolt and remove the arm from the shaft, noting the alignment punch marks **(see illustration)**. If no marks are visible, make your own before removing the arm so that it can be correctly aligned with the shaft on installation.

23 Wrap some insulating tape around the gearchange shaft splines to avoid damaging the oil seal as it is removed.

24 Note how the gearchange shaft centralising spring ends fit on each side of the locating pin in the crankcase and how the selector arm locates onto the pins on the stopper plate on the end of the selector drum. Grasp the end of the shaft and withdraw the shaft/arm assembly **(see illustration)**. If the

20.16b . . . then fit the assembly . . .

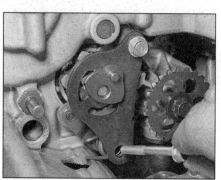

20.16c . . . and install the bolts

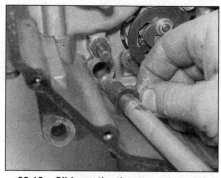

20.18a Slide on the thrust washer and insert the shaft . . .

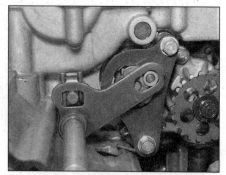

20.18b . . . making sure it locates as shown

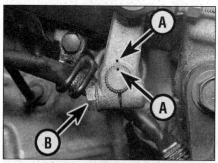

20.22 Note the alignment of the punch marks (A) then remove the bolt (B) and slide the arm off the shaft

20.24 Draw the shaft/arm off the selector drum and out of the engine

Engine, clutch and transmission 2•37

20.25 Stopper arm bolt (A), stopper plate bolt (B)

20.30a Lever out the old seal . . .

20.30b . . . and press or drive in a new one

thrust washer doesn't come away with the shaft, retrieve it from the crankcase.

25 Note how the stopper arm spring ends locate and how the arm itself locates in the neutral detent on the stopper plate, then unscrew the stopper arm bolt and remove the arm and spring **(see illustration)**.

26 If required, unscrew the bolt securing the stopper plate to the selector drum and remove the plate **(see illustration 20.25)**. Note the locating pin between the plate and the selector drum and remove it for safe keeping if required.

Inspection

27 Check the condition of the selector arm centralising spring and stopper arm return spring and renew them if they are cracked, weakened or distorted.

28 Check the selector arm for cracks, distortion and wear of its pawls, and check for any corresponding wear on the selector pins on the stopper plate. Also check the stopper arm roller and the stopper plate for any wear or damage, and make sure the roller turns freely. Renew any components that are worn or damaged.

29 Check that the centralising spring locating pin in the crankcase is securely tightened. If it is loose, remove it and apply a non-permanent thread locking compound to its threads, then tighten it to the torque setting specified at the beginning of the Chapter.

30 Check the gearchange shaft for straightness and damage to the splines. If the shaft is bent you can attempt to straighten it, but if the splines are damaged the shaft must be renewed. Also check the condition of the shaft oil seal in the crankcase. If it is damaged, deteriorated or shows signs of leakage it must be renewed. Lever out the old seal and press a new one squarely into place, with its lip facing inward, using thumb pressure, a seal driver or suitable socket **(see illustrations)**.

Installation

31 If removed, fit the stopper plate locating pin into the selector drum. Align the cutout in the back of the stopper plate with the locating pin and install the plate, then apply a suitable non-permanent thread locking compound to the bolt and tighten it securely.

32 Install the stopper arm return spring, washer, arm and bolt, locating the arm onto the neutral detent on the stopper plate, and tighten the bolt securely **(see illustration)**. Make sure the spring is positioned correctly **(see illustration 20.25)**.

33 Check that the gearchange shaft centralising spring is correctly positioned, then fit the thrust washer onto the end of the shaft **(see illustration)**. Apply some grease to the lips of the gearchange shaft oil seal. Slide the shaft into place and push it all the way through the case until the splined end comes out the other side, lifting the selector arm slightly as it is inserted and engage it with the pins on the stopper plate **(see illustration 20.24)**. Make sure the centralising spring ends locate correctly on each side of the locating pin **(see illustration)**. Remove the tape from the shaft splines.

34 Install the gearchange linkage arm onto the end of the shaft on the left-hand side of the engine, aligning the punch marks, and check that the mechanism works correctly **(see illustrations 20.22)**. Tighten the pinch bolt.

35 Install the oil pump drive sprocket and chain, and the clutch (see Section 16).

21 Crankcase halves – separation and reassembly

Note 1: *To separate the crankcase halves, the engine must be removed from the frame.*
Note 2: *If the crankcases are being separated to inspect the crankshaft without removing it, or to remove the crankshaft without removing the connecting rods and pistons, or to inspect or remove the transmission shafts, the cylinder head can remain in situ. However, if removal of the connecting rod assemblies is intended, full disassembly of the top-end is necessary. The gearchange mechanism external*

20.32 Fit the arm, washer and spring onto the bolt, locating the spring end in the cutout in the arm (arrowed), and install the assembly

20.33a Check the spring ends are correctly positioned and the washer is on the shaft

20.33b Gearchange mechanism correctly installed

2

2•38 Engine, clutch and transmission

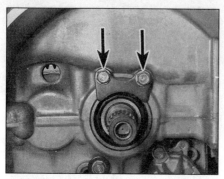

21.3 Unscrew the bolts (arrowed) and remove the plate

components can also remain in situ unless the selector drum and forks are being removed.

Separation

1 To access the pistons (L, N and R models only), connecting rods, crankshaft, bearings, transmission shafts and the selector drum and forks, the crankcase must be split into two parts.

2 To enable the crankcases to be separated, the engine must be removed from the frame (see Section 5). Before the crankcases can be separated the following components must be removed:

a) Valve cover (Section 8) (see **Note 2**).
b) Camshafts (Section 9) (see **Note 2**).
c) Cylinder head (Section 10) (see **Note 2**).
d) Camshaft drive gear assembly (Section 13) (see **Note 2**).
e) Cylinder block (J and K models) (Section 14) (see **Note 2**).
f) Starter clutch (Section 15).
g) Clutch (Section 16).
h) Water pump (Chapter 3).
i) Gearchange mechanism external components (Section 20) (see **Note 2**).
j) Oil sump (Section 18).
k) Oil pump (Section 19).
l) Alternator rotor (Chapter 9).

3 Unscrew the two bolts securing the input shaft bearing retainer plate to the right-hand side of the crankcase and remove the plate **(see illustration)**.

4 Unscrew the six (J and K models) or seven (L, N and R models) 6 mm upper crankcase bolts, noting their different lengths **(see illustrations)**. Unscrew the bolts evenly, a little at a time and in a **reverse** of the numerical tightening sequence until they are finger-tight, then remove them. **Note:** As each bolt is removed, store it in its relative position in a cardboard template of the crankcase halves. This will ensure all bolts are installed in their correct locations on reassembly. Also note the washers fitted with certain bolts, whose location should be identified by a triangle cast into the crankcase.

5 Turn the engine upside down.

6 Unscrew the single 8 mm bolt from the rear left-hand corner of the crankcase (the bolt will actually be on your right as you are looking down on the underside of the engine) **(see illustration 21.4a or b)**.

7 Unscrew the thirteen 6 mm lower crankcase bolts evenly, a little at a time and in a **reverse** of the numerical tightening sequence, until they are finger-tight, then

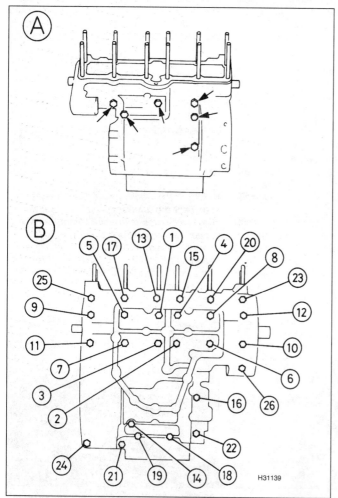

21.4a Upper crankcase bolts (A), lower crankcase bolts (B) – J and K models

Numbers indicate the tightening sequence

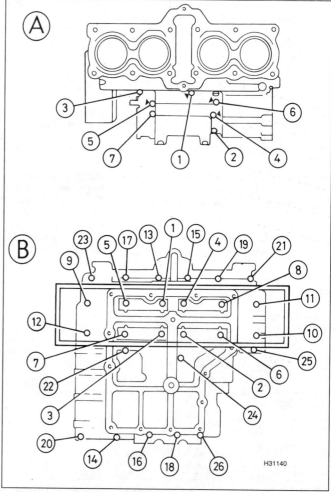

21.4b Upper crankcase bolts (A), lower crankcase bolts (B) – L, N and R models

Numbers indicate the tightening sequence

Engine, clutch and transmission 2•39

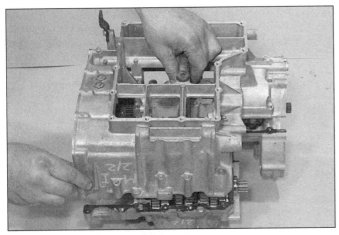

21.9 Carefully separate the halves

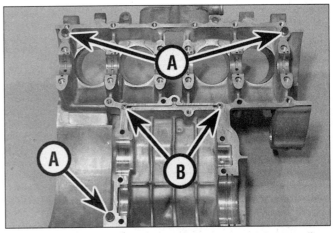

21.10 Remove the three dowels (A) if loose, and the two oil nozzles (B)

remove them **(see illustration 21.4a or b)**. **Note:** *As each bolt is removed, store it in its relative position in a cardboard template of the crankcase halves. This will ensure all bolts are installed in their correct locations on reassembly.*

8 Unscrew the twelve 8 mm lower crankcase (crankshaft journal) bolts evenly, a little at a time and in a **reverse** of the numerical tightening sequence, until they are finger-tight, then remove them, along with the washers **(see illustration 21.4a or b, bolts numbered 1 to 12)**. **Note:** *As each bolt is removed, store it in its relative position in a cardboard template of the crankcase halves. This will ensure all bolts are installed in their correct locations on reassembly.*

9 Carefully lift the lower crankcase half off the upper half, using a soft-faced hammer to tap around the joint to initially separate the halves if necessary **(see illustration)**. **Note:** *If the halves do not separate easily, make sure all fasteners have been removed. Do not try and separate the halves by levering against the crankcase mating surfaces as they are easily scored and will leak oil. Tap around the joint faces with a soft-faced mallet.* The lower crankcase half will come away with the gearchange mechanism external components (if not already removed) and the selector drum and forks, leaving the crankshaft and transmission shafts in the upper crankcase half.

10 If required, remove the three locating dowels from the crankcase if they are loose (they could be in either crankcase half), noting their locations **(see illustration)**. Similarly remove the two oil nozzles, noting which way up they fit.

11 Refer to Sections 22 to 30 for the removal, inspection and installation of the components housed within the crankcases.

Reassembly

12 Remove all traces of sealant from the crankcase mating surfaces.
13 Ensure that all components and their bearings are in place in the upper and lower crankcase halves. If the transmission shafts have not been removed, check the condition of the output shaft oil seal on the left-hand end of the shaft and renew it if it is damaged or deteriorated **(see illustration 28.4a)**; it is sound practice to renew this seal anyway. Apply some grease to the lips of the new seal on installation **(see illustration 28.4b)**.

14 Generously lubricate the crankshaft and transmission shafts, particularly around the bearings, with clean engine oil, then use a rag soaked in high flash-point solvent to wipe over the mating surfaces of both crankcase halves to remove all traces of oil.

15 If removed, install the three locating dowels and the two oil nozzles in the upper crankcase half **(see illustration 21.10)**. The oil nozzles must be fitted with the wider aperture facing out.

16 Apply a small amount of suitable sealant to the outer mating surface of the upper crankcase half.

Caution: *Do not apply an excessive amount of sealant as it will ooze out when the case halves are assembled and may obstruct oil passages. Do not apply the sealant on or too close to any of the bearing inserts or surfaces.*

17 Check again that all components are in position, particularly that the bearing shells are still correctly located in the lower crankcase half. Carefully install the lower crankcase half down onto the upper crankcase half, making sure each selector fork locates correctly in the groove in its pinion, and the dowels all locate correctly into the lower crankcase half **(see illustration 21.9)**.

18 Check that the lower crankcase half is correctly seated. **Note:** *The crankcase halves should fit together without being forced. If the casings are not correctly seated, remove the lower crankcase half and investigate the problem. Do not attempt to pull them together using the crankcase bolts as the casing will crack and be ruined.*

19 Clean the threads of the twelve 8 mm lower crankcase (crankshaft journal) bolts and insert them with their washers in their original locations **(see illustration 21.4a or b)**. Secure all bolts finger-tight at first, then tighten them evenly and a little at a time in the correct numerical sequence to the torque setting specified at the beginning of the Chapter.

20 Clean the threads of the thirteen 6 mm lower crankcase bolts and insert them in their original locations **(see illustration 21.4a or b)**. Secure all bolts finger-tight at first, then tighten them evenly a little at a time in the correct numerical sequence to the torque setting specified at the beginning of the Chapter.

21 Clean the threads of the single 8 mm lower crankcase bolt and insert it into the rear left-hand corner of the crankcase **(see illustration 21.4a or b)**. Tighten the bolt to the specified torque.

22 Turn the engine over. Clean the threads of the six (J and K models) or seven (L, N and R models) 6 mm upper crankcase bolts and insert them in their original locations, not forgetting the washers with the bolts whose locations are identified by a triangle cast into the crankcase **(see illustration 21.4a or b)**. Secure all bolts finger-tight at first, then tighten them evenly a little at a time in the numerical sequence to the torque setting specified at the beginning of the Chapter.

23 With all crankcase fasteners tightened, check that the crankshaft and transmission shafts rotate smoothly and easily. Check that the transmission shafts rotate freely and independently in neutral, then rotate the selector drum by hand and select each gear in turn whilst rotating the input shaft. Check that all gears can be selected and that the shafts rotate freely in every gear. If there are any signs of undue stiffness, tight or rough spots, or of any other problem, the fault must be rectified before proceeding further.

24 Install the transmission input shaft bearing retainer plate onto the right-hand side of the crankcase. Apply a suitable non-permanent

thread-locking compound to the threads of the bolts and tighten them to the specified torque setting **(see illustration 21.3)**.
25 Install all other removed assemblies in the reverse of the sequence given in Step 2.

22 Crankcase halves and cylinder bores – inspection and servicing

Crankcase halves

1 After the crankcases have been separated, remove the crankshaft, connecting rods, transmission shafts, selector drum and forks, and, if required, the oil pressure switch, neutral switch and cooling system union, referring to the relevant Sections of this Chapter, to Chapter 9 for the oil pressure switch and neutral switch, and to Chapter 3 for the coolant union. On J and K models, if required, unscrew the bolt securing the transmission input shaft needle bearing cage retainer in the bore in the crankcase, then remove the retainer and draw out the cage.
2 The crankcases should be cleaned thoroughly with new solvent and dried with compressed air. All oil passages should be blown out with compressed air.
3 All traces of old gasket sealant should be removed from the mating surfaces. Minor damage to the surfaces can be cleaned up with a fine sharpening stone or grindstone.
Caution: Be very careful not to nick or gouge the crankcase mating surfaces or oil leaks will result. Check both crankcase halves very carefully for cracks and other damage.
4 Small cracks or holes in aluminium castings may be repaired with an epoxy resin adhesive as a temporary measure. Permanent repairs can only be effected by argon-arc welding, and only a specialist in this process is in a position to advise on the economy or practical aspect of such a repair. Note that there are, however, kits available for low temperature welding. If any damage is found that can't be repaired, renew the crankcase halves as a set.
5 Damaged threads can be economically reclaimed using a thread insert, which is easily fitted after drilling and re-tapping the affected thread. There are a few types of thread insert available, of varying quality and cost – consult your dealer for his recommendation.
6 Sheared studs or bolts can usually be removed with stud or screw extractors. A stud extractor should be used if the stud is above the surface. Otherwise use a screw extractor, which consists of a tapered, left thread screw of very hard steel. These are inserted into a hole pre-drilled centrally in the stud, and usually succeed in dislodging the most stubborn stud or screw.

Refer to Tools and Workshop Tips for details of installing a thread insert and using screw extractors.

7 Install all components and assemblies, referring to the relevant Sections of this Chapter and to Chapters 9 and 3, before reassembling the crankcase halves.

Cylinder bores

8 Do not attempt to separate the cylinder liners from the cylinder block.
9 Check the cylinder walls carefully for scratches and score marks.
10 Using a precision straight-edge and a feeler gauge set to the warpage limit listed in the specifications at the beginning of the Chapter, check the block gasket mating surface for warpage. Refer to *Tools and Workshop Tips* in the Reference section for details of how to use the straight-edge. If warpage is excessive the block/crankcase must be renewed.
11 Using telescoping gauges and a micrometer (see *Tools and Workshop Tips*), check the dimensions of each cylinder to assess the amount of wear, taper and ovality. Measure near the top (but below the level of the top piston ring at TDC), centre and bottom (but above the level of the oil ring at BDC) of the bore, both parallel to and across the crankshaft axis **(see illustration)**. Compare the results to the specifications at the beginning of the Chapter. If the cylinders are worn, oval or tapered beyond the service limit, or badly scratched, scuffed or scored, the cylinder block (J and K models) or crankcases (L, N and R models) must be renewed – oversize pistons are not available, precluding the cylinders being rebored.
12 If the precision measuring tools are not available, take the block (J and K models) or upper crankcase (L, N and R models) to a dealer or motorcycle engineer for assessment and advice.
13 If the block and cylinders are in good condition and the piston-to-bore clearance is within specifications (see Section 25), the cylinders should be honed (de-glazed). To perform this operation you will need the proper size flexible hone with fine stones, or a bottle-brush type hone, plenty of light oil or honing oil, some clean rags and an electric drill motor.
14 Clamp the block/crankcase securely so that the bores are horizontal rather than vertical. Mount the hone in the drill motor, compress the stones and insert the hone into the cylinder. Thoroughly lubricate the cylinder, then turn on the drill and move the hone up and down in the cylinder at a pace which produces a fine cross-hatch pattern on the cylinder wall with the lines intersecting at an angle of approximately 60°. Be sure to use plenty of lubricant and do not take off any more material than is necessary to produce the desired effect. Do not withdraw the hone from the cylinder while it is still turning. Switch off the drill and continue to move it up and down in the cylinder until it has stopped turning, then compress the stones and withdraw the hone. Wipe the oil from the

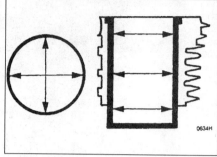

22.11 Measure the cylinder bore in the directions shown with a telescoping gauge, then measure the gauge with a micrometer

cylinder and repeat the procedure on the other cylinder. Remember, do not take too much material from the cylinder wall.
15 Wash the cylinders thoroughly with warm soapy water to remove all traces of the abrasive grit produced during the honing operation. Be sure to run a brush through the oil and coolant passages and flush them with running water. After rinsing, dry the cylinders thoroughly and apply a thin coat of light, rust-preventative oil to all machined surfaces.
16 If you do not have the equipment or desire to perform the honing operation, take the block to a dealer or motorcycle engineer.

23 Main and connecting rod bearings – general information

1 Even though main and connecting rod bearings are generally renewed during the engine overhaul, the old bearings should be retained for close examination as they may reveal valuable information about the condition of the engine.
2 Bearing failure occurs mainly because of lack of lubrication, the presence of dirt or other foreign particles, overloading the engine and/or corrosion. Regardless of the cause of bearing failure, it must be corrected before the engine is reassembled to prevent it from happening again.
3 When examining the connecting rod bearings, remove them from the connecting rods and caps and lay them out on a clean surface in the same general position as their location on the crankshaft journals. This will enable you to match any noted bearing problems with the corresponding crankshaft journal.
4 Dirt and other foreign particles get into the engine in a variety of ways. It may be left in the engine during assembly or it may pass through filters or breathers. It may get into the oil and from there into the bearings. Metal chips from machining operations and normal engine wear are often present. Abrasives are sometimes left in engine components after reconditioning operations, especially when parts are not thoroughly cleaned using the

proper cleaning methods. Whatever the source, these foreign objects often end up imbedded in the soft bearing material and are easily recognised. Large particles will not imbed in the bearing and will score or gouge the bearing and journal. The best prevention for this cause of bearing failure is to clean all parts thoroughly and keep everything spotlessly clean during engine reassembly. Frequent and regular oil and filter changes are also recommended.

5 Lack of lubrication or lubrication breakdown has a number of interrelated causes. Excessive heat (which thins the oil), overloading (which squeezes the oil from the bearing face) and oil leakage or throw off (from excessive bearing clearances, worn oil pump or high engine speeds) all contribute to lubrication breakdown. Blocked oil passages will also starve a bearing and destroy it. When lack of lubrication is the cause of bearing failure, the bearing material is wiped or extruded from the steel backing of the bearing. Temperatures may increase to the point where the steel backing and the journal turn blue from overheating.

 Refer to Tools and Workshop Tips for bearing fault finding.

6 Riding habits can have a definite effect on bearing life. Full throttle low speed operation, or labouring the engine, puts very high loads on bearings, which tend to squeeze out the oil film. These loads cause the bearings to flex, which produces fine cracks in the bearing face (fatigue failure). Eventually the bearing material will loosen in pieces and tear away from the steel backing. Short trip riding leads to corrosion of bearings, as insufficient engine heat is produced to drive off the condensed water and corrosive gases produced. These products collect in the engine oil, forming acid and sludge. As the oil is carried to the engine bearings, the acid attacks and corrodes the bearing material.

7 Incorrect bearing installation during engine assembly will lead to bearing failure as well. Tight fitting bearings which leave insufficient bearing oil clearances result in oil starvation. Dirt or foreign particles trapped behind a bearing insert result in high spots on the bearing which lead to failure.

8 To avoid bearing problems, clean all parts thoroughly before reassembly, double check all bearing clearance measurements and lubricate the new bearings with clean engine oil during installation.

24 Connecting rods – removal, inspection and installation

Note: *To remove the connecting rods the engine must be removed from the frame and the crankcases separated.*

Removal

1 Remove the engine from the frame (see Section 5). On J and K models, remove the cylinder block (see Section 14), and if required the pistons (see Section 25). On all models, separate the crankcase halves (see Section 21).

2 Before removing the rods from the crankshaft, measure the big-end side clearance with a feeler gauge **(see illustration)**. If the clearance between any rod is greater than the service limit listed in this Chapter's Specifications, renew that rod.

3 Using paint or a felt marker pen, mark the relevant cylinder identity on each connecting rod and cap. Mark across the cap-to-connecting rod join and note which side of the rod faces the front of the engine to ensure that the cap and rod are fitted the correct way around on reassembly. Note that the number already across the rod and cap indicates rod size grade, not cylinder number, and the letter indicates rod weight grade. The oil hole in the big-end of each connecting rod should face the back of the engine.

4 On J and K models, unscrew the big-end cap nuts and separate the caps from the crankpins **(see illustration 24.5)**. Detach the connecting rods from the crankshaft and remove them from the top of the crankcase. Keep the rods, caps, nuts and (if they are to be reused) the bearing shells together in their correct positions to ensure correct installation. Do not remove the bolts from the connecting rods.

5 On L, N and R models, unscrew the big-end cap nuts and separate the caps from the crankpins **(see illustration)**. Detach the connecting rods from the crankshaft and lift the crankshaft out of the upper crankcase half, taking care not to dislodge the main bearing shells **(see illustration 27.3)**. Turn the crankcase on its side, then push each piston/connecting rod assembly up and remove it from the top of the bore, making sure the connecting rod does not mark the cylinder bore walls. Keep the rods, caps, nuts and (if they are to be reused) the bearing shells together in their correct positions to ensure correct installation. Do not remove the bolts from the connecting rods.

 On L, N and R models, to ease removal of the pistons, remove any ridge of carbon built up on the top of each cylinder bore. If there is a pronounced wear ridge, remove it using a ridge reamer.

Caution: Do not try to remove the piston/connecting rod from the bottom of the cylinder bore. The piston will not pass the crankcase main bearing webs. If the piston is pulled right to the bottom of the bore the oil control ring will expand and lock the piston in position. If this happens it is likely the ring will be broken.

6 Immediately install the relevant bearing shells (if removed), bearing cap, and nuts on each connecting rod assembly so that they are all kept together as a matched set.

7 If required, and if not already done on J and K models, remove the pistons from the connecting rods (see Section 25).

Inspection

8 Check the connecting rods for cracks and other obvious damage.

9 Apply clean engine oil to the piston pin, insert it into the connecting rod small-end and check for any freeplay between the two **(see illustration)**. Measure the pin external

24.2 Measure the connecting rod side clearance using a feeler gauge

24.5 Unscrew the nuts (arrowed) and remove the connecting rod cap

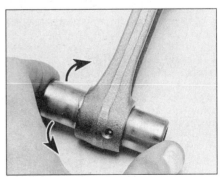

24.9a Slip the piston pin into the rod's small-end and rock it back and forth to check for looseness

2•42 Engine, clutch and transmission

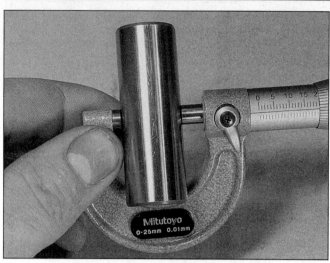

24.9b Measure the external diameter of the pin . . .

24.9c . . . and the internal diameter of the connecting rod small-end

diameter and the small-end bore diameter, then calculate the difference to obtain the small-end-to-piston pin clearance **(see illustrations)**. Compare the result to the specifications at the beginning of the Chapter. If the clearance is greater than specified, renew the components that are worn beyond their specified limits.

10 Refer to Section 23 and examine the connecting rod bearing shells. If they are scored, badly scuffed or appear to have seized, new shells must be installed. Always renew the shells in the connecting rods as a set. If they are badly damaged, check the corresponding crankpin. Evidence of extreme heat, such as discoloration, indicates that lubrication failure has occurred. Be sure to thoroughly check the oil pump and pressure regulator as well as all oil holes and passages before reassembling the engine.

11 Have the rods checked for twist and bend by a dealer if you are in doubt about their straightness.

Oil clearance check

12 Whether new bearing shells are being fitted or the original ones are being re-used, the connecting rod bearing oil clearance should be checked prior to reassembly. If not already done, remove the crankshaft from the crankcase **(see illustration 27.3)**.

13 Clean the backs of the bearing shells and the bearing locations in both the connecting rod and cap.

14 Press the bearing shells into their locations, ensuring that the tab on each shell engages the notch in the connecting rod/cap **(see illustration)**. Make sure the bearings are fitted in the correct locations and take care not to touch any shell's bearing surface with your fingers.

15 Cut a length of the appropriate size Plastigauge (it should be slightly shorter than the width of the crankpin). Place a strand of Plastigauge on the (cleaned) crankpin journal and fit the (clean) connecting rod, shells and cap. Make sure the cap is fitted the correct way around so the previously made markings align, and that the rod is facing the right way (see Step 3). Tighten the cap nuts evenly, in two or three stages, to the torque setting specified at the beginning of the Chapter, whilst ensuring that the connecting rod does not rotate on the crankshaft. Slacken the cap nuts and remove the connecting rod, again taking great care not to rotate the rod or crankshaft.

16 Compare the width of the crushed Plastigauge on the crankpin to the scale printed on the Plastigauge envelope to obtain the connecting rod bearing oil clearance **(see illustration 27.18)**. Compare the reading to the specifications at the beginning of the Chapter.

17 On completion carefully scrape away all traces of the Plastigauge material from the crankpin and bearing shells using a fingernail or other object which is unlikely to score the shells.

18 If the clearance is within the range listed in this Chapter's Specifications and the bearings are in perfect condition, they can be reused. If the clearance is beyond the service limit, renew the bearing shells (see Steps 21 and 22). Check the oil clearance once again (the new shells may be thick enough to bring bearing clearance within the specified range). Always renew all of the inserts at the same time.

19 If the clearance is still greater than the service limit listed in this Chapter's Specifications, the crankpin is worn and the crankshaft should be renewed.

20 Repeat the bearing selection procedure for the remaining connecting rods.

Bearing shell selection

21 New bearing shells for the big-end bearings are supplied on a selected fit basis. Code letters and numbers stamped on various components are used to identify the correct new bearings. The crankpin journal size letters are stamped on the outside of the left-hand crankshaft web and will be either an A or a B **(see illustration)**. From left-to-right, the letters represent cylinders 1 to 4 respectively. The connecting rod size code number is marked on the flat face of the

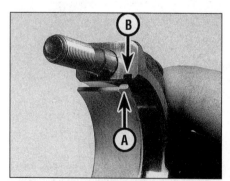

24.14 Make sure the tab (A) locates in the notch (B)

24.21a Crankpin journal size letters (arrowed)

Engine, clutch and transmission 2•43

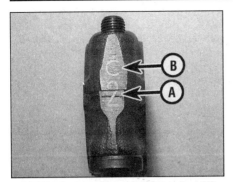

24.21b Connecting rod size number (A) and weight code (B)

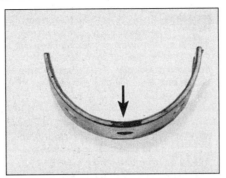

24.22 Bearing shell colour code location (arrowed)

24.26a Clamp the rings in position by tightening the bands on the compressor . . .

Connecting rod selection

23 If a connecting rod needs to be renewed, the weight of the new rod needs to be matched to the rod it is replacing. The connecting rod weight code is marked on the flat face of the connecting rod and cap and will be either an A, B, or C **(see illustration 24.21b)**.

Installation

24 On L, N and R models, and if required on J and K models (it can be done later), install the pistons onto the connecting rods (see Section 25).

25 Install the bearing shells in the connecting rods and caps, aligning the notch in the bearing with the groove in the rod or cap **(see illustration 24.14)**.

26 On L, N and R models, lubricate the pistons, rings and cylinder bore with clean engine oil. Insert the piston/connecting rod assembly into the top of its bore, taking care not to allow the connecting rod to mark the bore. Make sure the IN mark on the piston crown is on the inlet side of the bore and the connecting rod is the right way round (see Step 3), then carefully compress and feed

each piston ring into the bore until the piston crown is flush with the top of the bore. If available, a piston ring compressor makes installation a lot easier **(see illustrations)**. Turn the crankcase upside down. Lubricate the shells with molybdenum disulphide oil (a 50/50 mixture of molybdenum disulphide grease and clean engine oil), then lower the crankshaft into position in the upper crankcase, making sure all bearings remain in place **(see illustration 27.23 and 27.3)**.

27 Ensure that the connecting rod bearing insert is still correctly installed. Liberally lubricate the crankpin with molybdenum disulphide oil (a 50/50 mixture of molybdenum disulphide grease and clean engine oil). Fit the connecting rod onto the crankpin, on L, N and R models taking care not to mark the cylinder bores.

28 Fit the bearing cap with its shell onto the connecting rod **(see illustration)**. Make sure the cap is fitted the correct way around so the connecting rod and bearing cap weight/size markings are correctly aligned.

29 Lubricate the threads and seats of the cap nuts with clean engine oil. Fit the nuts to the connecting rod and tighten them evenly, in two or three stages, to the specified torque setting **(see illustration)**.

30 Check that the crankshaft is free to rotate easily. Check to make sure that all components have been returned to their original locations using the marks made on disassembly.

24.26b . . . then lower the connecting rod and piston into the bore . . .

connecting rod and cap and will be either a 1 or a 2 **(see illustration)**.

22 A range of bearing shells is available. To select the correct bearing for a particular big-end, using the table below cross-refer the crankpin journal size letter (stamped on the web) with the connecting rod size number (stamped on the rod) to determine the colour code of the bearing required. For example, if the connecting rod size is 2, and the crankpin size is A, then the bearing required is Green. The colour is marked on the side of the shell **(see illustration)**.

	Connecting rod code	
Crankpin code	1 – (32.992 to 33.000 mm)	2 – (33.001 to 33.008 mm)
A – (29.992 to 30.000 mm)	C – Yellow	B – Green
B – (29.984 to 29.991 mm)	B – Green	A – Brown

24.26c . . . and tap the top of the piston while holding the compressor

24.28 Fit the connecting rod cap . . .

24.29 . . . and tighten the nuts as described to the specified torque

31 Check that the rods rotate smoothly and freely on the crankpin. If there are any signs of roughness or tightness, remove the rods and re-check the bearing clearance. Sometimes tapping the bottom of the connecting rod cap will relieve tightness, but if in doubt, recheck the clearances.

32 Reassemble the crankcase halves (see Section 21). On J and K models, if not already done, install the pistons (see Section 25), then install the cylinder block (see Section 14).

25 Pistons – removal, inspection and installation

Note: *To remove the pistons the engine must be removed from the frame.*

Removal

1 On J and K models, remove the engine from the frame (see Section 5) and remove the cylinder block (see Section 14).

2 On L, N and R models, remove the engine from the frame (see Section 5) and separate the crankcase halves (see Section 21). Remove the piston/connecting rod assemblies (see Section 24).

3 Before removing the piston from the connecting rod, use a sharp scriber or felt marker pen to write the cylinder identity on the crown of each piston (or on the inside of the skirt if the piston is dirty and going to be cleaned). Each piston should also have an IN mark on its crown which should face the inlet side of the bore **(see illustration)**. If this is not visible, mark the piston accordingly so that it can be installed the correct way round.

4 Carefully prise out the circlip on one side of the piston using needle-nose pliers or a small flat-bladed screwdriver inserted into the notch **(see illustration)**. Push the piston pin out from the other side to free the piston from the connecting rod **(see illustration)**. Remove the other circlip and discard them as new ones must be used. When the piston has been removed, install its pin back into its bore so that related parts do not get mixed up.

 HAYNES HINT *To prevent the circlip from pinging away, pass a rod or screwdriver, whose diameter is greater than the gap between the circlip ends, through the piston pin. This will trap the circlip if it springs out.*

 HAYNES HINT *If a piston pin is a tight fit in the piston bosses, soak a rag in boiling water then wring it out and wrap it around the piston – this will expand the alloy piston sufficiently to release its grip on the pin. If the piston pin is particularly stubborn, extract it using a drawbolt tool, but be careful to protect the piston's working surfaces.*

Inspection

5 Before the inspection process can be carried out, the pistons must be cleaned and the old piston rings removed.

6 Using your thumbs or a piston ring removal and installation tool, carefully remove the rings from the pistons **(see illustration)**. Do not nick or gouge the pistons in the process. Carefully note which way up each ring fits and in which groove as they must be installed in their original positions if being re-used. The upper surface of each ring is usually marked at one end.

7 Scrape all traces of carbon from the tops of the pistons. A hand-held wire brush or a piece of fine emery cloth can be used once most of the deposits have been scraped away. Do not, under any circumstances, use a wire brush mounted in a drill motor to remove deposits from the pistons; the piston material is soft and will be eroded away by the wire brush.

8 Use a piston ring groove cleaning tool to remove any carbon deposits from the ring grooves. If a tool is not available, a piece broken off an old ring will do the job. Be very careful to remove only the carbon deposits. Do not remove any metal and do not nick or gouge the sides of the ring grooves.

25.3 Note the IN mark which faces the inlet side of the cylinder

9 Once the deposits have been removed, clean the pistons with solvent and dry them thoroughly. If the identification previously marked on the piston is cleaned off, be sure to re-mark it with the correct identity. Make sure the oil return holes below the oil ring groove are clear.

10 Carefully inspect each piston for cracks around the skirt, at the pin bosses and at the ring lands. Normal piston wear appears as even, vertical wear on the thrust surfaces of the piston and slight looseness of the top ring in its groove. If the skirt is scored or scuffed, the engine may have been suffering from overheating and/or abnormal combustion, which caused excessively high operating temperatures. The oil pump should be checked thoroughly. Also check that the circlip grooves are not damaged.

11 A hole in the piston crown, an extreme to be sure, is an indication that abnormal combustion (pre-ignition) was occurring. Burned areas at the edge of the piston crown are usually evidence of spark knock (detonation). If any of the above problems exist, the causes must be corrected or the damage will occur again.

12 Measure the piston ring-to-groove clearance by laying each piston ring in its groove and slipping a feeler gauge in beside it **(see illustration)**. Make sure you have the correct ring for the groove. Check the clearance at three or four locations around the groove. If the clearance is greater than

25.4a Prise out the circlip . . .

25.4b . . . then push out the pin and remove the piston

25.6 Removing the piston rings using a ring removal and installation tool

Engine, clutch and transmission 2•45

25.12 Measure the piston ring-to-groove clearance with a feeler gauge

25.13 Measure the piston diameter with a micrometer at the specified distance from the bottom of the skirt

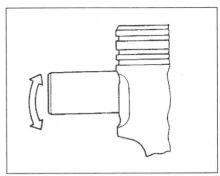

25.14a Slip the pin into the piston and try to rock it back and forth. If it's loose, renew the piston and pin

specified, renew both the piston and rings as a set. If new rings are being used, measure the clearance using the new rings. If the clearance is greater than that specified, the piston is worn and must be renewed.

13 Check the piston-to-bore clearance by measuring the bore (see Section 22) and the piston diameter. Make sure each piston is matched to its correct cylinder. Measure the piston at the specified distance (see Specifications) up from the bottom of the skirt and at 90° to the piston pin axis **(see illustration)**. Subtract the piston diameter from the bore diameter to obtain the clearance. If it is greater than the specified figure, the piston must be renewed (assuming the bore itself is within limits).

14 If not already done (see Section 24), apply clean engine oil to the piston pin, insert it into the piston and check for any freeplay between the two **(see illustration)**. Measure the pin external diameter **(see illustration 24.9b)**, and the pin bore in the piston **(see illustration)**. Calculate the difference to obtain the piston pin-to-piston pin bore clearance. Compare the result to the specifications at the beginning of the Chapter. If the clearance is greater than specified, renew the components that are worn beyond their specified limits. Repeat the measurements between the pin and the connecting rod small-end **(see illustration 24.9c)**.

Installation

15 Inspect and install the piston rings (see Section 26).

16 Lubricate the piston pin, the piston pin bore and the connecting rod small-end bore with molybdenum disulphide oil (a 50/50 mixture of molybdenum disulphide grease and clean engine oil).

17 When installing the pistons onto the connecting rods, make sure that the IN mark is on the same side as the oil hole in the connecting rod. Install a new circlip in one side of the piston (do not re-use old circlips). Line up the piston on its correct connecting rod, and insert the piston pin from the other side **(see illustration 25.4b)**. Secure the pin with the other new circlip. When installing the circlips, compress them only just enough to fit them in the piston, and make sure they are properly seated in their grooves with the open end away from the removal notch **(see illustration)**.

18 On J and K models, install the cylinder block (see Section 14), then fit engine into the frame (see Section 5).

19 On L, N and R models, install the piston/connecting rod assemblies (see Section 24). Reassemble the crankcase halves (see Section 21), then fit the engine into the frame (see Section 5).

26 Piston rings – inspection and installation

1 It is good practice to renew the piston rings when an engine is being overhauled. Before installing the new piston rings, the ring end gaps must be checked with the rings installed in the cylinder.

2 Lay out the pistons and the new ring sets so the rings will be matched with the same piston and cylinder during the end gap measurement procedure and engine assembly.

3 To measure the installed ring end gap, insert the top ring into the top of the first cylinder and square it up with the cylinder walls by pushing it in with the top of the piston. The ring should be about 20 mm below the top edge of the cylinder. To measure the end gap, slip a feeler gauge between the ends of the ring and compare the measurement to the specifications at the beginning of the Chapter **(see illustration)**.

4 If the gap is larger or smaller than specified, double check to make sure that you have the correct rings before proceeding.

5 If the gap is too small, it must be enlarged or the ring ends may come in contact with each other during engine operation, which

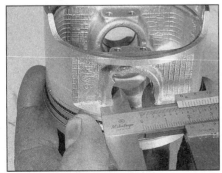

25.14b Measure the internal diameter of the bore in the piston

25.17 Do not over-compress the circlip when fitting it into the piston

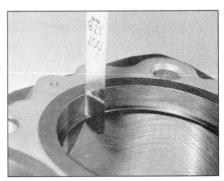

26.3 Measuring piston ring installed end gap

26.5 Ring end gap can be enlarged by clamping a file in a vice and filing the ring ends

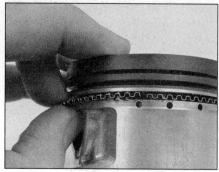

26.9a Install the oil ring expander in its groove . . .

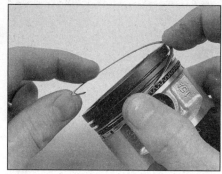

26.9b . . . and fit the side rails each side of it. The oil ring must be installed by hand

can cause serious damage. The end gap can be increased by filing the ring ends very carefully with a fine file. When performing this operation, file only from the outside in **(see illustration)**.

6 Excess end gap is not critical unless it exceeds the service limit. Again, double-check to make sure you have the correct rings for your engine and check that the bore is not worn.

7 Repeat the procedure for each ring that will be installed in the cylinders. Remember to keep the rings, pistons and cylinders matched up.

8 Once the ring end gaps have been checked/corrected, the rings can be installed on the pistons.

9 The oil control ring (lowest on the piston) is installed first. It is composed of three separate components, namely the expander and the upper and lower side rails. Slip the expander into the groove, then install the upper side rail.

Do not use a piston ring installation tool on the oil ring side rails as they may be damaged. Instead, place one end of the side rail into the groove between the expander and the ring land. Hold it firmly in place and slide a finger around the piston while pushing the rail into the groove. Next, install the lower side rail in the same manner **(see illustrations)**. Make sure the ends of the expander do not overlap.

10 After the three oil ring components have been installed, check to make sure that both the upper and lower side rails can be turned smoothly in the ring groove.

11 The upper surface of each compression ring should be marked at one end near the gap. Make sure that the identification mark or letter is facing up. Fit the middle ring into the middle groove in the piston. Do not expand the ring any more than is necessary to slide it into place. To avoid breaking the ring, use a piston ring installation tool.

12 Finally, install the top ring in the same manner into the top groove in the piston. Make sure the identification letter near the end gap is facing up.

13 Once the rings are correctly installed, check they move freely without snagging and stagger their end gaps as shown **(see illustration)**.

27 Crankshaft and main bearings – removal, inspection and installation

Note: *To remove the crankshaft the engine must be removed from the frame and the crankcase halves separated.*

Removal

1 Remove the engine from the frame (see Section 5) and separate the crankcase halves (see Section 21).

2 Separate the connecting rods from the crankshaft (see Section 24).

Note: *If no work is to be carried out on the piston/connecting rod assemblies there is no need to remove them from the bores, although the connecting rod bearing caps must be removed (see Section 24, Steps 3 and 4 or 5) and the pistons pushed up to the top of the bores so that the connecting rod ends are positioned clear of the crankshaft.*

3 Lift the crankshaft out of the upper crankcase half, taking care not to dislodge the main bearing shells **(see illustration)**.

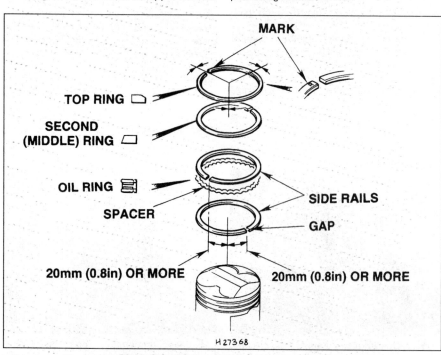

26.13 Stagger the ring end gaps as shown

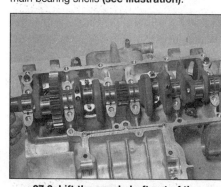

27.3 Lift the crankshaft out of the crankcase . . .

Engine, clutch and transmission 2•47

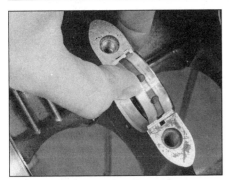

27.4 ... and remove the shells if required

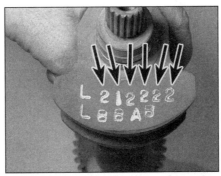

27.9a Main bearing journal size numbers (arrowed)

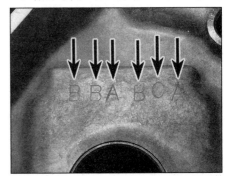

27.9b Main bearing housing size letters (arrowed)

4 The main bearing shells can be removed from the crankcase halves by pushing their centres to the side, then lifting them out **(see illustration)**. Keep the shells in order.

Inspection

5 Clean the crankshaft with solvent, using a rifle-cleaning brush to scrub out the oil passages. If available, blow the crank dry with compressed air, and also blow through the oil passages. Check the primary drive gear and the camshaft drive gear for wear or damage. If any of the gear teeth are excessively worn, chipped or broken, the crankshaft must be renewed. If wear or damage is found, also check the primary driven gear on the back of the clutch housing (see Section 16) and the camshaft drive gear assembly (see Section 13).

6 Refer to Section 23 and examine the main bearing shells. If they are scored, badly scuffed or appear to have been seized, new bearings must be installed. Always renew the main bearings as a set. If they are badly damaged, check the corresponding crankshaft journals. Evidence of extreme heat, such as discoloration, indicates that lubrication failure has occurred. Be sure to thoroughly check the oil pump and pressure regulator as well as all oil holes and passages before reassembling the engine.

7 Give the crankshaft journals a close visual examination, paying particular attention where damaged bearings have been discovered. If the journals are scored or pitted in any way a new crankshaft will be required. Note that undersizes are not available, precluding the option of re-grinding the crankshaft.

8 Place the crankshaft on V-blocks and check the runout at the main bearing journals using a dial gauge. Compare the reading to the maximum specified at the beginning of the Chapter. If the runout exceeds the limit, the crankshaft must be renewed.

Main bearing shell selection

9 New bearing shells for the main bearings are supplied on a selected fit basis. Code letters and numbers stamped on various components are used to identify the correct size new bearings. The crankshaft main bearing journal size numbers are stamped on the outside of the left-hand crankshaft web and will be either a 1 or a 2 **(see illustration)**. The corresponding main bearing housing size letters are stamped into the upper crankcase half and will be either an A or a B **(see illustration)**. The left-hand number or letter corresponds to the left-hand journal, and so on from left to right respectively.

10 A range of bearing shells is available. To select the correct bearing for a particular journal, using the table below cross-refer the main bearing journal size number (stamped on the crank web) with the main bearing housing size letter (stamped on the crankcase) to determine the colour code of the bearing required. For example, if the journal code is 1, and the housing code is B, then the bearing required is C (Yellow). The colour is marked on the side of the shell **(see illustration 24.22)**.

Oil clearance check

11 Whether new bearing shells are being fitted or the original ones are being re-used, the main bearing oil clearance should be checked before the engine is reassembled. Main bearing oil clearance is measured with a product known as Plastigauge.

12 Clean the backs of the bearing shells and the bearing housings in both crankcase halves.

13 Press the bearing shells into their cut-outs, ensuring that the tab on each shell engages in the notch in the crankcase **(see illustrations)**. Make sure the bearings are fitted in the correct locations and take care not to touch any shell's bearing surface with your fingers.

14 Ensure the shells and crankshaft are clean and dry. Lay the crankshaft in position in the upper crankcase **(see illustration 27.3)**.

15 Cut several lengths of the appropriate size Plastigauge (they should be slightly shorter than the width of the crankshaft journals). Place a strand of Plastigauge on each

	Main journal code	
Crankcase code	1 – (29.993 to 30.000 mm)	2 – (29.984 to 29.992 mm)
A – (33.000 to 33.008 mm)	D – Pink	C – Yellow
B – (33.009 to 33.016 mm)	C – Yellow	B – Green
C – (33.017 to 33.024 mm)	B – Green	A – Brown

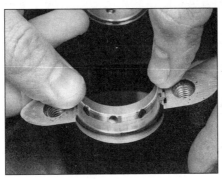

27.13a Fit the shells ...

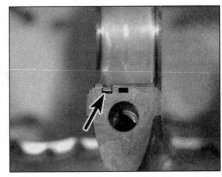

27.13b ... locating the tab in the notch (arrowed)

27.15 Lay a strip of Plastigauge on each journal parallel to the crankshaft centreline

27.18 Measure the width of the crushed Plastigauge

(cleaned) journal **(see illustration)**. Make sure the crankshaft is not rotated.

16 Carefully install the lower crankcase half on to the upper half **(see illustration 21.9)**. Check that the lower crankcase half is correctly seated. **Note:** *Do not tighten the crankcase bolts if the casing is not correctly seated.* Clean the threads of the twelve 8 mm lower crankcase (crankshaft journal) bolts and insert them with their washers in their original locations **(see illustration 21.4a or b)**. Secure all bolts finger-tight at first, then tighten them evenly and a little at a time in the correct numerical sequence to the torque setting specified at the beginning of the Chapter. Make sure that the crankshaft is not rotated as the bolts are tightened.

17 Slacken each bolt evenly a little at a time in a reverse of the numerical sequence until they are all finger-tight, then remove the bolts and washers. Carefully lift off the lower crankcase half, making sure the Plastigauge is not disturbed.

18 Compare the width of the crushed Plastigauge on each crankshaft journal to the scale printed on the Plastigauge envelope to obtain the main bearing oil clearance **(see illustration)**. Compare the reading to the specifications at the beginning of the Chapter.

19 On completion carefully scrape away all traces of the Plastigauge material from the crankshaft journal and bearing shells; use a fingernail or other object which is unlikely to score them.

20 If the oil clearance falls into the specified range, no bearing shell renewal is required (provided they are in good condition). If the clearance is beyond the service limit, refer to the marks on the case and the marks on the crankshaft and select new bearing shells (see Steps 9 and 10). Install the new shells and check the oil clearance once again (the new shells may bring bearing clearance within the specified range). Always renew all of the shells at the same time.

21 If the clearance is still greater than the service limit listed in this Chapter's Specifications (even with new shells), the crankshaft journal is worn and the crankshaft should be renewed. Measure the diameter of each journal and compare the measurements to the table above to confirm.

Installation

22 Clean the backs of the bearing shells and the bearing cut-outs in both crankcase halves. If new shells are being fitted, ensure that all traces of the protective grease are cleaned off using paraffin (kerosene). Wipe dry the shells and crankcase halves with a lint-free cloth. Make sure all the oil passages and holes are clear, and blow them through with compressed air if it is available.

23 Press the bearing shells into their cut-outs, ensuring that the tab on each shell engages in the notch in the crankcase **(see illustrations 27.13a and b)**. Make sure the bearings are fitted in the correct locations and take care not to touch any shell's bearing surface with your fingers. Lubricate each shell with molybdenum disulphide oil (a 50/50 mixture of molybdenum disulphide grease and clean engine oil) **(see illustration)**.

24 Lower the crankshaft into position in the upper crankcase, making sure all bearings remain in place **(see illustration 27.3)**. If the cylinder head was not removed, make sure the teeth of the drive gear on the crankshaft mesh with those on the camshaft drive gear assembly.

25 Fit the connecting rods onto the crankshaft (see Section 24).

26 Reassemble the crankcase halves (see Section 21).

28 Transmission shafts and bearings – removal and installation

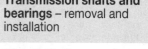

Note: *To remove the transmission shafts the engine must be removed from the frame. If the engine has already been removed, ignore the steps which do not apply.*

Removal

1 Remove the engine from the frame (see Section 5) and separate the crankcase halves (see Section 21).

2 On J and K models, lift the output shaft out of the casing, noting how it meshes with the input shaft. If it is stuck, use a soft-faced hammer and gently tap on the end of the shaft to free it. Remove the bearing half-ring retainer from the left-hand end of the output shaft; if it's not in its slot in the crankcase, remove it from the bearing on the shaft. Grasp the end of the shaft and draw it out of the crankcase. If the needle bearing on the left-hand end of the shaft does not come away with the shaft, remove it from its cage in the crankcase if required. If required, unscrew the bolt securing the cage retainer in the bore in the crankcase, then remove the retainer and draw out the cage. If necessary, the input shaft and output shaft can be disassembled and inspected for wear or damage (see Section 29).

3 On L, N and R models, lift the input shaft and output shaft out of the casing, noting how they mesh together **(see illustration)**. If they are stuck, use a soft-faced hammer and gently tap on the ends of the shafts to free them. Remove the dowel for the needle bearing on the left-hand end of the input shaft from the crankcase if it is loose, noting how it fits **(see illustrations 28.6a)**. If it is not in its hole in the crankcase, remove it from the bearing on the shaft. If necessary, the input shaft and output shaft can be disassembled and inspected for wear or damage (see Section 29).

4 Referring to *Tools and Workshop Tips* (Section 5) in the Reference Section, check the bearings. Renew the bearings if necessary, noting that the input shaft right-hand bearing and the output shaft left-hand

27.23 Generously lubricate all the bearing shells

28.3 Lift the input shaft (A) and the output shaft (B) out of the crankcase, noting how the retainer ring (C) and oil seal lip (D) locate in the grooves

Engine, clutch and transmission 2•49

28.4a Discard the oil seal if it is worn, damaged, or shows signs of leakage

28.4b Apply grease to the lips of the seal

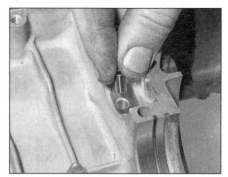

28.6a Fit the dowel . . .

bearing are not available separately and the shaft itself must be renewed. Also remove and check the condition of the output shaft oil seal and renew it if it is worn or damaged **(see illustration)**. Apply some grease to the lips of the seal on installation **(see illustration)**.

Installation

5 On J and K models, if removed, fit the needle bearing cage for the input shaft into its hole in the crankcase, then fit the retainer and tighten the bolt securely. Apply some engine oil to the bearing, then fit the bearing into the cage. Slide the shaft into the crankcase, making sure the shaft end locates into the needle bearing and that the shaft is slid fully home. Install the bearing half-ring retainer for the left-hand end of the output shaft into its slot in the upper crankcase half. Lower the output shaft into position in the crankcase half, making sure the groove in the bearing engages correctly with the bearing half-ring retainer.

6 On L, N and R models, if removed, fit the needle bearing dowel into its hole in the upper crankcase **(see illustration)**. Lower the input shaft into position, making sure it locates correctly onto the dowel **(see illustration)**. Lower the output shaft into position, making sure it meshes correctly with the input shaft, and the locating ring around the bearing and the oil seal lip fit in the slot in the crankcase **(see illustration 28.3)**.

7 Make sure both transmission shafts are correctly seated and their related pinions are correctly engaged.

Caution: If the bearing retainer or dowel are not correctly engaged with their bearings, the crankcase halves will not seat correctly.

8 Position the gears in the neutral position and check the shafts are free to rotate easily and independently (ie the input shaft can turn whilst the output shaft is held stationary) before proceeding further.

9 Reassemble the crankcase halves (see Section 21).

29 Transmission shafts – disassembly, inspection and reassembly

1 Remove the transmission shafts from the casing (see Section 28). Always disassemble the transmission shafts separately to avoid mixing up the components **(see illustrations)**.

> **HAYNES HiNT** When disassembling the transmission shafts, place the parts on a long rod or thread a wire through them to keep them in order and facing the proper direction.

28.6b . . . then install the input shaft, making sure the hole (arrowed) locates onto the dowel

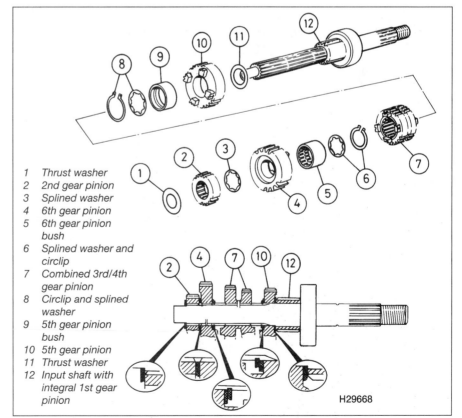

1 Thrust washer
2 2nd gear pinion
3 Splined washer
4 6th gear pinion
5 6th gear pinion bush
6 Splined washer and circlip
7 Combined 3rd/4th gear pinion
8 Circlip and splined washer
9 5th gear pinion bush
10 5th gear pinion
11 Thrust washer
12 Input shaft with integral 1st gear pinion

29.1a Transmission input shaft components

2•50 Engine, clutch and transmission

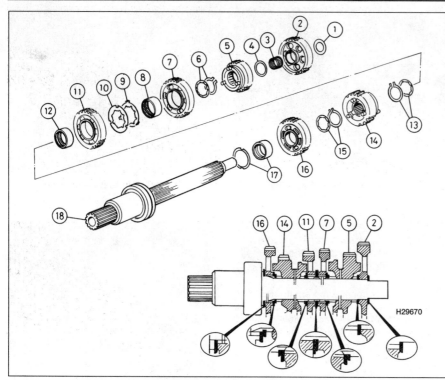

29.1b Transmission output shaft components

1. Thrust washer
2. 1st gear pinion
3. Needle bearing
4. Thrust washer
5. 5th gear pinion
6. Circlip and splined washer
7. 3rd gear pinion
8. 3rd gear pinion bush
9. Lockwasher
10. Splined washer
11. 4th gear pinion
12. 4th gear pinion bush
13. Splined washer and circlip
14. 6th gear pinion
15. Circlip and splined washer
16. 2nd gear pinion
17. 2nd gear pinion bush and thrust washer
18. Output shaft with integral bearing

Input shaft

Disassembly

2 On J and K models, if the needle bearing did not remain in its cage in the crankcase, slide it off the left-hand end of the shaft. On L, N and R models, slide the needle bearing cage and bearing off the left-hand end of the shaft **(see illustrations 29.18b and a)**.

3 Slide the thrust washer, the 2nd gear pinion, the splined washer, the 6th gear pinion and its bush, and the splined washer off the shaft **(see illustrations 29.17f, e, d, c, b and a)**.

4 Remove the circlip securing the combined 3rd/4th gear pinion, then slide the pinion off the shaft **(see illustrations 29.16b and a)**.

5 Remove the circlip securing the 5th gear pinion, then slide the splined washer, the pinion and its bush, and the thrust washer off the shaft **(see illustrations 29.15e, d, c, b and a)**.

6 The 1st gear pinion is integral with the shaft.

Inspection

7 Wash all of the components in clean solvent and dry them off.

8 Check the gear teeth for cracking, chipping, pitting and other obvious wear or damage. Any pinion that is damaged as such must be renewed.

9 Inspect the dogs and the dog holes in the gears for cracks, chips, and excessive wear especially in the form of rounded edges. Make sure mating gears engage properly. Renew the paired gears as a set if necessary.

10 Check for signs of scoring or bluing on the pinions, bushes and shaft. This could be caused by overheating due to inadequate lubrication. Check that all the oil holes and passages are clear. Renew any damaged pinions or bushes.

11 Check that each pinion moves freely on the shaft or bush but without undue freeplay. Check that each bush moves freely on the shaft but without undue freeplay. If the necessary measuring equipment is available, the gear, bush and shaft dimensions can be checked and compared with the specifications at the beginning of this chapter.

12 The shaft is unlikely to sustain damage unless the engine has seized, placing an unusually high loading on the transmission, or the machine has covered a very high mileage. Check the surface of the shaft, especially where a pinion turns on it, and renew the shaft if it has scored or picked up, or if there are any cracks. Damage of any kind can only be cured by renewal.

13 Check the washers and circlips and renew any that are bent or appear weakened or worn. Use new ones if in any doubt. It is good practice to renew the washers and circlips as a matter of course when overhauling the gearshaft.

Reassembly

14 During reassembly, apply engine oil to the mating surfaces of the shaft, pinions and bushes, and to the bearings. When installing the circlips, do not expand the ends any further than is necessary. Install the stamped circlips so that their chamfered side faces the pinion it secures (see *correct fitting of a stamped circlip* illustration in Tools and Workshop Tips of the Reference section) Also refer to illustration 29.1a.

15 Slide the thrust washer onto the left-hand end of the shaft, followed by the 5th gear pinion bush, aligning the oil hole in the bush with the hole in the shaft, and the 5th gear

29.15a Slide on the thrust washer . . .

29.15b . . . the 5th gear pinion bush . . .

29.15c . . . the 5th gear pinion . . .

Engine, clutch and transmission 2•51

29.15d ...and the splined washer...

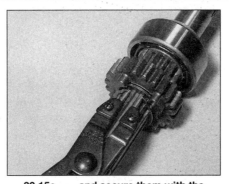

29.15e ...and secure them with the circlip

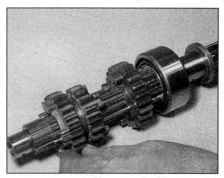

29.16a Slide on the 3rd/4th gear pinion...

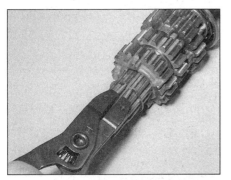

29.16b ...and secure it with the circlip

29.17a Slide on the splined washer...

29.17b ...the 6th gear pinion bush...

29.17c ...the 6th gear pinion...

29.17d ...the splined washer...

29.17e ...the 2nd gear pinion...

29.17f ...and the thrust washer

29.18a Fit the bearing...

pinion, with the pinion dogs facing away from the integral 1st gear **(see illustrations)**. Slide the splined washer onto the shaft, then fit the circlip, making sure that it locates correctly in the groove in the shaft **(see illustrations)**.

16 Slide the combined 3rd/4th gear pinion onto the shaft with the smaller 3rd gear pinion facing the 5th gear pinion **(see illustration)**. Fit the circlip, making sure it is locates correctly in its groove in the shaft **(see illustration)**.

17 Slide the splined washer onto the shaft, followed by the 6th gear pinion bush, aligning the oil hole in the bush with the hole in the shaft, and the 6th gear pinion, making sure the dogs in the pinion face the dog holes on the 3rd/4th gear pinion **(see illustrations)**. Slide the splined washer, the 2nd gear pinion and the thrust washer onto the end of the shaft **(see illustrations)**.

18 On L, N and R models, fit the needle roller bearing and its cage over the end of the shaft **(see illustrations)**.

2•52 Engine, clutch and transmission

29.18b ... and its cage

29.19 The complete input shaft should be as shown

29.29a Slide on the thrust washer ...

19 Check that all components have been correctly installed (see illustration).

Output shaft

Disassembly

20 Remove the caged ball bearing from the right-hand end of the shaft, referring to *Tools and Workshop Tips* (Section 5) in the Reference Section if required (see illustration 29.36).
21 Slide the thrust washer off the shaft, followed by the 1st gear pinion and its needle roller bearing, the thrust washer and the 5th gear pinion (see illustrations 29.35c, b and a, and 29.34b and a).
22 Remove the circlip securing the 3rd gear pinion, then slide the splined washer, the pinion and its splined bush off the shaft (see illustrations 29.33d, c, b and a).
23 Slide the lockwasher and the splined washer off the shaft, noting how they fit together (see illustration 29.32d, c, b and a).
24 Slide the 4th gear pinion and its splined bush, followed by the splined washer, off the shaft (see illustration 29.31c, b and a).
25 Remove the circlip securing the 6th gear pinion, then slide the pinion off the shaft (see illustrations 29.30b and a).
26 Remove the circlip securing the 2nd gear pinion, then slide the splined washer, the pinion and its bush, and the thrust washer off the shaft (see illustrations 29.29e, d, c, b and a).

Inspection

27 Refer to Steps 7 to 13 above.

Reassembly

28 During reassembly, apply engine oil to the mating surfaces of the shaft, pinions and bushes, and to the bearings. When installing the circlips, do not expand the ends any further than is necessary, and install them so that the chamfered side faces the pinion it secures (see *correct fitting of a stamped circlip* illustration in Tools and Workshop Tips of the Reference section). Also refer to illustration 29.1b.
29 Slide the thrust washer onto the right-hand end of the shaft, followed by the 2nd gear pinion bush, aligning the oil hole in the bush with the hole in the shaft, the 2nd gear pinion and the splined washer, then fit the circlip, making sure it is locates correctly in its groove in the shaft (see illustrations).
30 Slide on the 6th gear pinion with its selector fork groove facing away from the 2nd gear pinion, then fit the circlip, making sure it is locates correctly in its groove in the shaft (see illustrations).
31 Slide the splined washer and the 4th gear pinion bush onto the shaft, making sure the oil hole in the bush aligns with the hole in the shaft, followed by the 4th gear pinion; the open side of the gear pinion faces the 6th gear pinion (see illustrations).

29.29b ... the 2nd gear pinion bush ...

29.29c ... the 2nd gear pinion ...

29.29d ... and the splined washer ...

29.29e ... and secure them with the circlip

29.30a Slide on the 6th gear pinion ...

Engine, clutch and transmission 2•53

29.30b ... and secure it with the circlip

29.31a Slide on the splined washer ...

29.31b ... the 4th gear pinion bush ...

32 Slide the slotted splined washer onto the shaft and locate it in its groove, then turn it in the groove so that the splines on the washer locate between the splines of the shaft and secure the washer in the groove (**see illustrations**). Slide the lockwasher onto the shaft, so that the tabs on the lockwasher face the left-hand end of the shaft and locate into the slots in the outer rim of the splined washer (**see illustrations**).

33 Slide the 3rd gear pinion bush onto the shaft, making sure the oil hole in the bush aligns with the hole in the shaft, followed by the 3rd gear pinion (more recessed side facing away from the 4th gear pinion – see illustration 29.1b) and the splined washer, then fit the circlip, making sure it is locates correctly in its groove in the shaft (**see illustrations**).

29.31c ... and the 4th gear pinion

29.32a Slide the slotted splined washer onto the shaft ...

29.32b ... and locate it as shown

29.32c Slide the lockwasher onto the shaft ...

29.32d ... and locate it as shown

29.33a Slide on the 3rd gear pinion bush ...

29.33b ... the 3rd gear pinion ...

29.33c ... and the splined washer ...

2•54 Engine, clutch and transmission

29.33d ... and secure them with the circlip

29.34a Slide on the 5th gear pinion ...

29.34b ... and the thrust washer

34 Slide the 5th gear pinion onto the shaft with its selector fork groove facing the 3rd gear pinion, followed by the thrust washer **(see illustrations)**.

35 Slide the 1st gear pinion needle roller bearing onto the shaft, followed by the 1st gear pinion (the open side of the gear faces the 5th gear pinion) and the thrust washer **(see illustrations)**.

36 Fit the caged ball bearing onto the right-hand end of the shaft, referring to *Tools and Workshop Tips* (Section 5) in the Reference Section if required **(see illustration)**.

37 Check that all components have been correctly installed **(see illustration)**.

30 Selector drum and forks – removal, inspection and installation

Note: *To remove the selector drum and forks the engine must be removed from the frame.*

Removal

1 Remove the engine (see Section 5) and separate the crankcase halves (see Section 21). The selector drum and forks are located in the lower crankcase half.

2 If not already done, remove the gearchange mechanism (see Section 20).

3 Before removing the selector forks, note that each fork is lettered for identification. The right-hand fork has an R, the centre fork a C, and the left-hand fork an L **(see illustration)**. These letters face the right-hand side of the engine. If no letters are visible, mark them yourself using a felt pen.

4 On J and K models, unscrew the bolt securing the centre selector fork to the shaft and remove the lockwasher. Support the selector forks and withdraw the shaft from the casing, then remove the forks. Once removed from the case, slide the forks back onto the

29.35a Slide on the needle bearing ...

29.35b ... the 1st gear pinion ...

29.35c ... and the thrust washer ...

29.36 ... then fit the bearing onto the end of the shaft

29.37 The assembled output shaft should be as shown

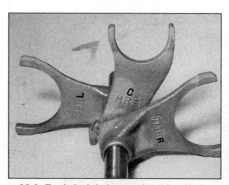

30.3 Each fork is lettered to identify its position on the shaft

Engine, clutch and transmission 2•55

30.4a Unscrew the bolt (A), then remove the dowel (B) and the plate (C)

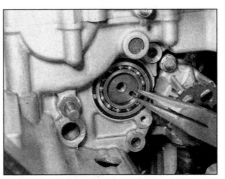

30.4b Remove the locating pin for safekeeping

30.5a Unscrew the bolt (arrowed) . . .

shaft in their correct order and way round. Unscrew the rear guide plate bolt and remove the dowel and the plate **(see illustration)**. If the drum selector cam has been removed, remove the locating pin for safekeeping **(see illustration)**. Slide the selector drum out of the crankcase.

5 On L, N and R models, unscrew the bolt retaining the fork shaft and remove the washer **(see illustration)**. Support the selector forks and withdraw the shaft from the casing, then remove the forks **(see illustration)**. Once removed from the case, slide the forks back onto the shaft in their correct order and way round. Unscrew the two bolts securing the selector drum bearing retainer plate and remove the plate **(see illustration)**. If the drum stopper plate has been removed, remove the locating pin for safekeeping **(see illustration 30.4b)**. Slide the selector drum out of the crankcase **(see illustration)**.

Inspection

6 Inspect the selector forks for any signs of wear or damage, especially around the fork ends where they engage with the groove in the pinion. Check that each fork fits correctly in its pinion groove. Check closely to see if the forks are bent. If the forks are in any way damaged they must be renewed.
7 Measure the thickness of the fork ends and compare the readings to the specifications

(see illustration). Renew the forks if they are worn beyond their specifications.
8 Check that the forks fit correctly on their shaft. They should move freely with a light fit but no appreciable freeplay. Measure the internal diameter of the fork bores and the corresponding diameter of the fork shaft. Renew the forks and/or shaft if they are worn beyond their specifications. Check that the fork shaft holes in the casing are not worn or damaged.
9 The selector fork shaft can be checked for trueness by rolling it along a flat surface. A bent rod will cause difficulty in selecting gears and make the gearchange action heavy. Renew the shaft if it is bent.
10 Inspect the selector drum grooves and selector fork guide pins for signs of wear or

damage. If either component shows signs of wear or damage the selector fork(s) and drum must be renewed.
11 Check that the selector drum bearing rotates freely and has no sign of freeplay between it and the casing. Renew the bearing if necessary (see *Tools and Workshop Tips* (Section 5) in the Reference Section) **(see illustration)**.

Installation

12 On J and K models, if the drum selector cam was not removed, locate the rear guide plate between it and the selector drum bearing before installing the drum. Slide the selector drum into position in the crankcase. Make sure the drum end locates into its bore in the casing, and position it so that the neutral

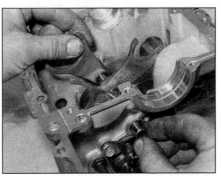

30.5b . . . and remove the shaft and the forks

30.5c Unscrew the two bolts (arrowed) and remove the plate . . .

30.5d . . . then draw the selector drum out

30.7 Measure the fork end thickness as shown

30.11 Renew the bearing if necessary

2

2•56 Engine, clutch and transmission

30.13a Slide in the drum, locating the shaft end in its bore (arrowed) . . .

30.13b . . . and the neutral contact against the switch (arrowed)

contact is against the neutral switch. If the drum selector cam was removed, now locate the rear guide plate. Fit the dowel into the top hole in the plate, then apply a suitable non-permanent thread locking compound to the bolt and tighten it securely **(see illustration 30.4a)**. Fit the locating pin into its hole in the selector drum **(see illustration 30.4b)**.

13 On L, N and R models, slide the selector drum into position in the crankcase **(see illustration)**. Make sure the drum end locates into its bore in the casing, and position it so that the neutral contact is against the neutral switch **(see illustration)**. Apply a suitable non-permanent thread-locking compound to the selector drum bearing retainer plate bolts, then fit the plate and tighten the bolts to the torque setting specified at the beginning of the Chapter **(see illustrations)**.

14 Refer to Step 3 for the correct location of each fork **(see illustration 30.3)**. Lubricate the selector fork shaft with clean engine oil and slide it through the crankcase and each fork in turn, and into its bore, locating the guide pin on the end of each fork into its groove in the drum as you do **(see illustrations)**.

15 On J and K models, rotate the shaft so that the threads for the centre fork bolt are correctly positioned, then fit the centre fork lockwasher, using a new one if it is fatigued or deformed, and tighten the bolt securely.

16 On L, N and R models, clean the threads of the fork shaft bolt, then apply a suitable non-permanent thread-locking compound **(see illustration)**. Locate the washer and tighten the bolt to the torque setting specified at the beginning of the Chapter **(see illustration 30.5a)**.

17 Reassemble the crankcase halves (see Section 21).

31 Initial start-up after overhaul

1 Make sure the engine oil level and coolant level are correct (see *Daily (pre-ride) checks*). Turn the fuel tap to the OFF position.

2 Pull the plug caps off the spark plugs and insert a spare spark plug into each cap. Position the spare plugs so that their bodies are earthed (grounded) against the engine. Turn on the ignition switch and crank the engine over with the starter until the oil pressure warning light goes off (which indicates that oil pressure exists). Turn off the ignition. Remove the spare spark plugs and reconnect the plug caps.

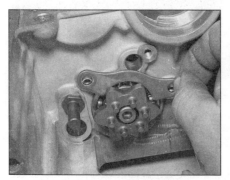

30.13c Fit the retainer plate . . .

30.13d . . . and tighten the bolts to the specified torque

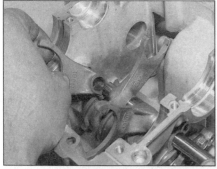

30.14a Slide the shaft through each fork in turn . . .

30.14b . . . locating the guide pins in the grooves (arrowed)

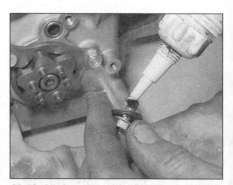

30.16 Apply a thread-lock to the fork shaft bolt

3 Make sure there is fuel in the tank, then turn the fuel tap to the ON or RES position as required, and set the choke.
4 Start the engine and allow it to run at a moderately fast idle until it reaches operating temperature.

 Warning: If the oil pressure indicator light doesn't go off, or it comes on while the engine is running, stop the engine immediately.

5 Check carefully for oil and coolant leaks and make sure the transmission and controls, especially the brakes, function properly before road testing the machine. Refer to Section 32 for the recommended running-in procedure.

6 Upon completion of the road test, and after the engine has cooled down completely, recheck the valve clearances (see Chapter 1) and check the engine oil and coolant levels (see *Daily (pre-ride) checks*).

32 Recommended running-in procedure

1 Treat the machine gently for the first few miles to make sure oil has circulated throughout the engine and any new parts installed have started to seat.

2 Even greater care is necessary if new pistons/rings or new crankshaft or connecting rod bearings have been installed. This means greater use of the transmission and a restraining hand on the throttle until at least 600 miles (1000 km) have been covered. There's no point in keeping to any set speed limit – the main idea is to keep from labouring the engine and to gradually increase performance up to the 600 mile (1000 km) mark. These recommendations can be lessened to an extent when only a new crankshaft is installed. Experience is the best guide, since it's easy to tell when an engine is running freely. The accompanying table showing maximum engine speed limitations, which Honda provide for new motorcycles, can be used as a guide.

3 If a lubrication failure is suspected, stop the engine immediately and try to find the cause. If an engine is run without oil, even for a short period of time, severe damage will occur.

Up to 600 miles (1000 km)	6000 rpm max	Vary throttle position/speed
600 to 1000 miles (1000 to 1600 km)	8000 rpm max	Vary throttle position/speed. Use full throttle for short bursts
Over 1000 miles (1600 km)	14,500 rpm max	Do not exceed tachometer red line

Notes

Chapter 3
Cooling system

Contents

Coolant hoses – removal and installation 9
Coolant level check .... see Daily (pre-ride) checks
Coolant reservoir – removal and installation 3
Coolant temperature gauge and sensor – check and
  replacement 5
Cooling fan and thermostatic switch – check and replacement 4
Cooling system checks .... see Chapter 1
Cooling system draining, flushing and refilling .... see Chapter 1
General information 1
Radiator – removal and installation 7
Radiator pressure cap – check 2
Thermostat and thermostat housing – removal,
  check and installation 6
Water pump – check, removal and installation 8

Degrees of difficulty

| Easy, suitable for novice with little experience | Fairly easy, suitable for beginner with some experience | Fairly difficult, suitable for competent DIY mechanic | Difficult, suitable for experienced DIY mechanic | Very difficult, suitable for expert DIY or professional 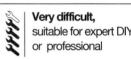 |

Specifications

Note: *Models are identified by their production code letter – refer to 'Identification numbers' at the front of this manual for details.*

Coolant
Mixture type and capacity .... see Chapter 1

Pressure cap
Cap valve opening pressure .... 13.5 to 17.7 psi (0.93 to 1.22 Bar)

Fan switch
Cooling fan cut-in temperature .... 98 to 102°C
Cooling fan cut-out temperature .... 93 to 97°C

Coolant temperature sensor
Resistance
  J and K models
    @ 50°C .... 130 to 180 ohms
    @ 80°C .... 26 to 30 ohms
  L, N and R models
    @ 80°C .... 47 to 57 ohms
    @ 120°C .... 14 to 18 ohms

Thermostat
Opening temperature .... 80 to 84°C
Valve lift .... 8 mm (min) @ 95°C

Torque settings
Note: *Where a specified setting is not given for a particular bolt, the general settings listed at the beginning apply. The dimension given applies to the diameter of the thread, not the head.*

5 mm bolt/nut .... 5 Nm
6 mm bolt/nut .... 10 Nm
8 mm bolt/nut .... 22 Nm
10 mm bolt/nut .... 35 Nm
12 mm bolt/nut .... 55 Nm
6 mm flange bolt with 8 mm head .... 9 Nm
6 mm flange bolt/nut with 10 mm head .... 12 Nm
8 mm flange bolt/nut .... 27 Nm
10 mm flange bolt/nut .... 40 Nm
Cooling fan switch .... 18 Nm
Coolant temperature sender .... 10 Nm

3•2 Cooling system

1 General information

The cooling system uses a water/antifreeze coolant to carry away excess energy in the form of heat. The cylinders are surrounded by a water jacket from which the heated coolant is circulated by thermo-syphonic action in conjunction with a water pump, driven by the oil pump. The hot coolant passes upwards to the thermostat and through to the radiator. The coolant then flows across the radiator core, where it is cooled by the passing air, to the water pump and back to the engine where the cycle is repeated.

A thermostat is fitted in the system to prevent the coolant flowing through the radiator when the engine is cold, therefore accelerating the speed at which the engine reaches normal operating temperature. A coolant temperature sender mounted in the thermostat housing transmits to the temperature gauge on the instrument panel. A thermostatically-controlled cooling fan is also fitted to aid cooling in extreme conditions.

The complete cooling system is partially sealed and pressurised, the pressure being controlled by a valve contained in the spring-loaded pressure cap. By pressurising the coolant the boiling point is raised, preventing premature boiling in adverse conditions. The overflow pipe from the system is connected to a reservoir into which excess coolant is expelled under pressure. The discharged coolant automatically returns to the radiator when the engine cools.

Coolant is routed around the oil filter adapter on the front of the crankcase to cool the oil passing through the filter. On J and K models, hoses route the coolant from the water pump cover to the filter adapter and back again. On L, N and R models, coolant is supplied to the filter adapter via a short hose running from the coolant union on the front of the engine and the hot coolant is returned via a hose to the water pump cover.

 Warning: Do not remove the pressure cap from the filler neck (J and K models) or radiator (L, N and R models) when the engine is hot. Scalding hot coolant and steam may be blown out under pressure, which could cause serious injury. When the engine has cooled, place a thick rag, like a towel over the pressure cap; slowly rotate the cap anti-clockwise to the first stop. This procedure allows any residual pressure to escape. When the steam has stopped escaping, press down on the cap while turning it anti-clockwise and remove it. Do not allow antifreeze to come in contact with your skin or painted surfaces of the motorcycle. Rinse off any spills immediately with plenty of water. Antifreeze is highly toxic if ingested. Never leave antifreeze lying around in an open container or in puddles on the floor; children and pets are attracted by its sweet smell and may drink it. Check with the local authorities about disposing of used antifreeze. Many communities will have collection centres which will see that antifreeze is disposed of safely.
Caution: At all times use the specified type of antifreeze, and always mix it with distilled water in the correct proportion. The antifreeze contains corrosion inhibitors which are essential to avoid damage to the cooling system. A lack of these inhibitors could lead to a build-up of corrosion which would block the coolant passages, resulting in overheating and severe engine damage. Distilled water must be used as opposed to tap water to avoid a build-up of scale which would also block the passages.

HAYNES HiNT *Models are identified by their production code letter – refer to 'Identification numbers' at the front of this manual for details.*

2 Pressure cap – check

1 If problems such as overheating or loss of coolant occur, check the entire system as described in Chapter 1. The filler neck (J and K models) or radiator (L, N and R models) cap opening pressure should be checked by a dealer with the special tester required to do the job. If the cap is defective renew it.

3 Coolant reservoir – removal and installation

Removal

1 The coolant reservoir is located under the seat, ahead of the rear wheel. Remove the rear wheel (see Chapter 7) and the rear shock absorber (see Chapter 6), and allow the swingarm to drop down for access.
2 Release the clamp securing the breather hose (coming out of the top of the reservoir) and detach the hose **(see illustration)**.
3 Place a suitable container underneath the reservoir, then release the clamp securing the radiator overflow hose to the base of the reservoir. Detach the hose and allow the coolant to drain into the container **(see illustration 3.2)**.
4 Unscrew the reservoir mounting bolt(s) **(see illustration 3.2)** and remove the reservoir; on L, N and R models note how the lug on the left-hand end locates in the hole **(see illustration)**.

Installation

5 Installation is the reverse of removal. Make sure the hoses are correctly installed and secured with their clamps. On completion refill the reservoir as described in Chapter 1.

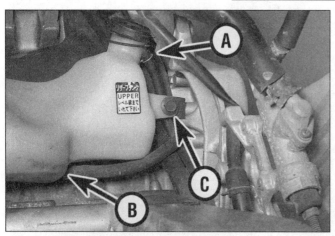

3.2 Detach the breather hose (A) and the overflow hose (B), then unscrew the bolt (C) and remove the reservoir . . .

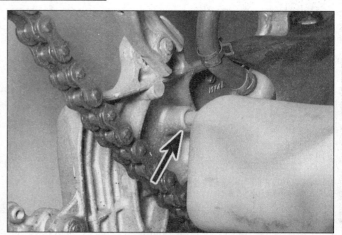

3.4 . . . noting how the lug (arrowed) locates

Cooling system 3•3

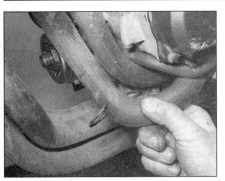

4.3a Release the hose from its clip . . .

4.3b . . . then unscrew the lower mounting bolt(s) and swing the radiator forward . . .

4.3c . . . and disconnect the fan wiring connector (arrowed)

4 Cooling fan and cooling fan switch – check and replacement

Cooling fan

Check

1 If the engine is overheating and the cooling fan isn't coming on, first check the cooling fan circuit fuse (see Chapter 9) and then the fan switch as described in Steps 8 to 13 below.

2 If the fan does not come on, (and the fan switch is good), the fault lies in either the cooling fan motor or the relevant wiring. Test all the wiring and connections as described in Chapter 9.

3 To test the cooling fan motor, remove the lower fairing (see Chapter 8). Release the radiator lower hose from its clip on the left-hand side of the engine, then unscrew the radiator lower mounting bolt(s) and swing the bottom of the radiator forward **(see illustrations)**. Disconnect the fan wiring connector **(see illustration)**. Using a 12 volt battery and two jumper wires, connect the battery positive (+ve) lead to the black/blue fan wire and the battery negative (-ve) lead to earth. Once connected the fan should operate. If it does not, and the wiring is all good, then the fan is faulty. Individual components are available for the fan assembly.

Replacement

> **Warning: The engine must be completely cool before carrying out this procedure.**

4 Remove the radiator (see Section 7).
5 Unscrew the three bolts securing the fan shroud and fan assembly to the radiator, noting that one bolt also secures the earth (ground) cable, and remove the fan **(see illustration)**.
6 If required, unscrew the three nuts on the front and separate the fan from the shroud.
7 Installation is the reverse of removal. Do not forget to attach the earth (ground) cable.
8 Install the radiator (see Section 7).

Cooling fan switch

Check

8 If the engine is overheating and the cooling fan isn't coming on, first check the cooling fan circuit fuse (see Chapter 9). If the fuse is blown, check the fan circuit for a short to earth (see the wiring diagrams at the end of this book).
9 If the fuse is good, remove the lower fairing (see Chapter 8), and disconnect the wiring connector from the fan switch on the left-hand side of the radiator **(see illustration)**. Using a jumper wire if necessary, connect the wire to earth (ground). The fan should come on. If it does, the fan switch is defective and must be renewed. If it does not come on, the fan should be tested (see Step 3).

10 If the fan stays on all the time, disconnect the wiring connector. The fan should stop. If it does, the switch is defective and must be renewed. If it doesn't, check the wiring between the switch and the fan for a short to earth, and the fan itself.
11 If the fan works but is suspected of cutting in at the wrong temperature, a more comprehensive test of the switch can be made as follows.
12 Remove the switch (see Steps 14 to 17). Fill a small heatproof container with coolant and place it on a stove. Connect the positive (+ve) probe of an ohmmeter to the terminal of the switch and the negative (-ve) probe to the switch body, and using some wire or other support suspend the switch in the coolant so that just the sensing portion and the threads are submerged **(see illustration)**. Also place a thermometer capable of reading temperatures up to 110°C in the coolant so that its bulb is close to the switch. **Note:** *None of the components should be allowed to directly touch the container.*
13 Initially the ohmmeter reading should be very high indicating that the switch is open (OFF). Heat the coolant, stirring it gently.

> **Warning: This must be done very carefully to avoid the risk of personal injury.**

When the temperature reaches around 98 to 102°C the meter reading should drop to around zero ohms, indicating that the switch

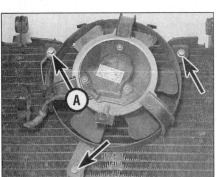

4.5 The fan assembly is secured by three bolts (arrowed). Note the fan switch earth lead secured by the bolt (A)

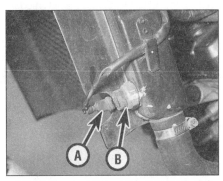

4.9 Disconnect the wiring connector (A) from the fan switch (B)

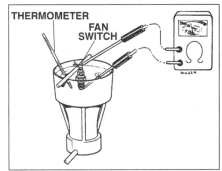

4.12 Fan switch testing set-up

has closed (ON). Now turn the heat off. As the temperature falls below 93 to 97°C the meter reading should show infinite (very high) resistance, indicating that the switch has opened (OFF). If the meter readings obtained are different, or they are obtained at different temperatures, then the fan switch is faulty and must be renewed.

Replacement

⚠️ **Warning: The engine must be completely cool before carrying out this procedure.**

14 Drain the cooling system (see Chapter 1).
15 Remove the lower fairing (see Chapter 8), and disconnect the wiring connector from the fan switch on the left-hand side of the radiator **(see illustration 4.9)**. Unscrew the switch and withdraw it from the radiator. Discard the O-ring as a new one must be used.
16 Apply a suitable sealant to the switch threads, then install the switch using a new O-ring and tighten it to the torque setting specified at the beginning of the Chapter. Take care not to overtighten the switch as the radiator could be damaged.
17 Reconnect the switch wiring and refill the cooling system (see Chapter 1). Install the lower fairing (see Chapter 8).

5 Coolant temperature gauge and sender – check and replacement

Coolant temperature gauge

Check

1 The circuit consists of the sender mounted in the thermostat housing and the gauge assembly mounted in the instrument panel. If the system malfunctions check first that the battery is fully charged and that the fuses are all good.
2 If the gauge is not working, remove the fuel tank, and on J and K models the air filter housing (see Chapter 4). On L, N and R models, displace the fuel pump from its mountings for improved access – do not disconnect any fuel hoses. Disconnect the wire from the sender and turn the ignition switch ON **(see illustrations)**. The temperature gauge needle should be on the 'C' on the gauge. Now earth the sender wire on the engine. The needle should swing immediately over to the 'H' on the gauge. If the needle moves as described, the sender is proven defective and must be renewed.
Caution: Do not earth the wire for any longer than is necessary to take the reading, or the gauge may be damaged.
3 If the needle movement is still faulty, or if it does not move at all, the fault lies in the wiring or the gauge itself. Check all the relevant wiring and wiring connectors (see Chapter 9). If all appears to be well, the gauge is defective and must be renewed.

Replacement

4 See Chapter 9.

Temperature gauge sender

Check

5 Remove the fuel tank, and on J and K models the air filter housing (see Chapter 4). On L, N and R models, displace the fuel pump from its mountings for improved access – do not disconnect any fuel hoses. The sender is mounted in the thermostat housing.
6 Disconnect the sender wiring connector **(see illustration 5.2a or b)**. Using a continuity tester, check for continuity between the sender body and earth (ground). There should be continuity. If there is no continuity, check that the thermostat mounting is secure.
7 Remove the sender (see Steps 9 to 11 below). Fill a small heatproof container with coolant and place it on a stove. Using an ohmmeter, connect the positive (+ve) probe of the meter to the terminal on the sender, and the negative (-ve) probe to the body of the sender **(see illustration 4.12)**. Using some wire or other support suspend the sender in the coolant so that just the sensing portion and the threads are submerged. Also place a thermometer capable of reading temperatures up to 120°C in the water so that its bulb is close to the sender. **Note:** *None of the components should be allowed to directly touch the container.*
8 Heat the coolant, stirring it gently.

⚠️ **Warning: This must be done very carefully to avoid the risk of personal injury.**

As the temperature of the coolant rises, the resistance of the sender will fall. Check that the correct resistance is obtained at the specified temperatures (see Specifications). If the meter readings obtained are different, or they are obtained at different temperatures, then the sender is faulty and must be renewed.

Replacement

⚠️ **Warning: The engine must be completely cool before carrying out this procedure.**

9 Drain the cooling system (see Chapter 1). Remove the fuel tank, and on J and K models

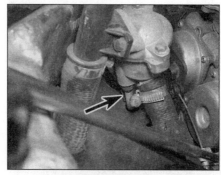

5.2a Temperature sender wiring connector (arrowed) – J and K models

the air filter housing (see Chapter 4). On L, N and R models, displace the fuel pump from its mountings for improved access – do not disconnect any fuel hoses. The sender is mounted in the thermostat housing.
10 Disconnect the sender wiring connector **(see illustration 5.2a or b)**. Unscrew the sender and remove it from the thermostat housing.
11 Apply a smear of sealant to the threads of the new sender, then install it into the thermostat housing and tighten it to the torque setting specified at the beginning of the Chapter. Connect the sender wiring.
12 Refill the cooling system (see Chapter 1).
13 Install the air filter housing (J and K models), the fuel pump (L, N and R models) and the fuel tank (see Chapter 4).

6 Thermostat and thermostat housing – removal, check and installation

Removal

Note: *On J and K models, the complete thermostat housing can be removed without removing the thermostat itself – ignore the points in Step 4 relating to cover and thermostat removal.*

⚠️ **Warning: The engine must be completely cool before carrying out this procedure.**

1 The thermostat is automatic in operation and should give many years service without requiring attention. In the event of a failure, the valve will probably jam open, in which case the engine will take much longer than normal to warm up. Conversely, if the valve jams shut, the coolant will be unable to circulate and the engine will overheat. Neither condition is acceptable, and the fault must be investigated promptly.
2 Drain the cooling system (see Chapter 1). Remove the fuel tank, and on J and K models the air filter housing (see Chapter 4). On L, N and R models, displace the fuel pump from its mounting lugs to improve access (see Chapter 4) – there is no need to disconnect the fuel hoses.

5.2b Temperature sender wiring connector (arrowed) – L, N and R models

Cooling system 3•5

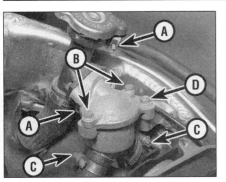

6.4 Filler neck hose clamps (A), cover bolts (B), housing hose clamps (C), mounting bolt (D)

6.5a Slacken the clamps (arrowed) and detach the hoses

6.5b Thermostat housing mounting bolt (arrowed)

3 The thermostat is located in the thermostat housing, which on J and K models is mounted to the right-hand frame spar, and on L, N and R models is mounted on the crankcase above the gearbox.

4 On J and K models, to remove the thermostat, slacken the clamp securing the overflow hose to the filler neck and detach the hose **(see illustration)**. Unscrew the two bolts securing the cover and separate it from the housing. Withdraw the thermostat, noting how it fits. Discard the cover O-ring as a new one must be used. To remove the thermostat housing, disconnect the temperature sender wiring connector **(see illustration 5.2a)**. Slacken the clamps securing the hoses to the filler neck and the thermostat housing and detach the hoses, noting which fits where. Unscrew the bolt securing the housing to the frame and remove the housing.

5 On L, N and R models, the thermostat housing must be removed to access the thermostat. Disconnect the temperature sender wiring connector **(see illustration 5.2b)**. Slacken the clamps securing the hoses to the thermostat housing and detach the hoses, noting which fits where **(see illustration)**. Unscrew the bolt securing the housing to the engine, accessing it from the left-hand side of the bike, and remove the housing **(see illustration)**. To remove the thermostat, unscrew the two bolts securing the cover and separate it from the housing **(see illustration)**. Withdraw the thermostat, noting how it fits. Discard the cover O-ring as a new one must be used.

Check

6 Examine the thermostat visually before carrying out the test. If it remains in the open position at room temperature, it should be renewed.

7 Suspend the thermostat by a piece of wire in a container of cold water. Place a thermometer in the water so that the bulb is close to the thermostat **(see illustration)**. Heat the water, noting the temperature when the thermostat opens, and compare the result with the specifications given at the beginning of the Chapter. Also check the amount the valve opens after it has been heated at 95°C for a few minutes and compare the measurement to the specifications. If the readings obtained differ from those given, the thermostat is faulty and must be renewed.

8 In the event of thermostat failure, as an emergency measure only, it can be removed and the machine used without it. **Note:** *Take care when starting the engine from cold as it will take much longer than usual to warm up.* Ensure that a new unit is installed as soon as possible.

Installation

9 Installation is the reverse of removal. Use a new O-ring **(see illustration)**. Make sure the thermostat seats correctly **(see illustration)**. On L, N and R models, make sure the locating hole in the hosing fits onto the locating pin **(see illustration)**. Tighten the bolts and the

6.5c Unscrew the bolts (arrowed), lift off the cover and remove the thermostat

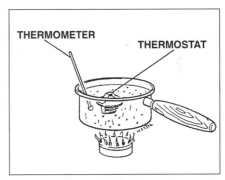

6.7 Thermostat testing set-up

6.9a Fit a new O-ring into the groove . . .

6.9b . . . then install the thermostat

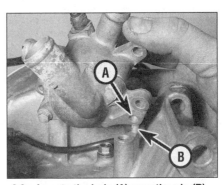

6.9c Locate the hole (A) over the pin (B) – L, N and R models shown

3

3•6 Cooling system

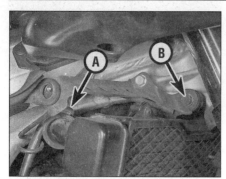

7.2a Radiator upper hose (A) and mounting bolt (B) (one on each side) . . .

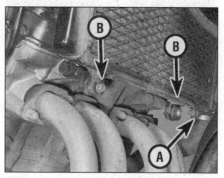

7.2b . . . and lower hose (A) and mounting bolts (B) – J and K models

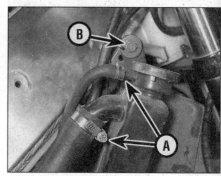

7.2c Hoses (A) and mounting bolts (B) – right-hand side . . .

hose clamps securely (see illustrations 6.4 or 6.5a).
10 Refill the cooling system (see Chapter 1).
11 Install the air filter housing (J and K models) and the fuel tank (see Chapter 4).

7 Radiator – removal and installation

Removal

 Warning: The engine must be completely cool before carrying out this procedure.

1 Remove the lower fairing (see Chapter 8) and drain the cooling system (see Chapter 1). On L, N and R models, also remove the fairing and the front trim panels (see Chapter 8).
2 Slacken the clamps securing the main hoses to the top right-hand side and bottom left-hand side of the radiator and detach the hoses (see illustrations). On L, N and R models, also detach the small hoses from the filler neck and the top left-hand side of the radiator, and unclip the rubber insulating pad from the top of the radiator (see illustrations). Unscrew the radiator lower mounting bolt(s) and swing the bottom of the radiator forward. Disconnect the fan wiring connector (see illustration 4.3c). Also disconnect the wiring connector from the fan switch in the radiator (see illustration 4.9).
3 Unscrew the radiator upper mounting bolts, noting the arrangement of the collars and rubber grommets, and remove the radiator (see illustrations 7.2a and b or c and d).
4 If necessary, separate the cooling fan (see Section 4) from the radiator.
5 Where fitted, remove the stone guard from the radiator. Check the stone guard and the radiator for signs of damage and clear any dirt or debris that might obstruct air flow and inhibit cooling. If the radiator fins are badly damaged or broken the radiator must be renewed. Also check the rubber mounting grommets, and renew them if necessary.

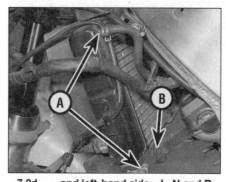

7.2d . . . and left-hand side – L, N and R models

Installation

6 Installation is the reverse of removal, noting the following.
a) Make sure the rubber grommets and collars are correctly installed with the mounting bolts.
b) Make sure that the fan wiring is correctly connected.
c) Ensure the coolant hoses are in good condition (see Chapter 1), and are securely retained by their clamps, using new ones if necessary.
d) On completion refill the cooling system as described in Chapter 1.

8 Water pump – check, removal and installation

Check

1 The water pump is located on the lower left-hand side of the engine. Visually check the area around the pump for signs of leakage.
2 To prevent leakage of water from the cooling system to the lubrication system and vice versa, two seals are fitted on the pump shaft. On the underside of the pump body there is also a drainage hole (see illustration). If either seal fails, this hole should allow the coolant or oil to escape and prevent the oil and coolant mixing.

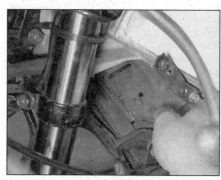

7.2e Also detach the insulating pad

3 The seal on the water pump side is of the mechanical type which bears on the rear face of the impeller. The second seal, which is mounted behind the mechanical seal is of the normal feathered lip type. However, neither seal is available as a separate item as the pump is sold as an assembly. Therefore, if on inspection the drainage hole shows signs of leakage, the pump must be removed and renewed.

Removal

4 Drain the coolant (see Chapter 1). Place a suitable container below the water pump to catch any residual oil as the water pump is removed.
5 To remove the pump cover, unscrew the

8.2 Check the drain hole (arrowed) for evidence of leakage

Cooling system 3•7

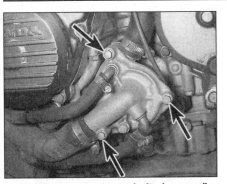

8.5a Water pump cover bolts (arrowed) and hoses – J and K models

8.5b Water pump cover bolts (arrowed) and hoses – L, N and R models

8.6 Check for freeplay between the impeller and the housing

three bolts securing the cover to the pump (one of which also secures the pump to the crankcase), and remove the cover **(see illustrations)**. There is no need to detach the hoses unless you want to. Note the position of each bolt as they are different lengths. Discard the cover O-ring as a new one must be used.

6 Wiggle the water pump impeller back-and-forth and in-and-out **(see illustration)**. If there is excessive movement the pump must be renewed. Also check for corrosion or a build-up of scale in the pump body and clean or renew the pump as necessary.

7 To remove the pump body, slacken the clamp securing the coolant hose to the pump body and detach the hose **(see illustration 8.8a)**. Unscrew the remaining bolt securing the pump to the crankcase, on L, N and R models noting the wiring clip **(see illustration 8.8b)**. Carefully draw the pump from the crankcase, noting how it fits **(see illustration 8.9b)**. Remove the O-ring from the rear of the pump body and discard it as a new one must be used.

8 To remove the whole pump as an assembly, slacken the clamps securing the coolant hoses to the pump cover and body, noting that access to the inner main hose is restricted (detach it from the engine if it proves stubborn), and detach the hoses, noting which fits where **(see illustrations 8.5a or b and 8.8a)**. Unscrew the two bolts securing the pump assembly to the crankcase, on L, N and R models noting the wiring clip, and carefully draw the pump out, noting how it fits **(see illustration and 8.9b)**. Note the position of each bolt as they are different lengths. Remove the O-ring from the rear of the pump body and discard it as a new one must be used **(see illustration 8.9a)**. Separate the cover from the pump if required.

Installation

9 Apply a smear of engine oil to the new pump body O-ring and install it onto the rear of the pump body **(see illustration)**. Install the pump into the crankcase, aligning the slot in the impeller shaft with the tab on the oil pump shaft **(see illustration)**.

10 If the pump was removed as a whole, install the two bolts securing the pump to the crankcase, on L, N and R models not forgetting the wiring clip **(see illustration 8.8b)**. Attach the coolant hoses to the pump and secure them with their clamps **(see illustrations 8.8a and 8.5a or b)**. Otherwise install the single bolt securing the body to the crankcase, on L, N and R models not forgetting the wiring clip, then attach the hose and secure it with its clamp **(see illustrations 8.8b and a)**.

11 To fit the cover, install a new O-ring into its groove in the pump body **(see illustration)**.

8.8a Access to the inner hose (arrowed) is restricted

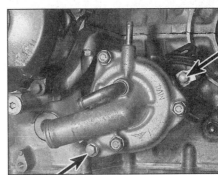

8.8b Pump mounting bolts (arrowed)

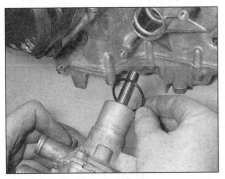

8.9a Fit a new O-ring . . .

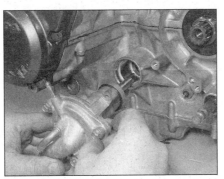

8.9b . . . and install the pump

8.11 Fit a new O-ring into the groove in the body . . .

12 Install the cover onto the pump, then install the bolts and tighten them securely **(see illustrations 8.5a and b)**. Make sure the different bolts are in their correct locations.
13 If detached, attach the coolant hoses to the pump cover and secure them with their clamps **(see illustrations 8.5a and b)**.
14 Refill the cooling system (see Chapter 1).

9 Coolant hoses – removal and installation

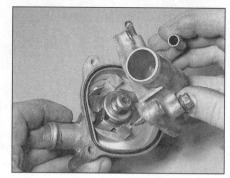

8.12 ... then fit the cover

Removal

1 Before removing a hose, drain the coolant (see Chapter 1).
2 Use a screwdriver to slacken the larger-bore hose clamps, then slide them back along the hose and clear of the union spigot. The smaller-bore hoses are secured by spring clamps which can be expanded by squeezing their ears together with pliers.
Caution: The radiator unions are fragile. Do not use excessive force when attempting to remove the hoses.

3 If a hose proves stubborn, release it by rotating it on its union before working it off. If all else fails, cut the hose with a sharp knife then slit it at each union so that it can be peeled off in two pieces. Whilst this means renewing the hose, it is preferable to buying a new radiator.
4 The water pipe inlet union to the cylinder block can be removed by unscrewing the retaining bolts **(see illustrations)**. On J and K models, remove the carburettors (see Chapter 4) for access. If the union is removed, the O-ring must be renewed. The outlet pipes are each secured by a single bolt **(see illustration)**. If they are removed, the O-rings must be renewed.

Installation

5 Slide the clips onto the hose and then work it on to its respective union.

> **HAYNES HINT**: If the hose is difficult to push on its union, it can be softened by soaking it in very hot water, or alternatively a little soapy water can be used as a lubricant.

6 Rotate the hose on its unions to settle it in position before sliding the clamps into place and tightening them securely.
7 If either the inlet union to the cylinder block or the outlet unions from the cylinder head have been removed, fit a new O-ring, then install the union and tighten the mounting bolts securely.

9.4a On J and K models, the union (arrowed) is on the back of the block ...

9.4b ... on L, N and R models, it is on the front (arrowed)

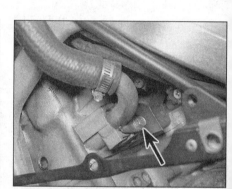

9.4c Each outlet pipe is secured by a bolt (arrowed)

Chapter 4
Fuel and exhaust systems

Contents

Air filter – cleaning and renewal .see Chapter 1	Fuel pump – check, removal and installation 14
Air filter housing – removal and installation 4	Fuel system – check . . . . .see Chapter 1
Carburettor overhaul – general information 6	Fuel tank – cleaning and repair 3
Carburettor synchronisation . . . . .see Chapter 1	Fuel tank and fuel tap – removal and installation 2
Carburettors – disassembly, cleaning and inspection 8	General information and precautions 1
Carburettors – reassembly, and float height check 10	Idle fuel/air mixture adjustment – general information 5
Carburettors – removal and installation 7	Idle speed – check . . . . .see Chapter 1
Carburettors – separation and joining 9	Throttle and choke cables – check and adjustment . . . .see Chapter 1
Choke cable – removal and installation 12	Throttle cables – removal and installation 11
Exhaust system – removal and installation 13	Throttle position sensor – check and replacement . . . . .see Chapter 5
Fuel hoses – check and renewal . . . . .see Chapter 1	

Degrees of difficulty

Easy, suitable for novice with little experience	Fairly easy, suitable for beginner with some experience	Fairly difficult, suitable for competent DIY mechanic	Difficult, suitable for experienced DIY mechanic	Very difficult, suitable for expert DIY or professional

Specifications

Note: *Models are identified by their production code letter – refer to 'Identification numbers' at the front of this manual for details.*

Fuel
Grade	Unleaded, minimum 91 RON (Research Octane Number)
Fuel tank capacity	15.0 litres

Carburettors
Type	CV
Pilot screw setting (turns out)	
J and K models	2 1/2 turns out
L, N and R models	2 1/4 turns out
Float height	
J and K models	7.0 mm
L, N and R models	13.7 mm
Idle speed	see Chapter 1

Carburettor jet sizes
Pilot jet	
J models	40
K, L, N and R models	35
Main jet	
J models	115
K models	105
L and N models	108
R models	105

4•2 Fuel and exhaust systems

Torque settings

Note: *Where a specified setting is not given for a particular bolt, the general settings listed at the beginning apply. The dimension given applies to the diameter of the thread, not the head.*

5 mm bolt/nut	5 Nm
6 mm bolt/nut	10 Nm
8 mm bolt/nut	22 Nm
10 mm bolt/nut	35 Nm
12 mm bolt/nut	55 Nm
6 mm flange bolt with 8 mm head	9 Nm
6 mm flange bolt/nut with 10 mm head	12 Nm
8 mm flange bolt/nut	27 Nm
10 mm flange bolt/nut	40 Nm
Exhaust downpipe nuts	12 Nm
Exhaust and silencer mounting bolts (J and K models)	27 Nm

1 General information and precautions

General information

The fuel system consists of the fuel tank, fuel tap, filters, fuel pump and relay, carburettors, fuel hoses and control cables.

There are two fuel filters, one is fitted inside the fuel tank and is part of the tap, and the other is in the fuel line to the pump.

The carburettors used on all models are CV types. On all models there is a carburettor for each cylinder. For cold starting, a choke knob is mounted on the top yoke, and is connected to the carburettors by a cable.

Air is drawn into the carburettors via an air filter which is housed under the fuel tank.

The exhaust system is a four-into-one design.

Many of the fuel system service procedures are considered routine maintenance items and for that reason are included in Chapter 1.

Precautions

 Warning: Petrol (gasoline) is extremely flammable, so take extra precautions when you work on any part of the fuel system. Don't smoke or allow open flames or bare light bulbs near the work area, and don't work in a garage where a natural gas-type appliance is present. If you spill any fuel on your skin, rinse it off immediately with soap and water. When you perform any kind of work on the fuel system, wear safety glasses and have a fire extinguisher suitable for a class B type fire (flammable liquids) on hand.

Always perform service procedures in a well-ventilated area to prevent a build-up of fumes.

Never work in a building containing a gas appliance with a pilot light, or any other form of naked flame. Ensure that there are no naked light bulbs or any sources of flame or sparks nearby.

Do not smoke (or allow anyone else to smoke) while in the vicinity of petrol (gasoline) or of components containing it. Remember the possible presence of vapour from these sources and move well clear before smoking.

Check all electrical equipment belonging to the house, garage or workshop where work is being undertaken (see the Safety first! section of this manual). Remember that certain electrical appliances such as drills, cutters etc. create sparks in the normal course of operation and must not be used near petrol (gasoline) or any component containing it. Again, remember the possible presence of fumes before using electrical equipment.

Always mop up any spilt fuel and safely dispose of the rag used.

Any stored fuel that is drained off during servicing work must be kept in sealed containers that are suitable for holding petrol (gasoline), and clearly marked as such; the containers themselves should be kept in a safe place. Note that this last point applies equally to the fuel tank if it is removed from the machine; also remember to keep its filler cap closed at all times.

Read the Safety first! section of this manual carefully before starting work.

 Models are identified by their production code letter – refer to 'Identification numbers' at the front of this manual for details.

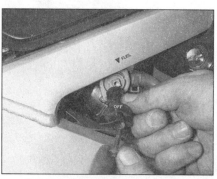

2.2 Remove the fuel tap knob . . .

2 Fuel tank and fuel tap – removal and installation

 Warning: Refer to the precautions given in Section 1 before starting work.

Fuel tank

Removal

1 Make sure the fuel filler cap is secure and fuel tap is in the OFF position. On J and K models, remove the seat cowling (see Chapter 8). On L, N and R models, remove the rider's seat (see Chapter 8).

2 Remove the screw from the centre of the fuel tap knob and remove the knob **(see illustration)**.

3 Place a rag under the fuel tap to catch any residual fuel as the hose is detached. Release the clamp securing the fuel hose to the tap and detach the hose **(see illustration)**.

4 Unscrew the bolt securing the front of the tank to the frame and carefully lift the tank away **(see illustrations)**.

5 Inspect the tank mounting rubbers for signs of damage or deterioration and renew them if necessary.

Installation

6 Check that the tank mounting rubbers and the collar for the front mounting are fitted, then carefully lower the fuel tank into position.

2.3 . . . then detach the fuel hose

Fuel and exhaust systems 4•3

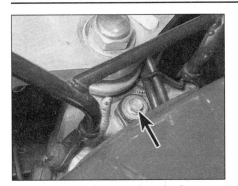

2.4a Unscrew the bolt at the front (arrowed) . . .

2.4b . . . and remove the tank

2.14 Unscrew the nut (arrowed) and remove the tap

7 Install the front mounting bolt and washer and tighten it securely **(see illustration 2.4)**.
8 Fit the fuel hose onto its union on the tap and secure it with the clamp **(see illustration 2.3)**.
9 Fit the fuel tap knob and tighten its screw **(see illustration 2.2)**.
10 Start the engine and check that there is no sign of fuel leakage, then shut if off.
11 On J and K models, install the seat cowling (see Chapter 8). On L, N and R models, install the rider's seat (see Chapter 8).

Fuel tap

Removal

12 The tap should not be removed unnecessarily from the tank to prevent the possibility of damaging the O-ring or the filter. If the fuel tap is leaking, tightening the retaining nut and the assembly screws may help. If leakage persists, remove the tap as described below and renew the O-ring. If the tap appears blocked, remove it and check the filter (see below). If a leakage or blockage cannot be cured, fit a new tap.
13 Remove the fuel tank as described above. Connect a drain hose to the fuel hose union on the tap and insert its end in a container suitable and large enough for storing the petrol. Refit the fuel tap knob and turn the tap to the RES position to allow the tank to drain.
14 Unscrew the nut securing the tap and withdraw it from the tank **(see illustration)**.

Check the condition of the O-ring. If it is in good condition it can be re-used, though it is better to use a new one. If it is in any way deteriorated or damaged it must be renewed.
15 Clean the gauze filter to remove all traces of dirt and fuel sediment. Check the gauze for holes. If any are found, a new one should be fitted.

Installation

16 Install the fuel tap into the tank, using a new O-ring if required, and tighten the nut securely **(see illustration 2.14)**.
17 Install the fuel tank (see above).

3 Fuel tank – cleaning and repair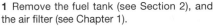

1 All repairs to the fuel tank should be carried out by a professional who has experience in this critical and potentially dangerous work. Even after cleaning and flushing of the fuel system, explosive fumes can remain and ignite during repair of the tank.
2 If the fuel tank is removed from the bike, it should not be placed in an area where sparks or open flames could ignite the fumes coming out of the tank. Be especially careful inside garages where a natural gas-type appliance is located, because the pilot light could cause an explosion.

4 Air filter housing – removal and installation

Removal

J and K models

1 Remove the fuel tank (see Section 2), and the air filter (see Chapter 1).
2 Remove the six screws securing the filter housing to the carburettor air duct holder and the bolt securing the front of the housing to the frame **(see illustration)**.
3 Lift the housing up off the carburettors, then release the clamp securing the engine breather hose to the union on the front left-hand corner of the housing and detach the hose **(see illustration)**.

L, N and R models

4 Remove the fuel tank (see Section 2).
5 Slacken the clamps securing the housing to the air inlet ducts at the front, then release the clamps securing the engine breather hoses to the unions on the front of the housing and detach the hoses **(see illustration)**. If access is too restricted, this can be done after the housing has been displaced from the carburettors.
6 Slacken the clamps securing the housing to the carburettor inlet ducts at the back (see

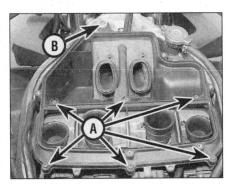

4.2 Remove the screws (A) and the bolt (B) . . .

4.3 . . . then lift the housing and detach the hose (arrowed)

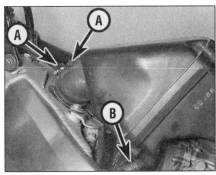

4.5 Slacken the clamps (A) and detach the hose (B) on each side . . .

4•4 Fuel and exhaust systems

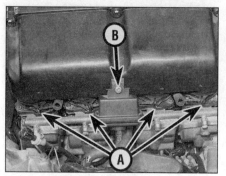

4.6 ... then slacken the four clamps (A) and detach the sub-air cleaner by removing the screw (B) ...

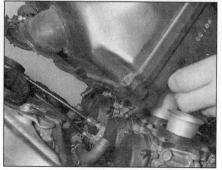

4.7 ... and remove the housing

4.8 On L, N and R models, locate the tabs on the inlet on each side of the lug on the air duct (arrowed)

illustration). Also remove the screw securing the sub-air cleaner to the rear of the housing and displace it.

7 Lift the housing up off the carburettors and remove it.

Installation

8 Installation is the reverse of removal. Check the condition of the breather hose(s) and clamp(s) and renew them if necessary. On J and K models, make sure the collar is in place in the front mounting rubber. On L, N and R models, make sure the inlets locate correctly onto the air ducts and tighten the clamps **(see illustration)**.

5 Idle fuel/air mixture adjustment – general information

1 The pilot screws **(see illustration 8.12a or 8.12b)** are set to their correct position by the manufacturer and should not be adjusted or removed unless it is necessary to do so during a carburettor overhaul. If the screws are to be removed, record the pilot screw's current setting by turning the screw it in until it seats lightly, counting the number of turns necessary to achieve this, then fully unscrew it. On installation, the screw is simply backed out the number of turns you've recorded.

2 If the engine runs extremely rough at idle or continually stalls, and if a carburettor overhaul

does not cure the problem, take the motorcycle to a dealer equipped with an exhaust gas analyser. They will be able to properly adjust the idle fuel/air mixture to achieve a smooth idle and restore low speed performance.

6 Carburettor overhaul – general information

1 Poor engine performance, hesitation, hard starting, stalling, flooding and backfiring are all signs that major carburettor maintenance may be required.

2 Keep in mind that many so-called carburettor problems are really not carburettor problems at all, but mechanical problems within the engine or ignition system malfunctions. Try to establish for certain that the carburettors are in need of maintenance before beginning a major overhaul.

3 Check the fuel tap and filters, the fuel hoses, the fuel pump and its relay, the inlet manifold joint clamps, the air filter, the ignition system, the spark plugs and carburettor synchronisation before assuming that a carburettor overhaul is required.

4 Most carburettor problems are caused by dirt particles, varnish and other deposits which build up in and block the fuel and air passages. Also, in time, gaskets and O-rings

shrink or deteriorate and cause fuel and air leaks which lead to poor performance.

5 When overhauling the carburettors, disassemble them completely and clean the parts thoroughly with a carburettor cleaning solvent and dry them with filtered, unlubricated compressed air. Blow through the fuel and air passages with compressed air to force out any dirt that may have been loosened but not removed by the solvent. Once the cleaning process is complete, reassemble the carburettor using new gaskets and O-rings.

6 Before disassembling the carburettors, make sure you have all necessary O-rings and other parts, some carburettor cleaner, a supply of clean rags, some means of blowing out the carburettor passages and a clean place to work. It is recommended that only one carburettor be overhauled at a time to avoid mixing up parts.

7 Carburettors – removal and installation

Warning: Refer to the precautions given in Section 1 before starting work.

Removal

1 Remove the fuel tank and the air filter housing (see Sections 2 and 4). On L, N and R models, remove the rear trim panels (see Chapter 8).

2 Detach the choke cable from the carburettors (see Section 12, Step 2).

3 Detach the throttle cables from the carburettors (see Section 11, Steps 3 and 4). If access is too restricted, detach them after the carburettors have been lifted off the cylinder head inlets.

4 On L, N and R models, disconnect the throttle position sensor wiring connector from the right-hand end of the carburettors **(see illustration)**. Also release the idle speed adjuster from its holder and feed it through to the base of the carburettors **(see illustration)**.

5 Release the clamp securing the fuel supply

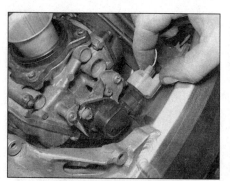

7.4a Disconnect the throttle position sensor wiring connector ...

7.4b ... and release the idle speed adjuster

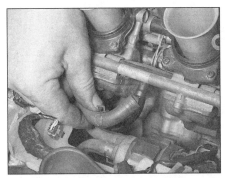

7.5 Release the clip and detach the fuel hose

7.7a Slacken the upper clamp on No. 1 cylinder

7.7b Access No. 2 via the hole in the left-hand side of the frame...

hose to the inlet union on the carburettors and detach the hose, being prepared for any residual fuel **(see illustration)**.

6 On J and K models, fully slacken the upper clamps on the cylinder head inlet rubbers, then ease the carburettors off the inlets, noting that they are quite a tight fit, and remove them. **Note:** *Keep the carburettors upright to prevent fuel spillage from the float chambers and the possibility of the piston diaphragms being damaged.*

7 On L, N and R models, fully slacken the upper clamps on the cylinder head inlet rubbers on cylinders 1 and 4, and the lower clamps on cylinders 2 and 3. Access the screws on cylinders 2, 3 and 4 using a long screwdriver inserted through the hole in each side of the frame **(see illustrations)**. Ease the carburettors off the inlets, noting that they are quite a tight fit, and remove them **(see illustration)**. **Note:** *Keep the carburettors upright to prevent fuel spillage from the float chambers and the possibility of the piston diaphragms being damaged.*
Caution: Stuff clean rag into each cylinder head inlet after removing the carburettors to prevent anything from falling in.

8 Place a suitable container below the float chambers, then slacken the drain screw on each chamber in turn and drain all the fuel from the carburettors **(see illustration)**. Discard the drain screw O-rings as new ones must be used. Once all the fuel has been drained, fit the new O-rings and tighten the drain screws securely.

9 If necessary, slacken the clamps securing the inlet rubbers to the cylinder head or carburettors and remove the rubbers, noting which way up and round they fit.

Installation

10 Installation is the reverse of removal, noting the following.
 a) Check for cracks or splits in the cylinder head inlet rubbers, and renew them if necessary.
 b) If removed, make sure the inlet rubbers are installed with the CARB marking facing out (towards the carburettor), and so that the arrow next to the UP mark points upwards and the slot in the base of the rubber locates over the lug on the inlet manifold. Also make sure the wire bracket on each clamp locates over the lug on the rubber for correct positioning of the clamps.
 c) Make sure the carburettors are fully

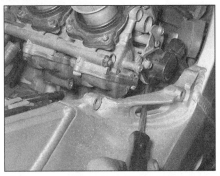

7.7c ...and Nos. 3 and 4 via the hole in the right-hand side

 engaged with the inlet rubbers and the clamps are securely tightened.
 d) Make sure all hoses are correctly routed and secured and not trapped or kinked.
 e) Refer to Section 11 for installation of the throttle cables and section 12 for the choke cable. Check the operation of the cables and adjust them as necessary (see Chapter 1).
 f) Check idle speed and carburettor synchronisation and adjust as necessary (see Chapter 1).

7.7d Carefully lift the carburettors off the inlets

7.8 Carburettor drain screws (arrowed)

8.3a Top cover screws (A) – J and K models. Remove the choke cable bracket (B) to access No. 3 carburettor

8.3b Top cover screws (arrowed) – L, N and R models

8 Carburettors – disassembly, cleaning and inspection

 Warning: *Refer to the precautions given in Section 1 before starting work.*

Disassembly

1 Remove the carburettors from the machine as described in the previous Section. **Note:** *Do not separate the carburettors unless absolutely necessary; each carburettor can be dismantled sufficiently for all normal cleaning and adjustments while in place on the mounting brackets. Dismantle the carburettors separately to avoid interchanging parts.*

2 On J and K models, if required, remove the air duct assembly from the carburettors (see Section 9, Step 3), but note that this is not necessary for carburettor disassembly and cleaning, unless they are being separated.

3 Unscrew and remove the top cover retaining screws **(see illustrations)**. Lift off the cover and remove the spring from inside the piston **(see illustration 10.14a)**. On J and K models, when working on No. 3 carburettor, it is necessary to remove the choke cable holder to access the screws.

4 Carefully peel the diaphragm away from its sealing groove in the carburettor and withdraw the diaphragm and piston assembly **(see illustration)**. Note how the tab on the diaphragm fits in the recess in the carburettor body.

Caution: *Do not use a sharp instrument to displace the diaphragm as it is easily damaged.*

5 On J and K models, if required, gently push down on the jet needle retainer using a Phillips screwdriver and rotate it until its tab is released from the protrusions inside the piston, then remove the retainer and spring **(see illustrations)**. Push the needle up from the bottom of the piston and withdraw it from the top **(see illustration)**. Take care not to lose the washer that fits between the head of the needle and the piston.

6 On L, N and R models, if required, thread a 4 mm screw into the top of the needle holder (one of the top cover retaining screws is ideal), then grasp the screw head using a pair of pliers and carefully draw the holder out of the piston **(see illustrations)**. Note the spring that

8.4 Peel the diaphragm off the carburettor and withdraw the diaphragm and piston assembly

8.5a Push down on the retainer and turn it to release it . . .

8.5b . . . then remove the retainer . . .

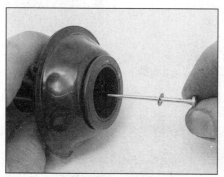

8.5c . . . and push the needle up from the bottom

8.6a Thread a screw into the retainer . . .

Fuel and exhaust systems 4•7

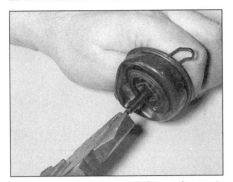

8.6b ... and use it to pull the retainer out

8.7a Float chamber screws (arrowed) – L, N and R models shown

8.7b The holder is secured by two screws (arrowed)

fits inside the holder. Push the needle up from the bottom of the piston and withdraw it from the top **(see illustration 10.12a)**. Note the washer that fits between the head of the needle and the piston. Discard the O-ring on the holder as a new one must be used.

7 Remove the screws securing the float chamber to the base of the carburettor and remove it **(see illustration)**. Remove the rubber gasket and discard it as a new one must be used. On J and K models, when working on Nos. 1 or 2 carburettors, first remove the idle speed adjuster holder **(see illustration)**.

8 Using a pair of thin-nose pliers, carefully withdraw the float pin **(see illustration)**. If necessary, displace the pin using a small punch or a nail. Remove the float and unhook the float needle valve, noting how it fits onto the tab on the float **(see illustration 10.8a)**. On J and K models, if required, unscrew and remove the float needle valve seat and its sealing washer **(see illustration 8.9a)**.

9 Unscrew and remove the pilot jet **(see illustrations)**.

10 Unscrew and remove the main jet from the base of the needle jet **(see illustration 8.9a or b)**.

11 Unscrew and remove the needle jet **(see illustration 8.9a or b)**.

12 The pilot screw can be removed if required, but note that its setting will be disturbed (see **Haynes Hint**). Unscrew and remove the pilot screw along with its spring, washer and O-ring **(see illustrations)**. Discard the O-ring as a new one must be used.

 HAYNES HINT To record the pilot screw's current setting, turn the screw in until it seats lightly, counting the number of turns necessary to achieve this, then fully unscrew it. On installation, the screw is simply backed out the number of turns you've recorded.

13 On J and K models, if you want to remove a choke plunger, slacken the screw on the plunger actuating arm, then lift the arm off the plunger **(see illustration)**. Unscrew the choke

8.8 Withdraw the float pin and remove the float assembly

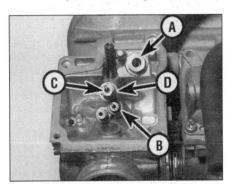

8.9a Float needle valve seat (A), pilot jet (B), main jet (C), needle jet (D) – J and K model

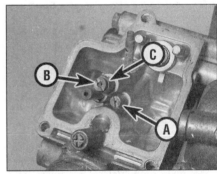

8.9b Pilot jet (A), main jet (B), needle jet (C)

8.12a Pilot screw (arrowed) – J and K models

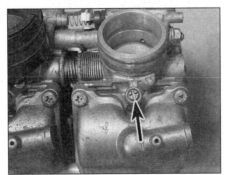

8.12 Pilot screw (arrowed) – L, N and R models

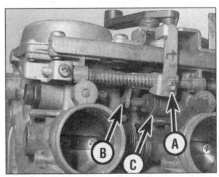

8.13 Slacken the screw on the arm (A), then lift the arm off the plunger (B) and unscrew the nut (C)

4•8 Fuel and exhaust systems

8.14a Remove the screws (arrowed) and washers and lift off the bar . . .

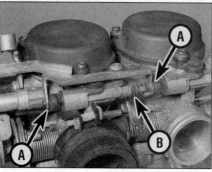

8.14b . . . noting how the arms locate on the plungers (A) and the return spring (B) fits

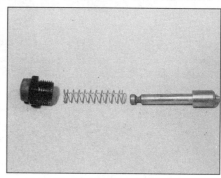

8.18 Choke plunger components – L, N and R models

plunger nut and withdraw the plunger assembly from the carburettor body, noting how it fits.

14 On L, N and R models, if you want to remove a choke plunger, remove the screws securing the choke plunger linkage bar to the carburettors, then remove the plastic washers **(see illustration)**. Lift off the bar and remove the return spring, noting how it fits, and remove the collars **(see illustration)**. Unscrew the choke plunger nut and withdraw the plunger and spring from the carburettor body, noting how they fit. Take care not to lose the spring when removing the nut.

Cleaning

Caution: Use only a petroleum based solvent for carburettor cleaning. Don't use caustic cleaners.

15 Submerge the metal components in the solvent for approximately thirty minutes (or longer, if the directions recommend it).
16 After the carburettor has soaked long enough for the cleaner to loosen and dissolve most of the varnish and other deposits, use a nylon-bristled brush to remove the stubborn deposits. Rinse it again, then dry it with compressed air.
17 Use a jet of compressed air to blow out all of the fuel and air passages in the main and upper body, not forgetting the air jets in the carburettor inlet.

Caution: Never clean the jets or passages with a piece of wire or a drill bit, as they will be enlarged, causing the fuel and air metering rates to be upset.

Inspection

18 Check the operation of the choke plunger. If it doesn't move smoothly, inspect the plunger assembly and linkage components and renew any that are worn, damaged or bent **(see illustration)**.
19 If removed from the carburettor, check the tapered portion of the pilot screw and the spring and O-ring for wear or damage **(see illustration)**. Renew them if necessary.
20 Check the carburettor body, float chamber and top cover for cracks, distorted sealing surfaces and other damage. If any defects are found, renew the faulty component, although renewal of the entire carburettor will probably be necessary (check with a dealer on the availability of separate components).
21 Check the piston diaphragm for splits, holes and general deterioration **(see illustration)**. Holding it up to a light will help to reveal problems of this nature.
22 Insert the piston in the carburettor body and check that the piston moves up-and-down smoothly. Check the surface of the piston for wear. If it's worn excessively or doesn't move smoothly in the guide, renew the components as necessary.
23 Check the jet needle for straightness by rolling it on a flat surface such as a piece of glass. Renew it if it's bent or if the tip is worn.
24 Check the tip of the float needle valve and the valve seat **(see illustration)**. If either has grooves or scratches in it, or is in any way worn, they must be renewed as a set. Gently push down on the rod on the top of the needle valve then release it – if it doesn't spring back, renew the valve. On J and K models, also check the condition of the valve seat sealing washer.
25 Operate the throttle shaft to make sure the throttle butterfly valve opens and closes smoothly. If it doesn't, cleaning the throttle linkage may help. Otherwise, renew the carburettor.
26 Check the float for damage. This will usually be apparent by the presence of fuel inside the float. If the float is damaged, it must be renewed.

9 Carburettors – separation and joining

 Warning: Refer to the precautions given in Section 1 before proceeding

Separation

1 The carburettors do not need to be separated for normal overhaul. If you need to separate them (to renew a carburettor body, for example), refer to the following procedure.
2 Remove the carburettors from the machine (see Section 7). Mark the body of each carburettor with its cylinder location to ensure that it is positioned correctly on reassembly **(see illustrations)**. On L, N and R models, if

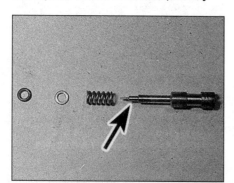

8.19 Check the tapered portion of the pilot screw (arrowed) for wear

8.21 Check the piston diaphragm for splits or other damage

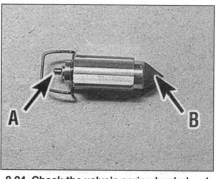

8.24 Check the valve's spring loaded rod (A) and tip (B) for wear or damage

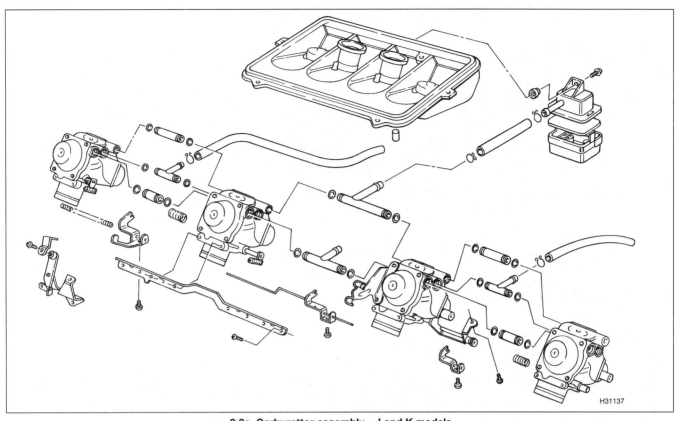

9.2a Carburettor assembly – J and K models

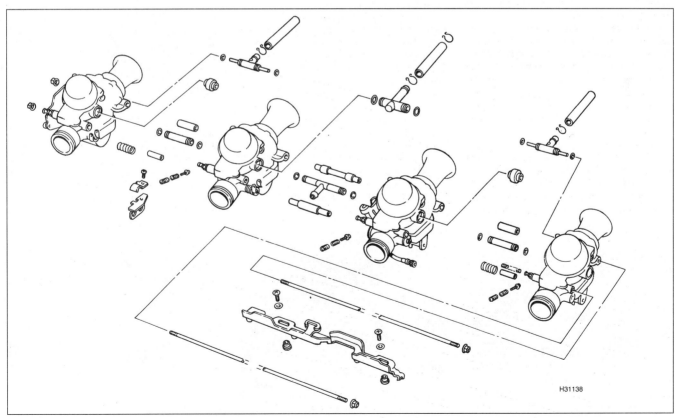

9.2b Carburettor assembly – L, N and R models

4•10 Fuel and exhaust systems

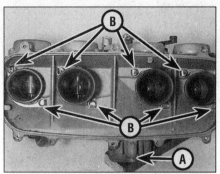

9.3a Remove the screw securing the sub-air cleaner (A), then bend back the tabs and remove the duct holder screws (B)

9.3b Remove the screws (arrowed) and the bracket

9.3c Slacken the screws (arrowed) and lift off the arms

required, remove the throttle position sensor (see Chapter 5).

3 On J and K models, remove the screw securing the sub-air cleaner to the air duct holder (see illustration). Bend back the tabs on the lockplates between each pair of screws securing the air duct holder to the carburettors, then remove the screws and lift off the duct holder, noting which way round it fits and how it locates onto the dowels. Also note how the rubber flanges locate around the air pipes. Remove the dowels from each carburettor inlet if they are loose. Discard the lockplates as new ones should be used. Remove the screws securing the idle speed adjuster bracket to the float chambers on Nos. 1 and 2 carburettors and detach the bracket (see illustration 8.7b). Remove the screws securing the choke actuating lever bracket to the top of the carburettors and remove the bracket (see illustration). Slacken all the choke plunger actuating arm screws and lift the arms off the plungers (see illustration). Remove the screws securing the joining plate to the top of the carburettors and remove the plate (see illustration).

4 On L, N and R models, remove the screws securing the choke plunger linkage bar to the carburettors, then remove the plastic washers (see illustration 8.14a). Lift off the bar and remove the return spring, noting how they fit, and remove the collars (see illus-

tration 8.14b). Unscrew the nut on one end of each threaded stud which joins the carburettors together, then withdraw the bars (see illustration).

5 Make a careful note of how the carburettor synchronisation springs and throttle linkage springs are arranged to ensure that they are fitted correctly on reassembly (see illustration). Also note the arrangement of the various hoses and their unions.

6 Carefully separate the carburettors, taking care not to damage the fuel and air vent joints between each carburettor or lose the screws from the throttle linkages. Keep a careful watch on all springs as the carburettors are separated; the synchronisation springs should stay with the adjusting screws, but if they don't, find them and install them so they aren't lost. On J and K models, note how the choke actuating arms, return spring and bar are fitted and come apart as the carburettors are separated as an aid to installation.

7 Pull out the fuel fittings and vent line fittings and the joining pieces and discard the O-rings where fitted as new ones must be used.

Joining

8 Assembly is the reverse of the disassembly procedure. Use new O-rings on the fuel and vent line fittings, and smear them with oil. Check the operation of both the choke and throttle linkages ensuring that both operate

smoothly and return quickly under spring pressure before installing the carburettors on the machine. Check carburettor synchronisation (see Chapter 1).

10 Carburettors – reassembly and float height check

Warning: Refer to the precautions given in Section 1 before proceeding.

Note: *When reassembling the carburettors, be sure to use new O-rings and seals. Do not overtighten the carburettor jets and screws as they are easily damaged.*

1 If removed, install the choke plunger assemblies into the carburettor bodies.

2 On J and K models, locate the actuating arms on the choke plungers (see illustration 8.13). Align the actuating arms so that the gap between them and each plunger is equidistant (so that each plunger opens simultaneously when the choke is activated), then tighten the clamp screws onto the bars.

3 On L, N and R models, fit the choke linkage bar collars, then locate the return spring and the bar, making sure the arms fit onto the plungers, then fit the plastic washers and secure the bar with the screws (see illustrations 8.14b and a).

9.3d Remove the screws (arrowed) and the joining plate

9.4 Unscrew the nuts (arrowed) and withdraw the studs

9.5 Synchronisation and linkage spring set-up – L, N and R models

Fuel and exhaust systems 4•11

10.5 Install the pilot jet . . .

10.6a . . . the needle jet . . .

10.6b . . . and the main jet

4 Install the pilot screw (if removed) along with its spring, washer and O-ring, turning it in until it seats lightly **(see illustrations 8.12a or b)**. Now, turn the screw out the number of turns previously recorded on disassembly.

5 Install the pilot jet into the carburettor **(see illustration)**.

6 Install the needle jet into the carburettor **(see illustration)**. Screw the main jet into the end of the needle jet **(see illustration)**.

7 On J and K models, install the float needle valve seat and its sealing washer, using a new one if the old is damaged or deformed **(see illustration 8.8.9a)**.

8 Fit the float needle valve onto the float, then position the float assembly in the carburettor, making sure the needle valve locates in its seat, and install the pin, making sure it is secure **(see illustrations)**.

9 To check the float height, hold the carburettor so the float hangs down, then tilt it back until the needle valve is just seated, but not so far that the needle's spring-loaded tip is compressed. Measure the distance between the gasket face (with the gasket removed) and the bottom of the float with an accurate ruler **(see illustration)**. The correct setting should be as given in the

10.8a Fit the needle valve onto the float . . .

Specifications at the beginning of the Chapter. On J and K models, if it is incorrect, adjust the float height by carefully bending the float tab a little at a time until the correct height is obtained. Repeat the procedure for all carburettors. On L, N and R models, the float height is not adjustable, so if it is incorrect the float may be damaged and should be renewed.

10 With the float height checked, fit a new rubber gasket onto the float chamber, making sure it is seated properly in its groove, and

10.8b . . . and install the float assembly – L, N and R models shown

install the chamber onto the carburettor **(see illustrations)**.

11 On J and K models, if removed, fit the washer onto the jet needle and insert the needle into the piston **(see illustration 8.5c)**. Fit the retainer and spring into the piston, then push down on it using a Phillips screwdriver and rotate it until its tab locks under the protrusion in the piston **(see illustrations 8.5b and a)**.

12 On L, N and R models, fit the washer onto the jet needle and insert the needle into the

10.9 Measuring float height

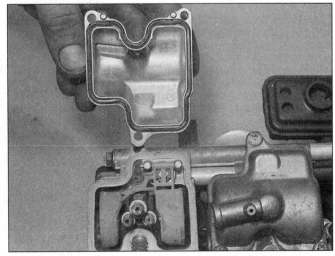

10.10 Fit a new gasket into the groove and install the float chamber

4•12 Fuel and exhaust systems

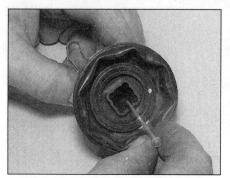

10.12a Insert the jet needle with its washer . . .

10.12b . . . then press in the holder

10.13a Install the diaphragm/piston assembly . . .

10.13b . . . making sure the diaphragm edge fits into the groove and the tab is correctly positioned around the air hole (arrow) – J and K models shown

10.14a Install the spring . . .

10.14b . . . then fit the top cover

piston **(see illustration)**. Fit a new O-ring into the groove in the needle holder, and make sure the spring is inside. Insert the holder into the centre of the piston and push it down until the O-ring is felt to locate in its groove in the piston **(see illustration)**. Remove the screw used on removal from the holder, if not already done.

13 Insert the piston assembly into the body and lightly push it down, ensuring the needle is correctly aligned with the needle jet **(see illustration)**. Align the tab on the diaphragm with the recess in the carburettor body, then press the diaphragm outer edge into its groove, making sure it is correctly seated and that the tab locates in the recess around the air hole **(see illustration)**. Check the diaphragm is not creased, and that the piston moves smoothly up and down in its guide.

14 Install the spring into the piston **(see illustration)**. Fit the top cover to the carburettor, making sure the top of the spring locates over or into (according to type) the raised section on the inside of the cover, and aligning the protrusion on the cover with the tab on the diaphragm and, then tighten the cover screws securely **(see illustration)**.

15 On J and K models, if removed, install the air duct assembly (see Section 9, Step 3).

16 Install the carburettors (see Section 7).

11 Throttle cables – removal and installation

> **Warning:** Refer to the precautions given in Section 1 before proceeding.

Removal

1 Remove the fuel tank and the air filter housing (see Sections 2 and 4). Whilst it is possible to detach the throttle cables with the carburettors in situ, there is a limited amount of space to work in and it can be tricky. If required, displace the carburettors to improve access (see Section 7).

2 Turn the handlebars onto full left lock to provide the maximum freeplay in the cables. Mark each cable according to its location. The accelerator cable is the upper cable in the bracket, the decelerator cable is the lower. Slacken the locknut on the upper cable adjuster and thread the adjuster fully in, then tighten the locknut against it **(see illustration 11.5a)**. This resets the adjuster to the start of its range.

3 At the carburettor end, slacken the upper (accelerator) cable adjuster locknut, then unscrew the adjuster until the captive nut is clear of the small lug on the bracket **(see illustration)**. Slip the adjuster out of the bracket and detach the cable nipple from the carburettor throttle cam **(see illustration)**.

11.3a Slacken the locknut (A), then unscrew the adjuster (B) until the captive nut (C) is free of its lug, then slip the adjuster out of the bracket . . .

11.3b . . . and the cable end out of the cam

Fuel and exhaust systems 4•13

11.4a Unscrew the holder until the captive nut is clear of its lug, then slip the cable out of the bracket . . .

11.4b . . . and the cable end out of the cam

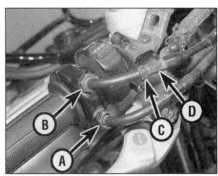

11.5a Unscrew the lower nut (A), and slacken the upper nut (B). Cable adjuster locknut (C) and adjuster (D)

4 Now unscrew the lower (decelerator) cable holder until the captive nut is free, then slip the holder out of the bracket and detach the cable nipple from the cam **(see illustrations)**. Withdraw the cables from the machine noting the correct routing of each cable.

5 Unscrew the lower (decelerator) cable elbow nut at the throttle pulley housing and slacken the upper (accelerator) cable locknut **(see illustration)**. Remove the housing screws and draw off the rear half **(see illustration)**. Detach the cable nipples from the pulley, then detach the front housing half from the handlebar and withdraw the lower elbow and cable from the housing **(see illustrations)**. Thread the housing off the upper elbow and withdraw the cable from the housing **(see illustration)**. Mark each cable to ensure it is connected correctly on installation.

Installation

6 Fit the accelerator cable elbow into the upper socket of the front half of the throttle pulley housing and thread the housing onto it **(see illustration)**. Lubricate the cable nipple with multi-purpose grease and install it into the throttle pulley **(see illustration)**. Fit the decelerator cable into the lower socket and tighten the nut, then lubricate the cable nipple with multi-purpose grease and install it into the pulley **(see illustration and 11.5c)**. Fit the front half of the housing over the pulley and onto the handlebar.

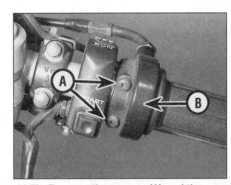

11.5b Remove the screws (A) and the rear half of the housing (B)

11.5c Detach the cable ends . . .

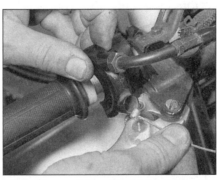

11.5d . . . and draw the lower cable out

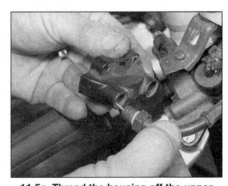

11.5e Thread the housing off the upper cable nut

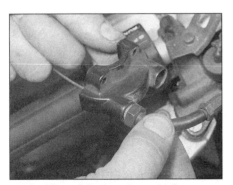

11.6a Thread the housing onto the upper cable nut . . .

11.6b . . . then fit the cable end into the pulley

11.6c Thread the lower cable nut into the housing

4

4•14 Fuel and exhaust systems

11.7 Locate the pin (arrowed) into the hole in the handlebar

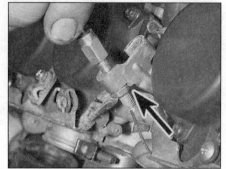

11.9 Locate the cable holder in the bracket so that the nut is against the lug (arrow) and fully tighten the adjuster

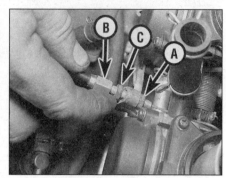

11.10 Locate the nut (A) so it is captive, then set the freeplay by turning the adjuster (B) and tighten the locknut (C) against the bracket

7 Fit the rear half of the housing onto the handlebar, making sure the pin locates in the hole in the handlebar, and install the screws, tightening them securely **(see illustration and 11.5b)**.
8 Feed the cables through to the carburettors, making sure they are correctly routed. The cables must not interfere with any other component and should not be kinked or bent sharply.
9 Lubricate the decelerator cable nipple with multi-purpose grease and fit it into the lower socket on the carburettor throttle cam **(see illustration 11.4b)**. Fit the cable holder into the lower bracket, locating the nut against the lug so that it is captive, then thread the holder into the nut until it is tight **(see illustration)**.
10 Lubricate the accelerator cable nipple with multi-purpose grease and fit it into the upper socket on the carburettor throttle cam **(see illustration 11.3b)**. Fit the accelerator cable adjuster into the upper bracket, then thread the lower nut up the adjuster **(see illustration 11.3a)**. Locate the nut against the lug so that it is captive, then thread the adjuster into the nut until the specified amount of cable freeplay is obtained **(see illustration)** (see Chapter 1). Tighten the locknut against the bracket.
11 Operate the throttle to check that it opens and closes freely.
12 Check and adjust the throttle cable freeplay if required (see Chapter 1). Turn the handlebars back and forth to make sure the cable doesn't cause the steering to bind.
13 Install the carburettors (if displaced), the air filter housing and the fuel tank (see Sections 7, 4 and 2).
14 Start the engine and check that the idle speed does not rise as the handlebars are turned. If it does, the throttle cables are routed incorrectly. Correct the problem before riding the motorcycle.

12 Choke cable – removal and installation

Removal

1 Remove the fuel tank and the air filter housing (see Sections 2 and 4).
2 Slacken the choke outer cable bracket screw and free the cable from the bracket on the front of the carburettors, then detach the inner cable nipple from the choke linkage lever **(see illustrations)**. Withdraw the cable from the machine noting the correct routing.
3 Unscrew the left-hand fork clamp bolt securing the choke knob to the top yoke and remove the cable **(see illustration)**.

Installation

4 Locate the choke knob bracket against the top yoke and tighten the bolt securely.
5 Feed the cable through to the carburettors, making sure it is correctly routed. The cable must not interfere with any other component and should not be kinked or bent sharply.
6 Lubricate the cable nipple with multi-purpose grease and attach it to the choke linkage lever on the carburettor **(see illustration 12.3b)**. Fit the outer cable into its bracket, making sure there is a small amount of freeplay in the inner cable, and tighten the screw **(see illustration 12.3a)**.
7 Check the operation of the choke cable (see Chapter 1).
8 Install the air filter housing and the fuel tank (see Sections 4 and 2).

13 Exhaust system – removal and installation

⚠️ **Warning:** *If the engine has been running the exhaust system will be very hot. Allow the system to cool before carrying out any work.*

Note: *The exhaust system can be removed as a complete assembly. The silencer and downpipe assembly can be separated if required, but this is best done after the complete system has been removed, rather than doing it in situ.*

12.2a Slacken the bracket screw (arrowed) and release the outer cable ...

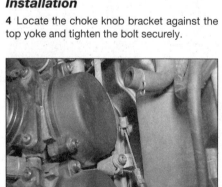

12.2b ... then free the inner cable nipple from the lever

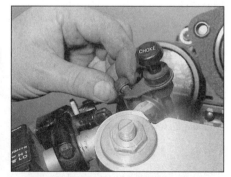

12.3 Unscrew the bolt and detach the choke knob bracket

Fuel and exhaust systems 4•15

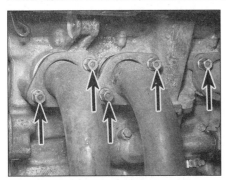

13.2 Unscrew the downpipe nuts (arrowed) . . .

13.3 . . . the silencer bolt . . .

13.4 . . . and the mounting bolt (arrowed)

13.5 The silencer end can is secured by three bolts

13.6a Lever out the old gaskets . . .

13.6b . . . and install new ones

Removal

1 Remove the lower fairing (see Chapter 8).
2 Unscrew the nuts securing the downpipes to the cylinder head **(see illustration)**.
3 Unscrew the nut and remove the bolt securing the silencer to the bracket **(see illustration)**.
4 Unscrew the nut that secures the exhaust system to the frame, but leave the bolt in place **(see illustration)**. Support the system, then withdraw the mounting bolt and manoeuvre the exhaust away from the bike.
5 If required, remove the three bolts to separate the silencer end can from the downpipe assembly **(see illustration)**. Discard the joint gasket as a new one must be used.

Installation

6 Installation is the reverse of removal, noting the following:
  a) Leave all fasteners loose until the entire system has been installed, making alignment easier.
  b) Use new gaskets in each cylinder head port and between the downpipe assembly and the silencer end can, if separated **(see illustrations)**. Use a dab of grease to hold them in place while installing the exhaust.
  c) Check the condition of the rubber mountings on the silencer end can bracket and the frame and renew them if they are damaged or deteriorated.
  d) Tighten the mounting nuts, and on J and K models the bolts to the torque settings specified at the beginning of the Chapter. Tighten the downpipe nuts first.
  e) Run the engine and check that there are no exhaust gas leaks.

14 Fuel pump – check, removal and installation

Warning: *Gasoline (petrol) is extremely flammable, so take extra precautions when you work on any part of the fuel system. Don't smoke or allow open flames or bare light bulbs near the work area, and don't work in a garage where a natural gas-type appliance (such as a water heater or clothes dryer) is present. If you spill any fuel on your skin, rinse it off immediately with soap and water. When you perform any kind of work on the fuel system, wear safety glasses and have a fire extinguisher suitable for a class B type fire (flammable liquids) on hand.*

Check

1 All models are fitted with a fuel pump. The fuel pump is located under the rider's seat **(see illustration)**. The fuel cut-off relay is also located under the seat on J and K models, but on L, N and R models it is on the right-hand side of the frame ahead of the rear master cylinder reservoir **(see illustrations)**. Remove

14.1a Fuel pump (arrowed) – J and K models shown

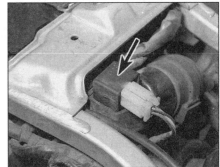

14.1b Cut-off relay (arrowed) – J and K models

4•16 Fuel and exhaust systems

14.1c Cut-off relay (arrowed) –
L, N and R models

14.6a Fuel pump wiring connector –
J and K models

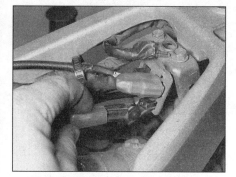

14.6b Fuel pump wiring connector –
L, N and R models

the seat cowling to access all areas (see Chapter 8).

2 The fuel pump is controlled through the fuel cut-off relay so that it runs whenever the ignition is switched ON and the ignition is operative (ie, only when the engine is turning over). As soon as the ignition is killed, the relay will cut off the fuel pump's electrical supply (so that there is no risk of fuel being sprayed out under pressure in the event of an accident).

3 It should be possible to hear or feel the fuel pump running whenever the engine is turning over – either place your ear close beside the pump or feel it with your fingertips. If you can't hear or feel anything, check the circuit fuse (see Chapter 9). If the fuse is good, check the pump and relay for loose or corroded connections or physical damage and rectify as necessary.

4 If the circuit is fine so far, switch the ignition OFF. Unplug the relay's wiring connector and connect across the relay's black and black/blue wire terminals with a short length of insulated jumper wire. Switch the ignition ON; the pump should operate. Turn OFF the ignition and disconnect the jumper wire when testing is complete.

5 If the pump operated when tested as described in Step 4 either the relay or its wiring is at fault.

6 If the pump still did not operate, trace the wiring from the pump and disconnect it at the black 2-pin wiring connector **(see illustrations)**. Using a fully charged 12 volt battery and two insulated jumper wires, connect the positive (+ve) terminal of the battery to the pump's black/blue wire terminal, and the negative (–ve) terminal of the battery to the pump's green wire terminal. The pump should operate. If the pump does not operate it must be renewed. Disconnect the battery and jumper wires.

7 Honda do not provide any test details for the relay, so the best way to determine whether it is faulty is by substituting a known good one. If the pump now works, the old relay is proven faulty. If the pump still does not work, check all the wiring and connectors between the pump and relay, and between the relay and the other components in the system, using the wiring diagrams at the end of Chapter 9. If this does not cure the problem, bear in mind that the ignition control unit could be faulty.

8 If the pump operates but is thought to be delivering an insufficient amount of fuel, first check that all fuel hoses are in good condition and not pinched or trapped. Check that the fuel filters in the fuel tank and fuel delivery hose are not blocked.

9 The fuel pump's output can be checked as follows: make sure the ignition switch is OFF.

10 Release the clamp securing the fuel supply hose to the inlet union on the carburettors and detach the hose, being prepared for any residual fuel **(see illustration 7.5)**. Place the end into a graduated beaker.

11 Disconnect the fuel cut-off relay wiring connector. Using a short length of insulated jumper wire, connect across the black and the black/blue wire terminals of the connector.

12 Turn the ignition switch ON and let fuel flow from the pump into the beaker for exactly 5 seconds, then switch the ignition OFF.

13 Measure the amount of fuel that has flowed into the beaker, then multiply that amount by 12 to determine the fuel pump flow rate per minute. The minimum flow rate required is 660 cc per minute on J and K models, and 700 cc per minute on L, N and R models. If the flow rate recorded is below the minimum required, then the fuel pump must be renewed.

Removal

14 Make sure both the ignition and the fuel tap are switched OFF. On L, N and R models, displace the carburettors from the inlet manifolds and position them clear to provide access to the fuel pump (see Section 7). There should be no need to disconnect the throttle cables, though care must be taken to properly support the carburettors. Remove them completely if required.

15 Trace the wiring from the fuel pump and disconnect it at the black 2 pin connector **(see illustrations 14.6a or b)**.

16 Make a note or sketch of which fuel hose fits where as an aid to installation. Using a rag to mop up any spilled fuel, disconnect the two fuel hoses from the fuel pump **(see illustration 14.1a)**. Displace the pump with its rubber mounting sleeve from the mounting bracket, then turn it to access the drain hose on the underside and detach the hose **(see illustrations)**. Remove the pump.

17 To remove the fuel cut-off relay, disconnect the relay wiring connector and remove the relay from its mounting lug.

Installation

18 Installation is a reverse of the removal procedure. Make sure the fuel hoses are correctly and securely fitted to the pump – the hose from the in-line filter attaches to the union marked INLET; the hose to the carburettors attaches to the other union. Start the engine and check carefully that there are no leaks at the pipe connections.

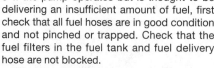
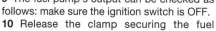
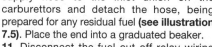

14.16a Pull the pump off its mounts . . .

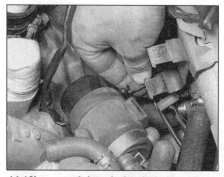

14.16b . . . and detach the drain hose from the underside

Chapter 5
Ignition system

Contents

General information 1	Neutral switch – check and replacement . . . . .see Chapter 9
Ignition control unit – check, removal and installation 5	Pulse generator coil assembly – check, removal
Ignition (main) switch – check, removal and installation .see Chapter 9	and installation 4
Ignition HT coils – check, removal and installation 3	Sidestand switch – check and replacement . . . . .see Chapter 9
Ignition system – check 2	Spark plugs – gap check and renewal . . . . .see Chapter 1
Ignition timing – general information and check 6	Throttle position sensor – check and replacement 7

Degrees of difficulty

Easy, suitable for novice with little experience		Fairly easy, suitable for beginner with some experience		Fairly difficult, suitable for competent DIY mechanic		Difficult, suitable for experienced DIY mechanic		Very difficult, suitable for expert DIY or professional	

Specifications

Note: *Models are identified by their production code letter – refer to 'Identification numbers' at the front of this manual for details.*

General information
Cylinder numbering (from left-hand to right-hand side of the bike)	1-2-3-4
Firing order	1-2-4-3
Spark plugs	see Chapter 1

Ignition timing
At idle	18° BTDC
Full advance	
J and K models	20° BTDC
L, N and R models	32° BTDC

Pulse generator coils
Resistance	340 to 420 ohms @ 20°C

Ignition HT coils
Primary winding resistance	
J and K models	2.6 to 3.2 ohms @ 20°C
L, N and R models	2.0 to 4.0 ohms @ 20°C
Secondary winding resistance	
with plug caps and leads	
J and K models	21.0 to 29.0 K-ohms @ 20°C
L, N and R models	23.0 to 27.0 K-ohms @ 20°C
without plug caps and leads	13.0 to 17.0 K-ohms @ 20°C

Torque settings
Note: *Where a specified setting is not given for a particular bolt, the general settings listed at the beginning apply. The dimension given applies to the diameter of the thread, not the head.*

5 mm bolt/nut	5 Nm
6 mm bolt/nut	10 Nm
8 mm bolt/nut	22 Nm
10 mm bolt/nut	35 Nm
12 mm bolt/nut	55 Nm
6 mm flange bolt with 8 mm head	9 Nm
6 mm flange bolt/nut with 10 mm head	12 Nm
8 mm flange bolt/nut	27 Nm
10 mm flange bolt/nut	40 Nm

5•2 Ignition system

1 General information

All models are fitted with a fully transistorised electronic ignition system, which due to its lack of mechanical parts is totally maintenance free. The system comprises a rotor, pulse generator coil(s), ignition control unit and ignition HT coils (refer to the wiring diagrams at the end of Chapter 9 for details). J and K models are fitted with two pulse generator coils, while L, N and R models have one.

The ignition triggers, which are on the rotor mounted on the left-hand end of the crankshaft, magnetically operate the pulse generator coil(s) as the crankshaft rotates. The pulse generator coil(s) sends a signal to the ignition control unit which then supplies the ignition HT coils with the power necessary to produce a spark at the plugs.

The system uses two coils, the left-hand one supplying nos. 1 and 4 cylinder spark plugs and the right-hand coil supplying nos. 2 and 3 cylinder plugs. On J and K models, the coils are mounted under the frame cross-member behind the air filter housing, while on L, N and R models they are mounted on the frame under the front of the air filter housing.

The ignition control unit incorporates an electronic advance system controlled by signals generated by the ignition triggers and the pulse generator coil.

On L, N and R models the system incorporates a safety interlock circuit which will cut the ignition if the sidestand is put down whilst the engine is running and in gear, or if a gear is selected whilst the engine is running and the sidestand is down. It also prevents the engine from being started if the sidestand is down and the engine is in gear unless the clutch lever is pulled in. J and K models have a simpler system, which does not include a sidestand switch, but prevents the engine from being started if it is in gear unless the clutch lever is pulled in, and it cuts the starter circuit rather than the ignition circuit.

Because of their nature, the individual ignition system components can be checked but not repaired. If ignition system troubles occur, and the faulty component can be isolated, the only cure for the problem is to renew the part. Keep in mind that most electrical parts, once purchased, cannot be returned. To avoid unnecessary expense, make very sure the faulty component has been positively identified before buying a new part.

Note that there is no provision for adjusting the ignition timing on these models.

> **HAYNES HiNT** *Models are identified by their production code letter – refer to 'Identification numbers' at the front of this manual for details.*

2 Ignition system – check

⚠️ **Warning:** *The energy levels in electronic systems can be very high. On no account should the ignition be switched on whilst the plugs or plug caps are being held. Shocks from the HT circuit can be most unpleasant. Secondly, it is vital that the engine is not turned over or run with any of the plug caps removed, and that the plugs are soundly earthed (grounded) when the system is checked for sparking. The ignition system components can be seriously damaged if the HT circuit becomes isolated.*

1 As no means of adjustment is available, any failure of the system can be traced to failure of a system component or a simple wiring fault. Of the two possibilities, the latter is by far the most likely. In the event of failure, check the system in a logical fashion, as described below.

2 Disconnect the HT leads from the spark plugs. Connect each lead to a spare spark plug and lay each plug on the engine with the threads contacting the engine **(see illustration)**. If necessary, hold each spark plug with an insulated tool.

⚠️ **Warning:** *Do not remove any of the spark plugs from the engine to perform this check – atomised fuel being pumped out of the open spark plug hole could ignite, causing severe injury!*

3 Having observed the above precautions, check that the kill switch is in the RUN position and the transmission is in neutral, then turn the ignition switch ON and turn the engine over on the starter motor. If the system is in good condition a regular, fat blue spark should be evident at each plug electrode. If the spark appears thin or yellowish, or is non-existent, further investigation will be necessary. Before proceeding further, turn the ignition OFF and remove the key as a safety measure.

4 The ignition system must be able to produce a spark which is capable of jumping a particular size gap. Honda do not provide a specification, but a healthy system should produce a spark capable of jumping at least 6 mm. A simple testing tool can be made to test the minimum gap across which the spark will jump (see **Tool Tip**) or alternatively it is possible to buy an ignition spark gap tester tool and some of these tools are adjustable to alter the spark gap.

5 Connect one of the spark plug HT leads from one coil to the protruding electrode on the test tool, and clip the tool to a good earth (ground) on the engine or frame **(see illustration)**. Check that the kill switch is in the RUN position, turn the ignition switch ON and turn the engine over on the starter motor. If the system is in good condition a regular, fat blue spark should be seen to jump the gap between the nail ends. Repeat the test for the other coil. If the test results are good the entire ignition system can be considered

2.2 Earth the spark plug and operate the starter – bright blue sparks should be visible

TOOL TiP

A simple spark gap testing tool can be made from a block of wood, a large alligator clip and two nails, one of which is fashioned so that a spark plug cap or bare HT lead end can be connected to its end. Make sure the gap between the two nail ends is the same as specified.

2.5 Connect the tester as shown – when the engine is cranked sparks should jump the gap between the nails

Ignition system 5•3

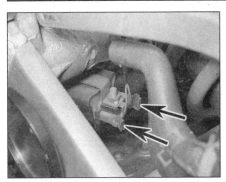

3.5a On J and K models, disconnect the primary circuit connectors from the coils (arrowed) . . .

3.5b . . . or at the connector (arrowed) if access is too restricted

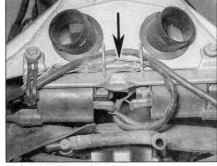

3.5c On L, N and R models, disconnect the bullet connectors (arrowed)

good. If the spark appears thin or yellowish, or is non-existent, further investigation will be necessary.

6 Ignition faults can be divided into two categories, namely those where the ignition system has failed completely, and those which are due to a partial failure. The likely faults are listed below, starting with the most probable source of failure. Work through the list systematically, referring to the subsequent sections for full details of the necessary checks and tests. **Note:** *Before checking the following items ensure that the battery is fully charged and that all fuses are in good condition.*

a) Loose, corroded or damaged wiring connections, broken or shorted wiring between any of the component parts of the ignition system (see Chapter 9).
b) Faulty HT lead or spark plug cap, faulty spark plug, dirty, worn or corroded plug electrodes, or incorrect gap between electrodes.
c) Faulty ignition (main) switch or engine kill switch (see Chapter 9).
d) Faulty neutral, clutch or sidestand switch (L, N and R models only) (see Chapter 9).
e) Faulty pulse generator coil or damaged rotor.
f) Faulty ignition HT coil(s).
g) Faulty ignition control unit.

7 If the above checks don't reveal the cause of the problem, have the ignition system tested by a dealer equipped with the special diagnostic tester.

3 Ignition HT coils – check, removal and installation

Check

1 In order to determine conclusively that the ignition coils are defective, they should be tested by a dealer equipped with the special diagnostic tester.
2 However, the coils can be checked visually (for cracks and other damage) and the primary and secondary coil resistance can be measured with a multimeter. If the coils are undamaged, and if the resistance readings are as specified at the beginning of the Chapter, they are probably capable of proper operation.
3 On J and K models, remove the fuel tank (see Chapter 4). On L, N and R models, remove the fuel tank and the air filter housing (see Chapter 4). Disconnect the battery negative (-ve) lead on all models.
4 On J and K models, the coils are mounted under the frame cross-member behind the air filter housing, while on L, N and R models they are mounted on the frame under the front of the air filter housing.
5 Disconnect the primary circuit electrical connectors from the coil being tested and the HT leads from the spark plugs. On J and K models, when testing the right-hand coil, it is easier to disconnect the two-pin connector in the bracket behind the air filter housing (black/white and blue/yellow wires) rather than disconnecting the wiring from the terminals on the coils **(see illustrations)**. On L, N and R models, it is easier to disconnect the bullet connectors in the primary circuit wiring on top of the coils rather than disconnecting the wiring from the terminals on the coils **(see illustration)**. If testing the left-hand coil, disconnect the yellow/blue and black/white connectors. If testing the right-hand coil, disconnect the blue/yellow and black/white connectors. Mark the locations of all wires and leads before disconnecting them.
6 Set the meter to the ohms x 1 scale and measure the resistance between the primary circuit terminals **(see illustration)**. This will give a resistance reading of the primary windings of the coil and should be consistent with the value given in the Specifications at the beginning of the Chapter.
7 To check the condition of the secondary windings, set the meter to the K-ohm scale. Connect one meter probe to one spark plug cap and the other probe to the other spark plug cap **(see illustration)**. If the reading obtained is not within the range shown in the Specifications, disconnect the HT leads from the coils and repeat the measurement between the lead sockets in the coil **(see illustration)**.

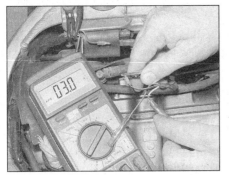

3.6 To test the coil primary resistance, connect the multimeter leads between the primary circuit terminals

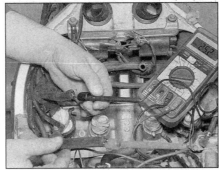

3.7a To test the coil secondary resistance, connect the multimeter leads between the spark plug caps

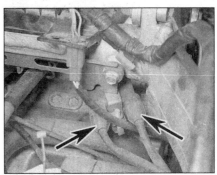

3.7b Disconnect the HT leads (arrowed) from the coils – J and K models shown

5•4 Ignition system

3.7c On L, N and R models, insert short lengths of metal (such as thick wire or welding rod) into the sockets as shown if access is too restricted for the meter probes

On L, N and R models, due to the positioning of the coils, it may be necessary to insert some metal rod into the secondary sockets on the coils as it is difficult to insert the meter probes **(see illustration)**; alternatively remove the coils for testing. If the reading is now as specified, renew the spark plug leads and caps. If the reading is still outside the specified range, it is likely that the coil is defective.

8 Should any of the above checks not produce the expected result, have your findings confirmed by a dealer (see Step 1). If the coil is confirmed to be faulty, it must be renewed; the coil is a sealed unit and cannot therefore be repaired.

Removal

9 On J and K models, remove the fuel tank (see Chapter 4). On L, N and R models, remove the fuel tank and the air filter housing, and detach the throttle cables from the carburettors (see Chapter 4). Disconnect the battery negative (-ve) lead.

10 On J and K models, the coils are mounted under the frame cross-member behind the air filter housing, while on L, N and R models they are mounted on the frame under the front of the air filter housing.

11 Disconnect the primary circuit electrical connectors from the coil and disconnect the HT leads from the spark plugs. On J and K models, when removing the right-hand coil, it

4.9 Unscrew the bolts (arrowed) and remove the cover

4.3a Pulse generator coil wiring connector (arrowed) – J and K models

is easier to disconnect the two-pin connector in the bracket behind the air filter housing (black/white and blue/yellow wires) rather than disconnecting the wiring from the terminals on the coils **(see illustrations 3.5a and b and 3.7b)**. On L, N and R models, it is easier to disconnect the bullet connectors in the primary circuit wiring on top of the coils rather than disconnecting the wiring from the terminals on the coils **(see illustration 3.5c)**. Mark the locations of all wires and leads before disconnecting them.

12 Unscrew the two bolts securing the coil mounting bracket, and remove the coil assembly. Note the routing of the HT leads. If required, separate the individual coils from the bracket, noting how they fit.

Installation

13 Installation is the reverse of removal. Make sure the wiring connectors and HT leads are securely connected.

4 Pulse generator coil assembly – check, removal and installation

Check

1 Remove the rider's seat (see Chapter 8) and disconnect the battery negative (-ve) lead.
2 Remove the lower fairing (see Chapter 8) and the fuel tank (see Chapter 4).

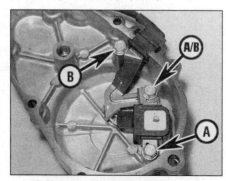

4.10 Unscrew the coil bolts (A) and the wiring guide bolts (B) – L, N and R models shown

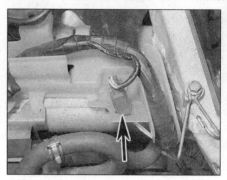

4.3b Pulse generator coil wiring connector (arrowed) – L, N and R models

3 Trace the pulse generator coil switch wiring back from the starter clutch cover on the left-hand side of the engine and disconnect it at the red 4-pin connector (J and K models) or the brown 2-pin connector (L, N and R models) **(see illustrations)**. Using a multimeter set to the ohms x 100 scale, measure the resistance between the white/yellow and yellow terminals on the pulse generator coil side of the connector. On J and K models with two coils, also measure the resistance between the white/blue and blue terminals.

4 Compare the reading obtained with that given in the Specifications at the beginning of this Chapter. The pulse generator coil(s) must be renewed if the reading obtained differs greatly from that given, particularly if the meter indicates a short circuit (no measurable resistance) or an open circuit (infinite, or very high resistance). On J and K models, even if only one of the coils is faulty both must be renewed as they come as an assembly.

5 If the pulse generator coil is thought to be faulty, first check that this is not due to a damaged or broken wire from the coil to the connector; pinched or broken wires can usually be repaired.

Removal

6 Remove the seat (see Chapter 8) and disconnect the battery negative (-ve) lead.
7 Remove the lower fairing (see Chapter 8) and the fuel tank (see Chapter 4).
8 Trace the pulse generator coil wiring back from the starter clutch cover on the left-hand side of the engine and disconnect it at the red 4-pin connector (J and K models) or the brown 2-pin connector (L, N and R models) **(see illustrations 4.3a or b)**. Feed the wiring through to the cover, noting its routing.
9 Unscrew the bolts securing the starter clutch cover and remove the cover, being prepared to catch any residue oil **(see illustration)**. Discard the gasket as a new one must be used. Remove the dowel from either the cover or the crankcase if it is loose.
10 Unscrew the bolts securing the pulse generator coil(s) and the wiring guide to the inside of the cover **(see illustration)**. Remove the rubber wiring grommet from its recess, then remove the coil(s).

Ignition system 5•5

4.13a Locate the new gasket onto the dowel (arrowed) . . .

4.13b . . . then install the cover

5.3a Ignition control unit – J and K models

11 Examine the triggers on the timing rotor for signs of damage and renew it if necessary (see Chapter 2, Section 15).

Installation

12 Install the pulse generator coil(s) into the cover and tighten the bolt(s) securely **(see illustration 4.10)**. Apply a suitable sealant to the wiring grommet and fit it into its recess.
13 If removed, insert the dowel in the crankcase, then install the starter clutch cover using a new gasket, making sure it locates correctly onto the dowel and the idle/reduction gear shaft **(see illustrations)**. Tighten the cover bolts evenly in a criss-cross sequence.
14 Feed the wiring up to its connector, making sure it is correctly routed, and reconnect it **(see illustrations 4.3a or b)**.
15 Install the lower fairing (see Chapter 8) and the fuel tank (see Chapter 4).
16 Reconnect the battery negative (-ve) lead and install the seat (see Chapter 8).

5 Ignition control unit – check, removal and installation

Check

1 If the tests shown in the preceding Sections have failed to isolate the cause of an ignition fault, it is possible that the ignition control unit itself is faulty. No test details are available with which the unit can be tested on home workshop equipment. Take the machine to a dealer for testing on the diagnostic tester.

Removal

2 Remove the rider's seat (see Chapter 8) and disconnect the battery negative (-ve) lead.
3 On J and K models the control unit is mounted on the rear mudguard, just in front of the tail light assembly **(see illustration)** – remove the passenger seat for access (see Chapter 8). On L, N and R models it is mounted on the left-hand side of the rear sub-frame **(see illustration)** – remove the seat cowling for access (see Chapter 8).
4 On J and K models, disconnect the ignition

5.3b Ignition control unit – L, N, and R models

control unit wiring connectors, which are housed inside the rubber boot **(see illustration)**. On L, N and R models, disconnect the wiring connector from the ignition control unit.
5 Remove the ignition control unit from its rubber sleeve, or lift the sleeve and unit together off the sleeve's mounting lugs, and remove the unit.

Installation

6 Installation is the reverse of removal. Make sure the wiring connector(s) are correctly and securely connected.

6 Ignition timing – general information and check

General information

1 Since no provision exists for adjusting the ignition timing and since no component is subject to mechanical wear, there is no need for regular checks; only if investigating a fault such as a loss of power or a misfire, should the ignition timing be checked.
2 The ignition timing is checked dynamically (engine running) using a stroboscopic lamp. The inexpensive neon lamps should be adequate in theory, but in practice may produce a pulse of such low intensity that the timing mark remains indistinct. If possible, one of the more precise xenon tube lamps

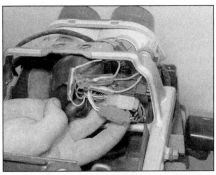

5.4 On J and K models, the wiring connectors are inside the rubber boot – trace the wiring from the unit and disconnect the relevant ones

should be used, powered by an external source of the appropriate voltage. **Note:** *Do not use the machine's own battery as an incorrect reading may result from stray impulses within the machine's electrical system.*

Check

3 Warm the engine up to normal operating temperature then stop it.
4 Unscrew the timing inspection plug from the alternator cover on the right-hand side of the engine **(see illustration)**. Discard the cover O-ring as a new one must be used.
5 The timing mark on the alternator rotor which indicates the firing point at idle speed for cylinders nos. 1 and 4 is a line with the

6.4 Unscrew the timing inspection plug (arrowed)

5•6 Ignition system

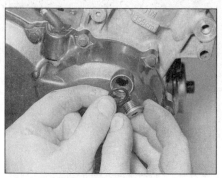

6.11 Install the plug using a new O-ring

letter F next to it. The static timing mark with which this should align is the notch in the timing inspection plug hole.

 The timing marks can be highlighted with white paint to make them more visible under the stroboscope light.

6 Connect the timing light to the no. 4 cylinder HT lead as described in the manufacturer's instructions.
7 Start the engine and aim the light at the static timing mark.
8 With the machine idling at the specified speed, the line next to the timing mark F should align with the static timing mark.
9 Slowly increase the engine speed whilst observing the timing mark. The timing mark should move anti-clockwise, increasing in relation to the engine speed until it reaches full advance (no identification mark).
10 As already stated, there is no means of adjusting the ignition timing on these machines. If the ignition timing is incorrect, or suspected of being incorrect, one of the ignition system components is at fault, and the system must be tested as described in the preceding Sections of this Chapter.
11 When the check is complete, install the timing inspection plug using a new O-ring and smear it and the cover threads with molybdenum disulphide oil (a 50/50 mixture of molybdenum disulphide grease and engine oil) **(see illustration)**.

7 Throttle position sensor – check and replacement

Check

1 The throttle position sensor is fitted to the carburettors on L, N and R models only. Remove the fuel tank (see Chapter 4) and the right-hand rear trim panel (see Chapter 8).
2 The throttle sensor is mounted on the outside of the right-hand (no. 4) carburettor. Disconnect the wiring connector from the sensor **(see illustration)**. Check the sensor visually for cracks and other damage.
3 Arrange a temporary fuel supply, either by using a small temporary tank or by using extra long fuel hoses to the now remote fuel tank on a nearby bench. Whichever method is used, connect the fuel supply hose to the inlet union on the in-line filter – do not connect it directly to the carburettors or you will by-pass the fuel pump.
4 Start the engine and increase speed to 3000 to 3500 rpm. Now connect the sensor wiring connector – as it is connected, the engine speed should rise. Stop the engine and turn the ignition OFF.
5 If the engine speed does not rise, disconnect the wring connector. Using a multi-meter set to the ohms x 10 scale, measure the resistance between the green/black and yellow/black terminals on the wiring loom side of the connector. A reading of 4 to 6 ohms should be recorded. Now set the meter to the dc volts scale and measure the voltage between the green/black and yellow/black terminals on the wiring loom side of the connector with the ignition switched ON. A reading of 5 volts should be recorded.
6 If the above readings are not as specified, disconnect the ignition control unit wiring connector (see Section 5). Check the sensor and control unit connectors for loose or corroded terminals. Using a multimeter set on the ohms range or a continuity tester, check the three wires which run from the throttle sensor connector to the ignition control unit connector for continuity. There should be continuity from one end of the wire to the other. If not, this is probably due to a damaged or broken wire between the connectors; pinched or broken wires can usually be repaired.
7 If the readings taken in Step 5 are as specified, disconnect the sensor wiring connector. Using a multimeter set on the ohms range, measure the resistance between each pair of terminals on the sensor in turn as the throttle is opened and closed. As the throttle is opened, the resistance should increase. As the throttle is closed, the resistance should decrease. If the readings obtained are not as specified, the sensor is faulty and should be renewed; the sensor is a sealed unit and cannot therefore be repaired.
8 If the readings taken are as specified, and the wiring is all good, it is possible that the ignition control unit is faulty.
9 If you are in doubt as to the results obtained, take the sensor to a dealer for testing.

Replacement

10 Remove the fuel tank (see Chapter 4) and the right-hand rear trim panel (see Chapter 8). Depending on the tools available, it may also be necessary to displace the carburettors to access the mounting plate screws (see Chapter 4).
11 The throttle sensor is mounted on the outside of the right-hand (no. 4) carburettor. Unscrew the sensor mounting plate screws and remove the sensor, noting how it fits **(see illustration)**. Do not separate the sensor from its mounting plate as its position has been pre-set.
12 Install the sensor, aligning the tab with the slot on the end of the throttle shaft, and tighten the mounting plate screws securely.
13 Install the carburettors (see Chapter 4).

7.2 Disconnect the throttle position sensor wiring connector

7.11 Remove the sensor mounting plate screws (arrowed) and remove the sensor

Chapter 6
Frame, suspension and final drive

Contents

Drive chain – removal, cleaning and installation 15
Drive chain and sprockets – check, adjustment
   and lubrication ....see Chapter 1
Footrests, brake pedal and gearchange lever –
   removal and installation 3
Forks – disassembly, inspection and reassembly 7
Forks – oil change ....see Chapter 1
Forks – removal and installation 6
Frame – inspection and repair 2
General information 1
Handlebars and levers – removal and installation 5
Handlebar switches – check ....see Chapter 9
Handlebar switches – removal and installation ....see Chapter 9
Rear shock absorber – removal, inspection and installation 10
Rear suspension linkage – removal, inspection and installation 11
Sidestand – check ....see Chapter 1
Sidestand – lubrication ....see Chapter 1
Sidestand – removal and installation 4
Sidestand switch – check and replacement ....see Chapter 9
Sprockets – check and renewal 16
Sprocket coupling/rubber damper – check and renewal 17
Steering head bearings – freeplay check and
   adjustment ....see Chapter 1
Steering head bearings – inspection and renewal 9
Steering head bearings – lubrication ....see Chapter 1
Steering stem – removal and installation 8
Suspension – adjustments 12
Suspension – check ....see Chapter 1
Swingarm – inspection and bearing renewal 14
Swingarm – removal and installation 13
Swingarm and suspension linkage bearings –
   lubrication ....see Chapter 1

Degrees of difficulty

Easy, suitable for novice with little experience	Fairly easy, suitable for beginner with some experience	Fairly difficult, suitable for competent DIY mechanic	Difficult, suitable for experienced DIY mechanic	Very difficult, suitable for expert DIY or professional

Specifications

Note: *Models are identified by their production code letter – refer to 'Identification numbers' at the front of this manual for details.*

Front forks
Fork oil type .... 10W fork oil
Fork oil capacity
   J and K models .... 490 cc
   L and N models .... 453 cc
   R models .... 408 cc
Fork oil level*
   J and K models .... 96 mm
   L and N models .... 105 mm
   R models .... 120 mm
Fork spring free length (min)
   J and K models
      Standard .... 308.0 mm
      Service limit .... 302.0 mm
   L and N models
      Standard .... 319.7 mm
      Service limit .... 313.0 mm
   R models
      Standard .... 295.5 mm
      Service limit .... 290.0 mm
Fork tube runout limit .... 0.2 mm
Fork air pressure (J and K models) .... 0 to 5.5 psi

*Oil level is measured from the top of the tube with the fork spring removed and the leg fully compressed.

Rear suspension

Shock absorber spring free length
   J and K models
      Standard .... 173.7 mm
      Service limit .... 170.2 mm
   L and N models
      Standard .... 154.9 mm
      Service limit .... 152.0 mm
   R models
      Standard .... 173.5 mm
      Service limit .... 170.0 mm

Final drive

Chain type
   J and K models .... RK525SM4 or DID525V8 (106 links, split)
   L and R models .... RK525SMOZ4 or DID525V8 (102 links, split)
   R models .... RK525SM5 or DID525V8 (102 links, split))
Joining link pin projection from side plate (unstaked) – R models
   RK type chain .... 1.2 to 1.4 mm
   DID type chain .... 1.15 to 1.55 mm
Joining link staked ends diameter – R models
   RK type chain .... 5.50 to 5.80 mm
   DID type chain .... 5.50 to 5.80 mm

Torque settings

Note: *Where a specified setting is not given for a particular bolt, the general settings listed at the beginning apply. The dimension given applies to the diameter of the thread, not the head.*

5 mm bolt/nut .... 5 Nm
6 mm bolt/nut .... 10 Nm
8 mm bolt/nut .... 22 Nm
10 mm bolt/nut .... 35 Nm
12 mm bolt/nut .... 55 Nm
6 mm flange bolt with 8 mm head .... 9 Nm
6 mm flange bolt/nut with 10 mm head .... 12 Nm
8 mm flange bolt/nut .... 27 Nm
10 mm flange bolt/nut .... 40 Nm
Rider's footrest bracket bolts .... 27 Nm
Rider's footrest holder bolt .... 37 Nm
Top yoke fork clamp bolts
   J and K models .... 11 Nm
   L, N and R models .... 23 Nm
Handlebar clamp bolts .... 27 Nm
Front brake master cylinder clamp bolts .... 12 Nm
Bottom yoke fork clamp bolts .... 35 Nm
Fork top bolt .... 23 Nm
Damper rod Allen bolt .... 20 Nm
Steering stem nut .... 105 Nm
Steering head bearing adjuster nut .... 22 Nm
Shock absorber mounting nuts/bolts .... 55 Nm
Suspension linkage mounting nuts/bolts .... 55 Nm
Swingarm adjuster bolt .... 15 Nm
Adjuster bolt locknut (special tool required – see text) .... 65 Nm
Pivot bolt nut
   J and K models .... 90 Nm
   L, N and R models .... 95 Nm
Front sprocket bolt .... 55 Nm
Rear sprocket bolts (J and K models) .... 75 Nm
Rear sprocket nuts (L, N and R models) .... 50 Nm

Frame, suspension and final drive 6•3

1 General information

All models use a twin spar box-section aluminium frame which uses the engine as a stressed member.

Front suspension is by a pair of oil-damped telescopic forks. On J and K models, the forks have a conventional damper system and are air-assisted. On L and N models, the forks have a conventional damper system and are adjustable for spring pre-load. On R models, the forks have a cartridge damper and are adjustable for spring pre-load and rebound damping.

At the rear, an alloy swingarm acts on a single shock absorber via a three-way linkage. The shock absorber is adjustable for spring pre-load on all models, and also for rebound damping on L, N and R models.

The drive to the rear wheel is by chain.

Haynes Hint: Models are identified by their production code letter – refer to 'Identification numbers' at the front of this manual for details.

2 Frame – inspection and repair

1 The frame should not require attention unless accident damage has occurred. In most cases, frame renewal is the only satisfactory remedy for such damage. A few frame specialists have the jigs and other equipment necessary for straightening the frame to the required standard of accuracy, but even then there is no simple way of assessing to what extent the frame may have been over stressed.

2 After the machine has accumulated a lot of miles, the frame should be examined closely for signs of cracking or splitting at the welded joints. Loose engine mount bolts can cause ovaling or fracturing of the mounting tabs. Minor damage can often be repaired by specialist welding, depending on the extent and nature of the damage.

3 Remember that a frame which is out of alignment will cause handling problems. If misalignment is suspected as the result of an accident, it will be necessary to strip the machine completely so the frame can be thoroughly checked.

3 Footrests, brake pedal and gearchange lever – removal and installation

Rider's footrests

Removal

1 Remove the split pin and washer from the bottom of the footrest pivot pin, then withdraw the pivot pin and remove the footrest, noting the fitting of the return spring **(see illustration)**.

Installation

2 Installation is the reverse of removal. Use a new split pin and bend its ends securely.

Passenger footrests

Removal

3 Remove the seat cowling (see Chapter 8).
4 Unscrew the two bolts securing the bracket to the frame and remove the footrest assembly, noting how it fits **(see illustration)**.

Installation

5 Installation is the reverse of removal. Prior to installation, clean the assembly and remove all the old grease, then apply some new grease to the release mechanism and pivot points.

Brake pedal

Removal

6 Unhook the brake pedal return spring and the brake light switch spring from the pin on the pedal **(see illustration)**.
7 Remove the split pin and washer from the clevis pin securing the brake pedal to the master cylinder pushrod **(see illustration)**. Remove the clevis pin and separate the pedal from the pushrod.
8 The pedal pivots on the footrest holder. Remove the bolt on the inside of the footrest bracket and remove the footrest and its holder **(see illustration 3.7)**. Remove the wave washer and slide the pedal off the holder.

Installation

9 Installation is the reverse of removal, noting the following:
a) Apply molybdenum disulphide grease to the brake pedal pivot.
b) Align the flat on the footrest holder with that in the bracket. Do not forget the wave washer.
c) Apply a suitable non-permanent thread locking compound to the brake pedal/footrest holder bolt and tighten it to the torque setting specified at the beginning of the Chapter.
d) Use a new split pin on the clevis pin securing the brake pedal to the master cylinder pushrod and bend the split pin ends securely.
e) Check the operation of the rear brake light switch (see Chapter 1).

Gearchange lever

Removal

10 Slacken the gearchange lever linkage rod locknuts, then unscrew the rod and separate

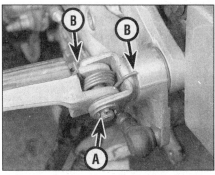
3.1 Remove the split pin and washer (A) and withdraw the pivot pin. Note how the spring ends (B) locate

3.4 Unscrew the two bolts (arrowed) and remove the assembly

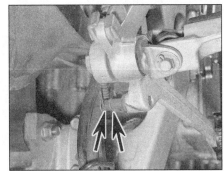
3.6 Unhook the springs (arrowed) from the pin

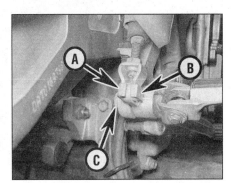
3.7 Remove the split pin (A) and withdraw the clevis pin (B). Brake pedal/footrest holder bolt (C)

6•4 Frame, suspension and final drive

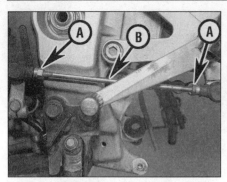

3.10 Slacken the locknuts (A), then unscrew the linkage rod (B)

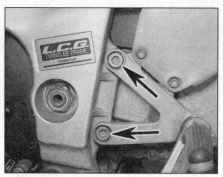

3.11a Unscrew the two bolts (arrowed) and displace the bracket . . .

3.11b . . . to access the holder bolt (arrowed)

it from the lever and the arm (the rod is reverse-threaded on one end and so will simultaneously unscrew from both lever and arm when turned in the one direction) **(see illustration)**. Note the how far the rod is threaded into the lever and arm as this determines the height of the lever relative to the footrest.

11 The lever pivots on the footrest holder. Unscrew the two bolts securing the footrest bracket to the frame and turn it round to access the holder bolt **(see illustrations)**. Unscrew the bolt and remove the footrest and its holder. Remove the wave washer and slide the lever off the holder.

4.3 Unhook the spring (A), then unscrew the nut (B) and remove the bolt

Installation

12 Installation is the reverse of removal, noting the following:
a) Apply molybdenum disulphide grease to the gear lever pivot.
b) Align the flat on the footrest holder with that in the bracket. Do not forget the wave washer.
c) Apply a suitable non-permanent thread locking compound to the footrest holder bolt and tighten it to the torque setting specified at the beginning of the Chapter.
d) Adjust the gear lever height as required by screwing the rod in or out of the lever and arm. Tighten the locknuts securely **(see illustration 3.10)**.

4 Sidestand – removal and installation

1 The sidestand is attached to a bracket on the frame. A spring anchored to the stand ensures that it is held in the retracted or extended position.
2 Support the bike using an auxiliary stand. On L, N and R models, remove the sidestand switch (see Chapter 9).
3 Unhook the stand spring and unscrew the nut securing the stand on the pivot bolt **(see**

illustration). Remove the pivot bolt from the inside of the bracket and remove the stand, and on J and K models the thrust washers.
4 If required, unscrew the bolts securing the sidestand bracket to the frame and remove the bracket.
5 On installation apply grease to the pivot bolt shank and tighten the nut securely. Reconnect the sidestand spring and check that it holds the stand securely up when not in use – an accident is almost certain to occur if the stand extends while the machine is in motion.
6 On L, N and R models, check the operation of the sidestand switch (see Chapter 1).

5 Handlebars and levers – removal and installation

Right-hand handlebar

Removal

1 Disconnect the brake light switch wires from the switch on the underside of the master cylinder assembly. Unscrew the two handlebar switch screws and free the switch from the handlebar **(see illustration)**. Unscrew the two throttle cable housing screws and release the throttle cables from the throttle pulley **(see illustration)**. Position

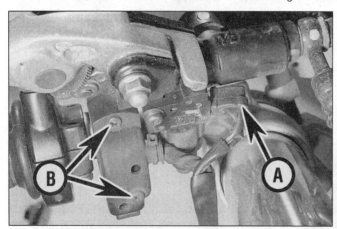

5.1a Disconnect the brake light switch wiring connectors (A) and unscrew the switch housing screws (B)

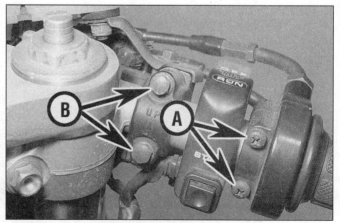

5.1b Throttle cable housing screws (A), master cylinder clamp bolts (B)

Frame, suspension and final drive

5.3a Slacken the right-hand clamp bolt (A) and remove the left-hand bolt (B), noting the choke knob bracket (C)

5.3b Remove the plug (where fitted) . . .

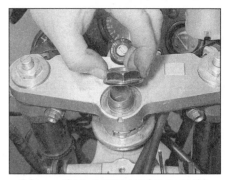

5.3c . . . then unscrew and remove the steering stem nut

the switch housing and the throttle housing away from the handlebar.

2 Unscrew the two master cylinder assembly clamp bolts and position the assembly clear of the handlebar, making sure no strain is placed on the hydraulic hose **(see illustration 5.1b)**. Keep the master cylinder reservoir upright to prevent possible fluid leakage.

3 Slacken the right-hand fork clamp bolt in the top yoke and remove the left-hand clamp bolt, noting how the choke knob bracket locates **(see illustration)**. On L, N and R models, prise the plug out of the steering stem nut **(see illustration)**. Unscrew and remove the nut using a 30 mm socket or spanner, and on J and K models remove the washer **(see illustration)**. Free the clutch cable from its guide on the top yoke. Gently ease the top yoke upwards off the fork tubes and position it clear, using a rag to protect other components **(see illustration)**.

4 Slacken the handlebar holder clamp bolt, then ease the handlebar up and off the fork **(see illustration 5.8a)**. If necessary, unscrew the handlebar end-weight retaining screw, then remove the weight from the end of the handlebar and slide off the throttle twistgrip **(see illustration 5.8b)**.

Left-hand handlebar

Removal

5 Disconnect the clutch switch wires from the switch in the clutch lever bracket **(see illustration)**. Unscrew the two handlebar switch screws and free the switch from the handlebar **(see illustration)**. Position the switch housing away from the handlebar.

6 Free the clutch cable from its guide on the top yoke. Unscrew the two clutch lever assembly clamp bolts and position the assembly clear of the handlebar, making sure the cable is not unduly bent or kinked **(see illustration)**.

7 Slacken the right-hand fork clamp bolt in the top yoke and remove the left-hand clamp bolt, noting how the choke knob bracket

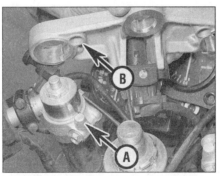

5.3d Ease the top yoke up off the forks. Note how the lug (A) locates in the hole (B)

locates **(see illustration 5.3a)**. On L, N and R models, prise the plug out of the steering stem nut **(see illustration 5.3b)**. Unscrew and remove the nut using a 30 mm socket or spanner, and on J and K models remove the washer **(see illustration 5.3c)**. Gently ease the top yoke upwards off the fork tubes and position it clear, using a rag to protect other components **(see illustration 5.3d)**.

8 Slacken the handlebar holder clamp bolt, then ease the handlebar up and off the fork **(see illustration)**. If necessary, unscrew the handlebar end-weight retaining screw, then remove the weight from the end of the handlebar and remove the grip **(see

5.5a Disconnect the clutch switch wiring connectors (arrowed) . . .

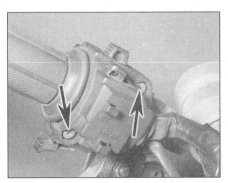

5.5b . . . and remove the switch housing screws (arrowed)

5.6 Clutch lever assembly clamp bolts (arrowed)

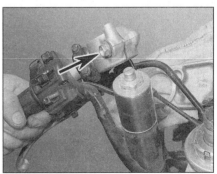

5.8a Slacken the clamp bolt (arrowed) and ease the handlebar up and off the fork

6•6 Frame, suspension and final drive

5.8b Handlebar end-weight screw (arrowed)

5.9 Tighten the specified bolts to their correct torque settings

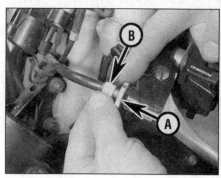

5.10a Slacken the lockring (A) and thread the adjuster (B) in . . .

illustration). It may be necessary to slit the grip open using a sharp blade in order to remove it as they are sometimes stuck in place. This will mean using a new grip on assembly.

Installation

9 Installation is the reverse of removal, noting the following.
  a) Do not tighten the handlebar clamp bolt until the top yoke has been installed, and align the lug on the top of each handlebar holder with its hole in the yoke, so that the handlebars are set in the correct position *(see illustration 5.3d)*.
  b) Do not forget to install the choke knob with the left-hand fork clamp bolt *(see illustration 5.3a)*.
  c) Refer to the Specifications at the beginning of the Chapter and tighten the fork clamp bolts, the handlebar holder clamp bolts, and the steering stem nut to the specified torque settings, in that order *(see illustration)*. When tightening the fork clamp bolts, push the top yoke down onto the handlebars. When tightening the handlebar bolts, push the bar forward.
  d) Make sure the front brake and clutch lever assembly clamps are installed with the UP mark facing up *(see illustrations 5.1b and 5.6)*. Tighten the brake master cylinder clamp bolts to the torque setting specified at the beginning of the Chapter.
  e) Make sure the pin in the lower half of each switch housing and in the rear half of the throttle cable housing locates in the hole in the handlebar.
  f) If removed, apply a suitable non-permanent locking compound to the handlebar end-weight retaining screws. If new grips are being fitted, secure them using a suitable adhesive to the handlebar (left-hand grip) or to the throttle twistgrip (right-hand grip).
  g) Do not forget to reconnect the front brake light switch and clutch switch wiring connectors.

Handlebar levers

Removal – clutch lever

10 Slacken the clutch cable adjuster lockring and thread the adjuster fully into the bracket to provide maximum freeplay in the cable *(see illustration)*. Unscrew the lever pivot bolt locknut, then withdraw the pivot bolt and remove the lever, detaching the cable nipple as you do so *(see illustration)*.

Removal – brake lever

11 Unscrew the lever pivot bolt locknut, then withdraw the pivot bolt and remove the lever *(see illustrations)*.

Installation

12 Installation is the reverse of removal. Apply grease to the pivot bolt shafts and the contact areas between the lever and its bracket, and to the clutch cable nipple. Adjust the clutch cable freeplay (see Chapter 1).

6 Forks – removal and installation

Removal

Caution: Although not strictly necessary, before removing the forks it is recommended that the fairing and fairing panels are removed (see Chapter 8). This will prevent accidental damage to the paintwork.

1 Remove the front wheel (see Chapter 7). Remove each fork individually.
2 Unscrew the bolt securing the brake hose/pipe union to the fork and detach the union *(see illustration)*. Remove the front mudguard (see Chapter 8).
3 Slacken the handlebar holder clamp bolt *(see illustration)* and the fork clamp bolt in

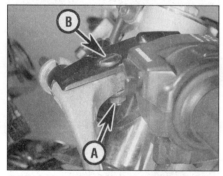

5.10b . . . then unscrew the nut (A) and the pivot bolt (B)

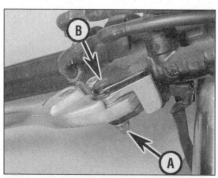

5.11 Unscrew the nut (A) and the pivot bolt (B)

6.2 Unscrew the bolt (arrowed) and detach the brake hose/pipe union

6.3a Slacken the handlebar holder clamp bolt (arrowed) . . .

Frame, suspension and final drive 6•7

6.3b ... and the fork clamp bolt (arrowed) – left-hand fork shown

6.3c If required, slacken the fork top bolt (arrowed) now

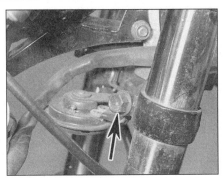

6.4a Slacken the clamp bolt (arrowed) in the bottom yoke and slide the fork down ...

the top yoke **(see illustration)**. Depending on the tools available, access to the right-hand bolt may be restricted by the front brake master cylinder. If this is the case, unscrew the two master cylinder assembly clamp bolts and position the assembly clear of the handlebar, making sure no strain is placed on the hydraulic hose **(see illustration 5.1b)**. Keep the master cylinder reservoir upright to prevent possible fluid leakage. If the forks are to be disassembled, or if the fork oil is being changed, it is advisable at this stage to release any air pressure in the forks on J and K models and adjust the spring pre-load to a minimum on L, N and R models (see Section 12), then slacken the fork top bolt. Where fitted, slacken the cable-tie securing the wiring against the left-hand fork. On J and K models, release the air pressure by removing the valve cap and depressing the pin in the centre of the valve.

4 Slacken but do not remove the fork clamp bolt in the bottom yoke, and remove the fork by twisting it and pulling it downwards **(see illustration)**. As the fork drops clear of the top yoke, draw the handlebar assembly, and on the left-hand fork the wiring tie, off the top, and remove the circlip, which supports the handlebar holder, from its groove in the fork **(see illustrations)**. If removing the right-hand fork, support the right handlebar so that no strain is placed on the brake master cylinder hose and the reservoir is upright.

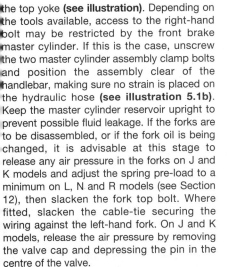

6.4b ... sliding off the handlebar ...

 HAYNES HiNT *If the fork legs are seized in the yokes, spray the area with penetrating oil and allow time for it to soak in before trying again.*

Installation

5 Remove all traces of corrosion from the fork tubes and the yokes. Install each fork individually. Slide the fork up through the bottom yoke, then install the circlip, wiring tie (left-hand fork) and the handlebar onto the fork **(see illustrations)**. Make sure the circlip locates properly into the groove in the fork **(see illustration 6.4c)**, and press the

6.4b ... and removing the circlip

handlebar holder down onto the circlip so that it locates up inside the base of the holder. Slide the fork up into the top yoke, aligning the handlebar holders so that the lug on the top locates into its slot in the top yoke **(see illustration 5.3d)**. Make sure the forks are pushed fully home, so that the handlebar holders are held securely between the top yoke and the circlip. Check that the amount of protrusion of the fork tube above the top yoke is equal on both sides.

6 Tighten the fork clamp bolts in the bottom yoke to the torque setting specified at the beginning of the Chapter **(see illustration)**. If the fork legs have been dismantled or if the fork oil has been changed, the fork top bolts should now be tightened to the specified

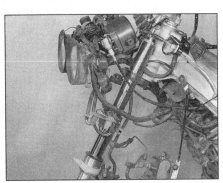

6.5a Slide the fork up through the bottom yoke ...

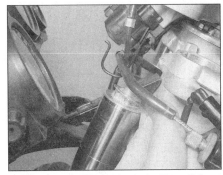

6.5b ... and fit the circlip and handlebar onto it

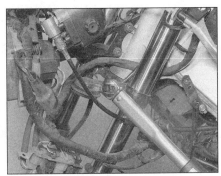

6.6a Tighten the bottom yoke clamp bolt ...

6•8 Frame, suspension and final drive

6.6b ... then the fork top bolt ...

6.6c ... and finally the top yoke and handlebar clamp bolts to their specified torque settings

torque setting **(see illustration)**. Now tighten the fork clamp bolts in the top yoke and the handlebar holder clamp bolts to the specified torque setting **(see illustration)**.

7 If displaced, make sure the front brake master cylinder clamp is installed with the UP mark facing up and tighten the clamp bolts to the torque setting specified at the beginning of the Chapter **(see illustration 5.1b)**.

8 Install the front wheel (see Chapter 7), and the front mudguard (see Chapter 8). Mount the front brake hose/pipe union on the fork and tighten its bolt **(see illustration 6.2)**.

9 Check the operation of the front forks and brakes before taking the machine out on the road.

7 Forks – disassembly, inspection and reassembly

J, K, L and N models

Disassembly

1 Always dismantle the fork legs separately to avoid interchanging parts and thus causing an accelerated rate of wear. Store all components in separate, clearly marked containers **(see illustration)**.

2 Before dismantling the fork, it is advised that the damper rod bolt be slackened at this stage. Compress the fork tube in the slider so that the spring exerts maximum pressure on the damper rod head, then have an assistant slacken the damper rod bolt in the base of the fork slider **(see illustration 7.31)**. If an assistant is not available, clamp the slider via its brake caliper mounting lugs in a soft-jawed vice to support the fork. If required, remove the circlip securing the mudguard holder to the top of the fork slider and remove the holder.

3 If the fork top bolt was not slackened with the fork in situ, release any air pressure in the forks on J and K models and adjust the spring pre-load to a minimum on L and N models (see Section 12), then carefully clamp the fork tube in a vice equipped with soft jaws, taking care not to overtighten or score its surface, and slacken the top bolt.

4 Unscrew the fork top bolt from the top of the fork tube **(see illustration)**.

⚠️ **Warning: The fork spring is pressing on the fork top bolt with considerable pressure. Unscrew the bolt very carefully, keeping a downward pressure on it and release it slowly as it is likely to spring clear. It is advisable to wear some form of eye and face protection when carrying out this operation.**

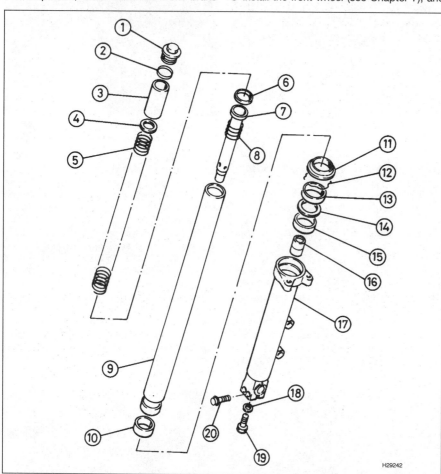

7.1 Front fork components – J, K, L and N models

1 Top bolt (L and N model shown)	7 Damper rod	14 Washer
2 O-ring	8 Rebound spring	15 Top bush
3 Spacer	9 Fork tube	16 Damper rod seat
4 Spring seat	10 Bottom bush	17 Fork slider
5 Spring	11 Dust seal	18 Sealing washer
6 Piston ring	12 Retaining clip	19 Damper rod bolt
	13 Oil seal	20 Axle clamp bolt

7.4 Unscrew and remove the fork top bolt (L and N model type shown)

Frame, suspension and final drive 6•9

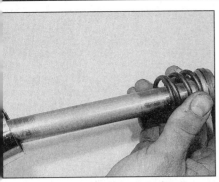

7.8 Withdraw the damper rod and rebound spring from the tube

7.9 Prise out the dust seal using a flat-bladed screwdriver

7.10 Prise out the retaining clip using a flat-bladed screwdriver

5 Slide the fork tube down into the slider and withdraw the spacer, spring seat and the spring from the tube **(see illustrations 7.26c, b and a)**. Note which way up the spring is fitted.

6 Invert the fork leg over a suitable container and pump the fork vigorously to expel as much fork oil as possible.

7 Remove the previously slackened damper rod bolt and its copper sealing washer from the bottom of the slider **(see illustration 7.37)**. Discard the sealing washer as a new one must be used on reassembly. If the damper rod bolt was not slackened before dismantling the fork, it may be necessary to re-install the spring, spring seat, spacer and top bolt to prevent the damper rod from turning. Alternatively, a long metal bar passed down through the fork tube and pressed hard into the damper rod head quite often suffices.

8 Invert the fork and withdraw the damper rod from inside the fork tube. Remove the rebound spring from the damper rod **(see illustration)**.

9 Carefully prise out the dust seal from the top of the slider to gain access to the oil seal retaining clip **(see illustration)**. Discard the dust seal as a new one must be used.

10 Carefully remove the retaining clip, taking care not to scratch the surface of the tube **(see illustration)**.

11 To separate the tube from the slider it is necessary to displace the top bush and oil seal. The bottom bush should not pass through the top bush, and this can be used to good effect. Push the tube gently inwards until it stops against the damper rod seat. Take care not to do this forcibly or the seat may be damaged. Then pull the tube sharply outwards until the bottom bush strikes the top bush. Repeat this operation until the top bush and seal are tapped out of the slider **(see illustration)**.

12 With the tube removed, slide off the oil seal and its washer, noting which way up they fit **(see illustration)**. Discard the oil seal as a new one must be used. The top bush can then also be slid off its upper end.

Caution: Do not remove the bottom bush from the tube unless it is to be renewed.

13 Tip the damper rod seat out of the slider, noting which way up it fits.

Inspection

14 Clean all parts in solvent and blow them dry with compressed air, if available. Check the fork tube for score marks, scratches, flaking of the chrome finish and excessive or abnormal wear. Look for dents in the tube and renew the tube in both forks if any are found. Check the fork seal seat for nicks, gouges and scratches. If damage is evident, leaks will occur. Also check the oil seal washer for damage or distortion and renew it if necessary.

15 Check the fork tube for runout using V-blocks and a dial gauge, or have it done by a dealer **(see illustration)**. If the amount of runout exceeds the service limit specified, the tube should be renewed.

⚠ **Warning: If the tube is bent or exceeds the runout limit, it should not be straightened; renew it.**

16 Check the spring for cracks and other damage. Measure the spring free length and compare the measurement to the specifications at the beginning of the Chapter. If it is defective or sagged below the service limit, renew the springs in both forks. Never renew only one spring. Also check the rebound spring.

17 Examine the working surfaces of the two bushes; if worn or scuffed they must be renewed. To remove the bottom bush from the fork tube, prise it apart at the slit using a flat-bladed screwdriver and slide it off **(see illustration)**. Make sure the new one seats properly.

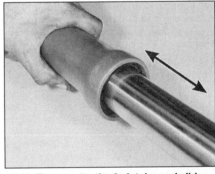

7.11 To separate the fork tube and slider, pull them apart firmly several times – the slide-hammer effect will pull the tubes apart

7.12 The oil seal (1), washer (2), top bush (3) and bottom bush (4) will come out with the fork tube

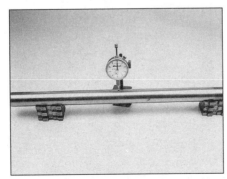

7.15 Check the fork tube for runout using V-blocks and a dial gauge

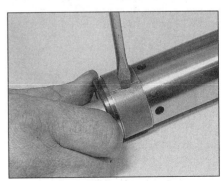

7.17 Prise off the bottom bush using a flat-bladed screwdriver

6•10 Frame, suspension and final drive

7.18 Renew the damper rod piston ring if it is worn or damaged

7.19a Slide the rebound spring onto the damper rod

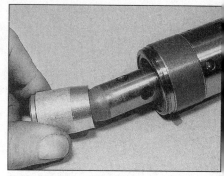

7.19b Fit the seat to the bottom of the rod

7.20a Slide the tube into the slider

7.20b Apply a thread-locking compound to the damper rod bolt and use a new sealing washer

7.21a Install the top bush . . .

18 Check the damper rod and its piston ring for damage and wear, and renew them if necessary (see illustration). Do not remove the ring from the piston unless it requires renewal

Reassembly

19 If removed, install the new piston ring into the groove in the damper rod head, then slide the rebound spring onto the rod (see illustration). Insert the damper rod into the fork tube and slide it into place so that it projects fully from the bottom of the tube, then install the seat on the bottom of the damper rod (see illustration).
20 Oil the fork tube and bottom bush with the specified fork oil and insert the assembly into the slider (see illustration). Fit a new copper sealing washer to the damper rod bolt and apply a few drops of a suitable non-permanent thread-locking compound, then install the bolt into the bottom of the slider (see illustration). Tighten the bolt to the specified torque setting. If the damper rod rotates inside the tube, temporarily install the fork spring and top bolt (see Steps 26 and 27) and compress the fork to hold the damper rod. Alternatively, a long metal bar pressed hard into the damper rod head quite often suffices.
21 Push the fork tube fully into the slider, then oil the top bush and slide it down over the tube (see illustration). Press the bush squarely into its recess in the slider as far as possible, then install the oil seal washer (see illustration). Either use the Honda service tool or a suitable piece of tubing to tap the bush fully into place; the tubing must be slightly larger in diameter than the fork tube and slightly smaller in diameter than the bush recess in the slider. Take care not to scratch the fork tube during this operation; it is best to make sure that the fork tube is pushed fully into the slider so that any accidental scratching is confined to the area above the oil seal.
22 When the bush is seated fully and squarely in its recess in the slider, (remove the washer to check, wipe the recess clean, then reinstall the washer), install the new oil seal.

HAYNES HiNT

Wrap some insulating tape around the circlip groove in the top of the fork tube – this will prevent the possibility of damage to the oil seal lips as it is slid down the tube.

7.21b . . . followed by the washer

Smear the seal's lips with fork oil and slide it over the tube so that its markings face upwards and drive the seal into place as described in Step 21 until the retaining clip groove is visible above the seal (see illustration).

7.22 Make sure the oil seal is the correct way up

Frame, suspension and final drive 6•11

7.23 Install the retaining clip . . .

7.24 . . . followed by the dust seal

7.25a Pour the oil into the top of the tube

7.25b Measure the oil level with the fork held vertical

7.26a Install the spring . . .

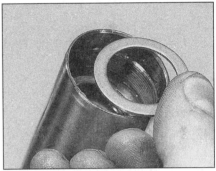

7.26b . . . followed by the spring seat . . .

23 Once the seal is correctly seated, fit the retaining clip, making sure it is correctly located in its groove **(see illustration)**.
24 Lubricate the lips of the new dust seal then slide it down the fork tube and press it into position **(see illustration)**.
25 Slowly pour in the correct quantity of the specified grade of fork oil and pump the fork at least ten times to distribute it evenly **(see illustration)**; the oil level should also be measured and adjustment made by adding or subtracting oil. Fully compress the fork tube into the slider and measure the fork oil level from the top of the tube **(see illustration)**. Add or subtract fork oil until it is at the level specified at the beginning of the Chapter.

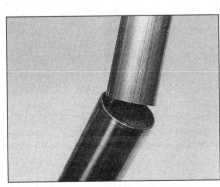

7.26c . . . and the spacer

26 Clamp the fork tube upright in the padded jaws of a vice, taking care not to overtighten it and damage the tube's surface. Install the spring with its closer-wound coils at the bottom, followed by the spring seat and the spacer **(see illustrations)**.
27 Fit a new O-ring to the fork top bolt and thread the bolt into the top of the fork tube **(see illustration 7.4)**.

 Warning: It will be necessary to compress the spring by pressing it down using the top

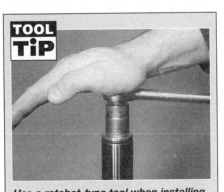

Use a ratchet-type tool when installing the fork top bolt. This makes it unnecessary to remove the tool from the bolt whilst threading it in making it easier to maintain a downward pressure on the spring.

bolt to engage the threads of the top bolt with the fork tube. This is a potentially dangerous operation and should be performed with care, using an assistant if necessary. Wipe off any excess oil before starting to prevent the possibility of slipping.

Keep the fork tube fully extended whilst pressing on the spring. Screw the top bolt carefully into the fork tube making sure it is not cross-threaded. **Note:** *The top bolt can be tightened to the specified torque setting at this stage if the tube is held between the padded jaws of a vice, but do not risk distorting the tube by doing so. A better method is to tighten the top bolt when the fork has been installed in the bike and is securely held in the bottom yoke.*

28 Remove the insulating tape from around the circlip groove in the fork tube. If removed, fit the mudguard holder onto the top of the slider, noting that each is marked L or R according to its side, and secure it with its circlip.
29 Install the forks (see Section 6). Set the air pressure or spring pre-load as required (see Section 12).

R models

Disassembly

30 Always dismantle the fork legs separately to avoid interchanging parts and thus causing an accelerated rate of wear. Store all

6•12 Frame, suspension and final drive

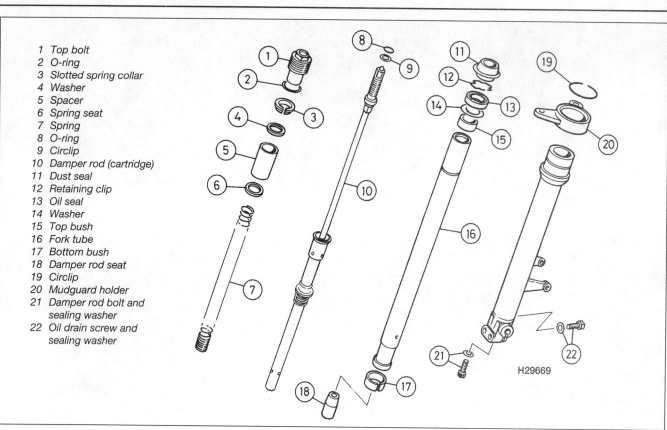

1 Top bolt
2 O-ring
3 Slotted spring collar
4 Washer
5 Spacer
6 Spring seat
7 Spring
8 O-ring
9 Circlip
10 Damper rod (cartridge)
11 Dust seal
12 Retaining clip
13 Oil seal
14 Washer
15 Top bush
16 Fork tube
17 Bottom bush
18 Damper rod seat
19 Circlip
20 Mudguard holder
21 Damper rod bolt and sealing washer
22 Oil drain screw and sealing washer

7.30 Front fork components – R models

components in separate, clearly marked containers (see illustration).
31 Before dismantling the fork, it is advised that the damper rod (cartridge) bolt be slackened at this stage. Compress the fork tube in the slider so that the spring exerts maximum pressure on the damper rod head, then have an assistant slacken the damper rod bolt in the base of the fork slider (see illustration). If required, remove the circlip securing the mudguard holder to the top of the fork slider and remove the holder.
32 Set the spring pre-load adjuster to its minimum setting (see Section 12). If the fork top bolt was not slackened with the fork in situ, carefully clamp the fork tube in a vice equipped with soft jaws, taking care not to overtighten or score its surface, and slacken the top bolt.
33 Unscrew the fork top bolt from the top of the fork tube (see illustration 7.59). The bolt will remain threaded on the damper rod.
34 Carefully clamp the fork slider in a vice and slide the fork tube down into the slider a little way (wrap a rag around the top of the tube to minimise oil spillage) while, with the aid of an assistant if necessary, keeping the damper rod fully extended. Counter-hold the pre-load adjuster and unscrew the fork top bolt from the damper rod (see illustration 7.58b).
35 Remove the slotted spring collar by slipping it out to the side (see illustration 7.57), then remove the washer, the spacer and the spring seat (see illustration 7.56d, c and b). Withdraw the spring from the tube, noting which way up it fits (see illustration 7.56a).
36 Invert the fork leg over a suitable container and pump the fork vigorously to expel as much fork oil as possible.
37 Remove the previously slackened damper rod bolt and its copper sealing washer from the bottom of the slider (see illustration). Discard the sealing washer as a new one must be used on reassembly.
38 Invert the fork and withdraw the damper rod (cartridge) from inside the fork tube (see illustration 7.49a).
39 Carefully prise out the dust seal from the top of the slider to gain access to the oil seal retaining clip (see illustration). Discard the dust seal as a new one must be used.

7.31 Slacken the damper rod bolt

7.37 Unscrew and remove the damper rod bolt

7.39 Prise out the dust seal using a flat-bladed screwdriver

Frame, suspension and final drive 6•13

7.40 Prise out the retaining clip using a flat-bladed screwdriver

7.49a Slide the damper rod (cartridge) into the tube . . .

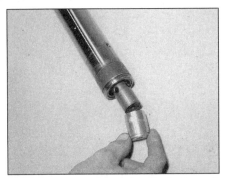

7.49b . . . and fit the seat to the bottom of the rod

40 Carefully remove the retaining clip, taking care not to scratch the surface of the tube **(see illustration)**.

41 To separate the tube from the slider it is necessary to displace the top bush and oil seal. The bottom bush should not pass through the top bush, and this can be used to good effect. Push the tube gently inwards until it stops against the damper rod seat. Take care not to do this forcibly or the seat may be damaged. Then pull the tube sharply outwards until the bottom bush strikes the top bush. Repeat this operation until the top bush and seal are tapped out of the slider **(see illustration 7.11)**.

42 With the tube removed, slide off the oil seal, washer and top bush, noting which way up they fit **(see illustration 7.12)**. Discard the oil seal as a new one must be used.

Caution: Do not remove the bottom bush from the tube unless it is to be renewed.

43 Tip the damper rod seat out of the slider, noting which way up it fits.

Inspection

44 Clean all parts in solvent and blow them dry with compressed air, if available. Check the fork tube for score marks, scratches, flaking of the chrome finish and excessive or abnormal wear. Look for dents in the tube and renew the tube in both forks if any are found. Check the fork seal seat for nicks, gouges and scratches. If damage is evident, leaks will occur. Also check the oil seal washer for damage or distortion and renew it if necessary.

45 Check the fork tube for runout using V-blocks and a dial gauge, or have it done by a dealer **(see illustration 7.15)**. If the amount of runout exceeds the service limit specified, the tube should be renewed.

 Warning: If the tube is bent or exceeds the runout limit, it should not be straightened; renew it.

46 Check the spring for cracks and other damage. Measure the spring free length and compare the measurement to the specifications at the beginning of the Chapter. If it is defective or sagged below the service limit, renew the springs in both forks. Never renew only one spring.

47 Examine the working surfaces of the two bushes; if worn or scuffed they must be renewed. To remove the bottom bush from the fork tube, prise it apart at the slit using a flat-bladed screwdriver and slide it off **(see illustration 7.17)**. Make sure the new one seats properly.

48 Check the damper rod (cartridge) assembly for damage and wear, and renew it if necessary. Holding the outside of the damper, pump the rod in and out of the damper. If the rod does not move smoothly in the damper it must be renewed.

Reassembly

49 Insert the damper rod (cartridge) into the fork tube and slide it into place so that it projects fully from the bottom of the tube, then install the seat on the bottom of the damper rod **(see illustrations)**.

50 Oil the fork tube and bottom bush with the specified fork oil and insert the assembly into the slider **(see illustration)**. Fit a new copper sealing washer to the damper rod bolt and apply a few drops of a suitable non-permanent thread locking compound, then install the bolt into the bottom of the slider **(see illustration)**. Tighten the bolt to the specified torque setting **(see illustration)**.

51 Push the fork tube fully into the slider, then oil the top bush and slide it down over the tube **(see illustration)**. Press the bush squarely into its recess in the slider as far as possible, then install the oil seal washer with

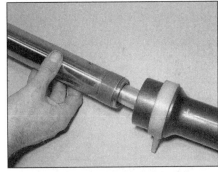

7.50a Slide the tube into the slider

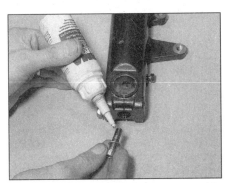

7.50b Apply a thread-locking compound to the damper rod bolt and use a new sealing washer . . .

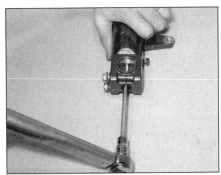

7.50c . . . and tighten the bolt to the specified torque

7.51a Install the top bush . . .

6•14 Frame, suspension and final drive

7.51b ... followed by the washer

7.52 Make sure the oil seal is the correct way up

7.53 Install the retaining clip ...

7.54 ... followed by the dust seal

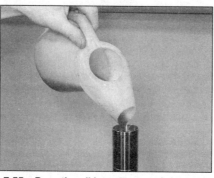
7.55a Pour the oil into the top of the tube

7.55b Measure the oil level with the fork held vertical

its flat side facing up **(see illustration)**. Either use the service tool or a suitable piece of tubing to tap the bush fully into place; the tubing must be slightly larger in diameter than the fork tube and slightly smaller in diameter than the bush recess in the slider. Take care not to scratch the fork tube during this operation; it is best to make sure that the fork tube is pushed fully into the slider so that any accidental scratching is confined to the area above the oil seal.

52 When the bush is seated fully and squarely in its recess in the slider, (remove the washer to check, wipe the recess clean, then reinstall the washer), install the new oil seal.

Protect the oil seal during installation by applying tape to the circlip groove in the fork tube (see **Haynes Hint**). Smear the seal's lips with fork oil and slide it over the tube so that its markings face upwards and drive the seal into place as described in Step 51 until the retaining clip groove is visible above the seal **(see illustration)**.

53 Once the seal is correctly seated, fit the retaining clip, making sure it is correctly located in its groove **(see illustration)**.

54 Lubricate the lips of the new dust seal then slide it down the fork tube and press it into position **(see illustration)**.

55 Slowly pour in the correct quantity of the specified grade of fork oil and pump the fork and damper rod at least ten times each to distribute it evenly **(see illustration)**; the oil level should also be measured and adjustment made by adding or subtracting oil. Fully compress the fork tube and damper rod into the slider and measure the fork oil level from the top of the tube **(see illustration)**. Add or subtract fork oil until it is at the level specified at the beginning of the Chapter.

56 Clamp the fork tube upright between the padded jaws of a vice, taking care not to overtighten the vice and damage the tube. Install the spring with its tapered coils at the bottom **(see illustration)**. Install the

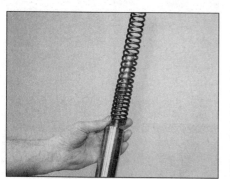
7.56a Install the spring ...

7.56b ... followed by the spring seat ...

7.56c ... the spacer ...

Frame, suspension and final drive 6•15

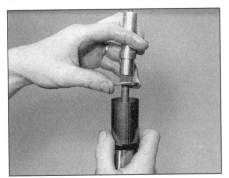

7.56d ... and the washer

7.57 Fit the slotted collar between the nut and the washer

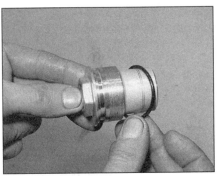

7.58a Fit a new O-ring ...

spring seat, the spacer and the washer over the damping adjuster rod **(see illustrations)**.

57 Slide the slotted spring collar into position between the washer and the nut on the damping adjuster rod **(see illustration)**.

58 Fit a new O-ring onto the fork top bolt, then thread the top bolt fully, but not tightly, onto the damping adjuster rod, using a spanner on either the nut or the pre-load adjuster to prevent the rod from turning **(see illustrations)**.

59 Withdraw the tube fully from the slider and carefully screw the top bolt into the fork tube making sure it is not cross-threaded **(see illustration)**. **Note:** *The top bolt can be tightened to the specified torque setting at this stage if the tube is held between the padded jaws of a vice, but do not risk distorting the tube by doing so. A better method is to tighten the top bolt when the fork leg has been installed and is securely held in the bottom yoke. See* **Tool Tip** *earlier in this section.*

60 Remove the insulating tape from around the circlip groove in the fork tube. If removed, fit the mudguard holder onto the top of the slider, noting that each is marked L or R according to its side, and secure it with its circlip.

61 Install the forks (see Section 6). Set the spring pre-load and damping as required (see Section 12).

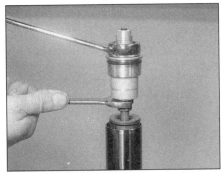

7.58b ... then thread the top bolt onto the damper rod

8 Steering stem – removal and installation

Removal

1 Remove the front forks (see Section 6) and the horn (see Chapter 9). Remove the left-hand fork clamp bolt in the top yoke to release the choke knob – the right-hand bolt can remain in place if the master cylinder was displaced when removing the forks, otherwise it must also be removed **(see illustration 5.3a)**. Unscrew the bolt securing the front brake hose clamp **(see illustration)**. Take care not to strain or knock the hoses when removing the steering stem.

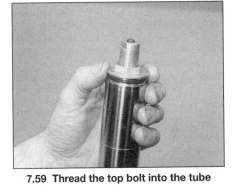

7.59 Thread the top bolt into the tube

2 On L, N and R models, prise the plug out of the steering stem nut **(see illustration 5.3b)**. Unscrew and remove the nut using a 30 mm socket or spanner, and on J and K models remove the washer **(see illustration 5.3c)**. Free the clutch cable from its guide on the top yoke. Lift the top yoke off the steering stem and place it aside, making sure no strain is placed on the ignition switch wiring **(see illustration 5.3d)**. Use a rag to protect other components. If required, remove the fairing (see Chapter 8), then trace the ignition switch wiring and disconnect it at the connector.

3 Prise the lockwasher tabs out of the notches in the locknut **(see illustration)**. Unscrew the locknut using either a C-spanner or a suitable drift located in one of the notches **(see illustration)**. Remove the lockwasher,

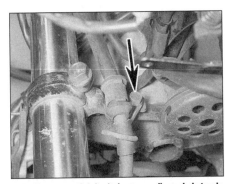

8.1 Remove the bolt (arrowed) and detach the brake hose clamp

8.3a Bend down the tabs ...

8.3b ... then unscrew the locknut ...

6•16 Frame, suspension and final drive

8.3c ... and remove the lockwasher

bending up the remaining tabs to release it from the adjuster nut if necessary **(see illustration)**. Inspect the tabs for cracks or signs of fatigue. If there are any, discard the lockwasher and use a new one; otherwise the old one can be reused.

4 Supporting the bottom yoke, unscrew the adjuster nut using either a C-spanner, a peg-spanner or a drift located in one of the notches, then remove the adjuster nut and the bearing cover from the steering stem **(see illustration)**.

5 Gently lower the bottom yoke and steering stem out of the frame.

6 Remove the upper bearing and its inner race from the top of the steering head **(see illustration)**. Remove all traces of old grease from the bearings and races and check them for wear or damage as described in Section 9.

Note: *Do not attempt to remove the races from the frame or the lower bearing from the steering stem unless they are to be renewed.*

Installation

7 Smear a liberal quantity of grease on the bearing races in the frame. Work the grease well into both the upper and lower bearings.

8 Carefully lift the steering stem/bottom yoke up through the frame. Install the upper bearing and its inner race in the top of the steering head **(see illustration 8.6)**, then install the bearing cover. Thread the adjuster nut onto the steering stem. Tighten the nut lightly to settle the bearings, then slacken if off so that it is finger-tight. Using either the C-spanner or drift **(see illustration)**, tighten the nut a little at

8.6 Remove the upper bearing

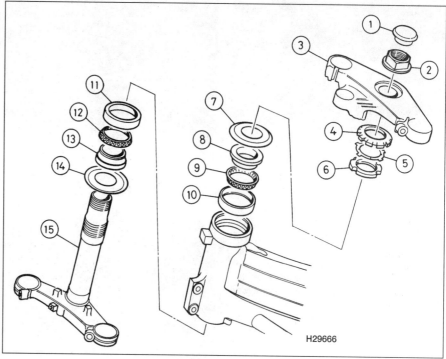

8.4 Steering stem components

1 Steering stem nut cap (where fitted)
2 Steering stem nut
3 Top yoke
4 Locknut
5 Lockwasher
6 Adjuster nut
7 Bearing cover
8 Upper bearing inner race
9 Upper bearing
10 Upper bearing outer race
11 Lower bearing outer race
12 Lower bearing
13 Lower bearing inner race
14 Dust seal
15 Bottom yoke and steering stem

a time until all freeplay is removed, yet the steering is able to move freely. If the Honda adapter is available you can apply the torque setting specified at the beginning of the Chapter. Now turn the steering from lock to lock five times to settle the bearings, then recheck the adjustment or the torque setting. The object is to set the adjuster nut so that the bearings are under a very light loading, just enough to remove any freeplay.

Caution: *Take great care not to apply excessive pressure because this will cause premature failure of the bearings. If the torque setting is applied and the bearings are too loose or tight, set them up according to feel.*

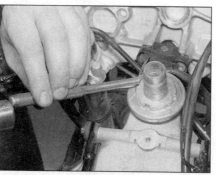

8.8 Tighten the adjuster nut as described

9 Attach the front brake hose guide to the bottom yoke **(see illustration 8.1)**. Feed the clutch cable into its guide.

10 Install the horn (see Chapter 9) and the front forks (see Section 6), but do not yet tighten the fork clamp bolts.

11 Carry out a check and adjustment of the steering head bearing freeplay as described in Chapter 1, and follow the procedure for the installation of the remaining disturbed components. Reconnect the ignition switch wiring connector if it was disconnected. If not already done, tighten the handlebar clamp bolts to the specified torque **(see illustration 5.9)**.

9 Steering head bearings – inspection and renewal

Inspection

1 Remove the steering stem (see Section 8).
2 Remove all traces of old grease from the bearings and races and check them for wear or damage.
3 The outer races should be polished and free from indentations. Inspect the bearing rollers for signs of wear, damage or discoloration, and examine the bearing ball

Frame, suspension and final drive 6•17

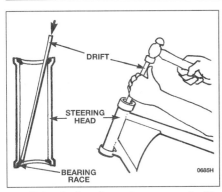

9.4 Drive the bearing outer races out with a brass drift as shown

retainer cage for signs of cracks or splits. Spin the bearings by hand. They should spin freely and smoothly. If there are any signs of wear on any of the above components both upper and lower bearing assemblies must be renewed as a set. Only remove the outer races from the frame headstock if they need to be renewed – do not re-use them once they have been removed.

Renewal

4 The outer races are an interference fit in the frame headstock and can be tapped from position with a suitable drift **(see illustration)**. Tap firmly and evenly around each race to ensure that it is driven out squarely. It may prove advantageous to curve the end of the drift slightly to improve access.

5 Alternatively, the races can be removed using a slide-hammer type bearing extractor; these can often be hired from tool shops.

6 The new outer races can be pressed into the head using a drawbolt arrangement **(see illustration)**, or by using a large diameter tubular drift. Ensure that the drawbolt washer or drift (as applicable) bears only on the outer edge of the race and does not contact the working surface. Alternatively, have the races installed by a dealer equipped.

 HAYNES HiNT *Installation of new bearing outer races is made much easier if the races are left overnight in the freezer. This causes them to contract slightly making them a looser fit.*

7 To remove the lower bearing inner race from the steering stem, use two screwdrivers placed on opposite sides of the race to work it free. If the bearing is firmly in place it will be necessary to use a bearing puller, or in extreme circumstances to split the bearing's inner section using an angle grinder **(see illustration)**. Take the steering stem to a dealer if required. Check the condition of the dust seal that fits under the inner race and renew it if it is worn, damaged or deteriorated.

8 Fit the new lower bearing inner race onto the steering stem. A length of tubing with an

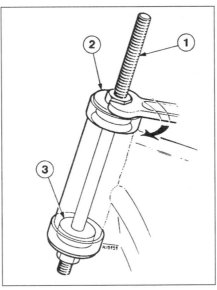

9.6 Drawbolt arrangement for fitting steering stem bearing outer races

1. Long bolt or threaded bar
2. Thick washer
3. Guide for lower race

internal diameter slightly larger than the steering stem will be needed to tap the new bearing into position **(see illustration)**. Ensure that the drift bears only on the inner edge of the bearing race and does not contact its working surface.

9 Install the steering stem (see Section 8).

10 Rear shock absorber – removal, inspection and installation

Removal

1 Place the machine on an auxiliary stand. Position a support under the rear wheel so that it does not drop when the shock absorber is removed, but also making sure that the weight of the machine is off the rear suspension so that the shock is not compressed.

2 Remove the seat cowling (see Chapter 8).

10.2 The heel plates are secured to the footrest brackets by two bolts (arrowed)

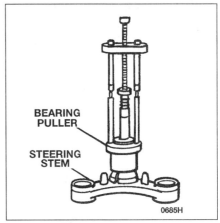

9.7 It is best to remove the lower bearing using a puller

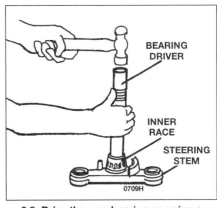

9.8 Drive the new bearing on using a suitable bearing driver or a length of pipe

On J and K models, unscrew the bolts securing both rider's heel plates and remove them **(see illustration)**. On L, N and R models, unscrew the bolts securing the rider's left-hand heel plate and remove it.

3 On J and K models, using socket extensions inserted through the holes in the swingarm, counter-hold the shock absorber lower mounting bolt and unscrew the nut, then withdraw the bolt **(see illustration)**. On L, N and R models, using a socket extension inserted through the hole in the left-hand side

10.3a Shock absorber lower mounting bolt (arrowed) – J and K models

6•18 Frame, suspension and final drive

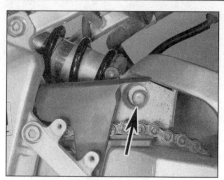

10.3b Shock absorber lower mounting bolt (arrowed) – L, N and R models

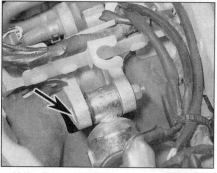

10.4a Remove the upper mounting bolt (arrowed) . . .

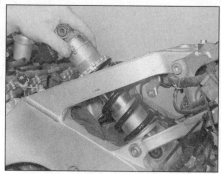

10.4b . . . and lift the shock out of the frame

of the swingarm, unscrew the shock absorber lower mounting bolt (see illustration).

4 Support the shock absorber, then unscrew the upper mounting nut (where fitted) and bolt and lift the shock out of the top of the frame, noting the routing of the drain hose (see illustrations).

Inspection

5 Individual components for the rear shock absorber are available for all except R models. Check with your dealer for availability. On R models, the entire unit must be renewed if it is worn or damaged.

6 Inspect the shock absorber for obvious physical damage and oil leakage, and the coil spring for looseness, cracks or signs of fatigue (see illustration).

10.6 Check for signs of oil leaks

7 Inspect the pivot hardware at the top and bottom of the shock for wear or damage.

8 On L, N and R models, withdraw the inner sleeve from the shock absorber lower mounting (see illustration). Inspect the sleeve, bearing and dust seals for signs of wear and renew worn components as necessary. The bearing is a press-fit and can be removed and installed using a drawbolt arrangement – see *Tools and Workshop Tips* in the Reference section.

9 To measure the shock absorber spring free length, the unit must be disassembled, which requires the use of a suitable spring compressor. If the tools or expertise are not available, take the unit to a dealer. Otherwise, on J and K models, compress the spring using a coil spring compressor by just enough to access the spring stopper ring. Remove the ring and the spring seat, then carefully release the compressor until the spring is relaxed. On L and N models, compress the spring using a coil spring compressor by just enough to access the spring seat halves, then remove them, noting how they fit. Remove the spring sleeve, then carefully release the compressor until the spring is relaxed. Remove the spring, noting which way up it fits. Measure the free length of the spring and compare it to the specifications. If the spring has sagged below its service limit, it must be renewed. Individual components for the rear shock absorber are available for all except R models. Check with your dealer for availability. On R models, the entire unit must be renewed if it is worn or damaged.

Installation

10 Installation is the reverse of removal, noting the following.

a) Apply molybdenum disulphide grease to the pivot points.
b) Make sure the drain hose is correctly routed.
c) On L, N and R models, clean the threads of the lower mounting bolt and apply a suitable non-permanent thread locking compound (see illustration).
d) Install the bolts and nut(s) finger-tight only until both are in position, then tighten them to the torque settings specified at the beginning of the Chapter (see illustration).
e) Adjust the suspension as required (see Section 12).

11 Rear suspension linkage – removal, inspection and installation

Removal

1 Place the machine on an auxiliary stand. Position a support under the rear wheel so that it does not drop when the shock absorber

10.8 Withdraw the sleeve and check the bearing

10.10a Apply a thread-lock to the bolt

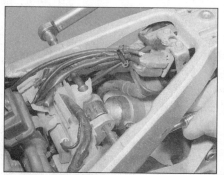

10.10b Tightening the upper mounting bolt on L, N and R models

Frame, suspension and final drive 6•19

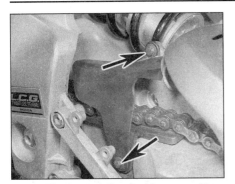

11.2 The front chainguard is secured by two bolts (arrowed)

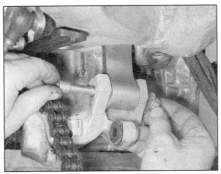

11.4 Remove the linkage-to-frame bolt ...

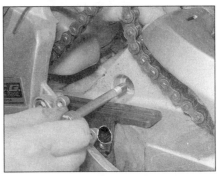

11.5a ... and the linkage-to-swingarm bolt ...

lower mounting bolt is removed, but also making sure that the weight of the machine is off the rear suspension so that the shock is not compressed.

2 Remove the seat cowling (see Chapter 8). On J and K models, unscrew the bolts securing the rider's right-hand footrest bracket and displace it, making sure no strain is placed on the brake hose. On L, N and R models, unscrew the bolts securing the rider's left-hand heel plate **(see illustration 10.2)** and the front chainguard and remove them **(see illustration)**.

3 On J and K models, remove the shock absorber (see Section 10). On L, N and R models using a socket extension inserted through the hole in the left-hand side of the swingarm, unscrew the shock absorber lower mounting bolt **(see illustration 10.3b)**.

4 Unscrew and remove the nut and bolt securing the linkage to the frame **(see illustration)**.

5 Unscrew and remove the bolt securing the linkage to the swingarm, on J and K models accessing it from the right, and on L, N and R models accessing it from the left using a socket extension inserted through the hole in the swingarm **(see illustration)**. Remove the linkage from the top of the frame on J and K models, and from the bottom on L, N and R models **(see illustration)**.

Inspection

6 Undo the nut and remove the pivot bolt and separate the two linkage components.

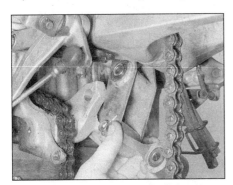

11.5b ... and remove the linkage assembly – L, N and R models shown

Withdraw the inner sleeves from both links **(see illustration)**.

7 Thoroughly clean all components, removing all traces of dirt, corrosion and grease.

8 Inspect all components closely, looking for obvious signs of wear such as heavy scoring, or for damage such as cracks or distortion.

9 Carefully lever out the dust seals, using a flat-bladed screwdriver, and check them for signs of wear or damage; renew them if necessary.

10 Worn bearings or bushes can be drifted out of their bores, but note that removal will destroy them; new components should be obtained before work commences. On J and K models, a short bearing is fitted on each side of each pivot, while on L, N and R models, a single long bearing is fitted for each pivot. Honda do not list the specific bearings as being available separately from the linkage components themselves, so check with a dealer or a bearing specialist before removing them – if they are not available, new linkage components must be installed. If obtainable, the new bearings should be pressed or drawn into their bores rather than driven into position. In the absence of a press, a suitable drawbolt arrangement can be made up as described in *Tools and Workshop Tips* in the Reference section.

11 Lubricate the needle roller bearings and the inner sleeves with molybdenum disulphide grease and install the sleeves **(see illustration 11.6)**.

12 Check the condition of the dust seals and

11.6 Linkage assembly components – L, N and R models shown

renew them if they are damaged or deteriorated. Press the seals squarely into place.

13 Couple the two suspension links. On J and K models, make sure the rear link is fitted so that the centre hole is at the bottom – if it is at the top, turn it upside down. On L, N and R models, make sure the 'F MV4' mark on the rear link is positioned on the left side with the arrow pointing forward. Apply some grease to the bolt shank. On J and K models, install the bolt and tighten the nut to the specified torque setting. On L, N and R models, clean the threads of the pivot bolt and apply a suitable non-permanent thread locking compound. Tighten the bolt to the specified torque setting.

Installation

14 If not already done, lubricate the seals, needle roller bearings, inner sleeves and the pivot bolts with molybdenum disulphide grease.

15 Manoeuvre the linkage assembly into position making sure that the 'F MV4' or 'FR MAL' mark on the rear link is on the left side **(see illustration 11.5b)**.

16 Install the linkage assembly and the shock absorber in a reverse of the removal procedure. Install the bolts and nuts finger-tight only until all components are in position, then tighten them to the torque setting specified at the beginning of the Chapter **(see illustrations 11.5a and 11.4)**.

17 Install any other disturbed components (see Step 2).

18 Check the operation of the rear suspension before taking the machine on the road. Adjust the suspension as required (see Section 12).

12 Suspension – adjustments

Front forks

1 On J and K models, the forks are air assisted. Remove the cap from the valve in

12.1 Remove the cap to access the air valve

12.2 Spring pre-load adjuster (arrowed) – L and N models

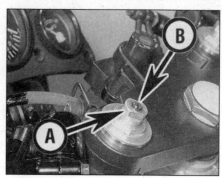

12.3 Spring pre-load adjuster (A) and damping adjuster (B) – R models

the centre of the fork top bolt **(see illustration)**. Use a pressure gauge and pump designed for suspension systems – on no account use a filling station air line; combined gauge and pump units are available for motorcycle suspension systems. Adjust the air pressure as required and according to the Specifications at the beginning of the Chapter. To decrease the pressure, depress the pin in the centre of the valve. Ensure that the air pressure is the same in each fork.

Caution: Do not use a high pressure source of air or you will blow the seals.

2 On L and N models, the forks are adjustable for spring pre-load. Spring pre-load is adjusted using a suitable screwdriver in the slot in the top of the adjuster in the centre of the fork top bolt **(see illustration)**. Turn it clockwise to increase pre-load and anti-clockwise to decrease it. The amount of pre-load is indicated by lines on the body of the adjuster. Always make sure both adjusters are set equally.

3 On R models, the forks are adjustable for spring pre-load and rebound damping. Spring pre-load is adjusted using a suitable spanner on the adjuster flats in the centre of the fork top bolt **(see illustration)**. Turn it clockwise to increase pre-load and anti-clockwise to decrease it. The amount of pre-load is indicated by lines on the adjuster. Always make sure both adjusters are set equally. Damping is adjusted using a screwdriver in the slot in the top of the damper rod protruding from the pre-load adjuster. Turn it clockwise to increase damping and anti-clockwise to decrease it.

Rear shock absorber

4 On J and K models, the rear shock absorber is adjustable for spring pre-load. Adjustment is made using a suitable C-spanner (one is provided in the toolkit) to turn the spring seat on the bottom of the shock absorber **(see illustration)**. There are seven positions. Position 1 is the softest setting, position 7 is the hardest. Align the setting required with the adjustment stopper. To increase the pre-load, turn the spring seat clockwise. To decrease the pre-load, turn the spring seat anti-clockwise.

5 On L, N and R models, the rear shock absorber is adjustable for spring pre-load and rebound damping. Pre-load adjustment is made by turning the adjuster nut on the threads on the shock absorber body **(see illustration)**. Slacken the locknut, then turn the adjuster nut towards the spring to increase pre-load and away from the spring to decrease it. Tighten the locknut securely after adjustment.

6 Damping adjustment is made by turning the adjuster on the bottom of the shock absorber using a flat-bladed screwdriver **(see illustration)**. To increase the damping, turn the adjuster clockwise. To decrease the damping, turn the adjuster anti-clockwise.

13 Swingarm – removal and installation

Note: *The swingarm can be removed with the shock absorber and linkage assembly attached if required. Otherwise, the linkage assembly must either be removed completely (see Section 11), or it can remain attached to the swingarm and removed with it after detaching it from the shock absorber and frame.*

Removal

1 Remove the rear wheel (see Chapter 7). Displace the brake caliper bracket from the swingarm, noting how it fits, and tie it to the top of the frame, making sure no strain is placed on the hose **(see illustration 13.19)**.

2 Unscrew the gearchange lever linkage arm pinch bolt and remove the arm from the shaft, noting the alignment punch marks **(see illustration 16.1a)**. If no marks are visible, make your own before removing the arm so that it can be correctly aligned with the shaft on installation. If required, detach the speedometer cable from the front sprocket cover. Unscrew the bolts securing the sprocket cover and remove it, noting the guide plate **(see illustration 16.1b)**. Let the cover hang by the cable, if it wasn't detached. On L, N and R models, note the wiring clip

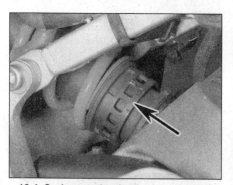

12.4 Spring pre-load adjuster – J and K models

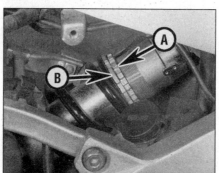

12.5 Slacken the locknut (A) and turn the adjusting ring (B) as required

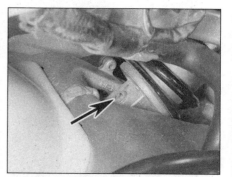

12.6 Damping adjuster (arrowed) – L, N and R models

Frame, suspension and final drive 6•21

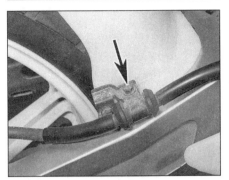

13.3 Unscrew the bolt (arrowed) and detach the brake hose guide

13.5 Unscrew the swingarm nut (arrowed)

13.6a Unscrew the locknut...

secured by the lower bolt. Slip the drive chain off the front sprocket.

3 Unscrew the bolt securing the rear brake hose guide to the swingarm (see illustration).
4 Either remove the suspension linkage (see Section 11), or remove the shock absorber upper mounting bolt and the linkage assembly to frame bolt, or remove the shock absorber lower mounting bolt and the linkage to frame bolt, as required (see **Note** above), referring to the relevant Steps of Sections 10 and 11.
5 Unscrew the nut on the left-hand end of the swingarm pivot bolt (see illustration).

6 Unscrew the locknut on the adjuster bolt on the right-hand end of the pivot bolt (see illustration). This requires the use of a Honda service tool (Pt. No. 07908-4690001), which is a special wrench that fits the locknut (see illustration). Although it is possible to fabricate a peg spanner from an old socket which will enable the locknut to be unscrewed, the special tool provides a means of tightening the locknut to the correct torque on installation and is therefore considered essential.
7 The swingarm pivot bolt fits into the head of

the adjuster bolt (actually a threaded sleeve). Using an Allen key, unscrew the pivot bolt which will also unscrew the adjuster bolt (see illustration). It is not necessary to completely remove the adjuster bolt – it can remain loosely threaded in.
8 When the adjuster bolt is fully unscrewed, support the swingarm then withdraw the pivot bolt and remove the swingarm (see illustration).
9 Remove the chain slider, chainguards and mudguard from the swingarm if necessary, noting how they fit (see illustrations). If the

13.6b ... using the special tool. This tool MUST be used for the swingarm tightening procedure

13.7 Unscrew the pivot bolt and the adjuster bolt together

13.8 Withdraw the pivot bolt and remove the swingarm

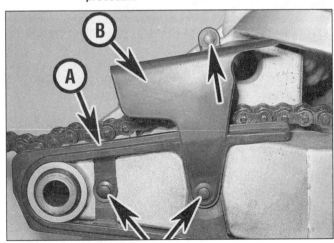

13.9a Unscrew the bolts (arrowed) and remove the chain slider (A), the front chainguard (B), ...

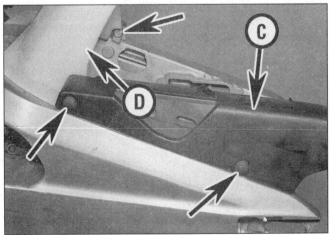

13.9b ... the rear chainguard (C) and the mudguard (D)

6•22 Frame, suspension and final drive

13.13a Thread in the adjuster bolt a couple of turns

13.13b Tighten the adjuster bolt as shown to the specified torque

13.14 Thread on the pivot bolt nut and tighten it finger-tight

chain slider is badly worn or damaged, it should be renewed. If required, split the drive chain at its joining link to separate the chain from the swingarm (see Section 15).

10 If attached, and if required, separate the suspension linkage from the swingarm (see Section 11). Inspect all components for wear or damage as described in Section 14.

Installation

11 If removed, install the mudguard, chainguards and chain slider (see illustrations 13.9a and b).
12 Lubricate the dust seals, bearings, sleeve or collar, and the pivot bolt with grease.
13 If removed, thread the adjuster bolt a couple of turns only into the swingarm (see

13.15a Thread on the locknut . . .

illustration). Offer up the swingarm and have an assistant hold it in place. Install the pivot bolt through the swingarm and engage its flats with those of the adjuster bolt (see illustration 13.8). Tighten the adjuster bolt to the torque setting specified at the beginning of the Chapter by turning the pivot bolt using a suitable Allen key (see illustration).
14 Thread the pivot bolt nut finger-tight onto the left-hand end of the bolt (see illustration).
15 Install the adjuster bolt locknut and tighten it as much as possible by hand (see illustration). Tighten the locknut further using the service tool as described in Step 6, and using an Allen key applied through its middle to counter-hold the pivot bolt and adjuster bolt and prevent them from turning. Then, using a torque wrench applied to the hole in the arm of the service tool, tighten the locknut to the specified torque setting (see illustration). Note: *The specified torque setting takes into account the extra leverage provided by the service tool and cannot be duplicated without it.*
16 Tighten the pivot bolt nut to the specified torque setting, again using an Allen key to counter-hold the pivot bolt and adjuster bolt and prevent them from turning (see illustration).
17 Install the rear shock absorber and linkage assembly as required by your removal procedure (see Sections 10 and 11).

18 Fit the drive chain around the front sprocket, then install the sprocket cover with its guide plate, making sure the speedometer cable drive socket on the inside of the cover engages correctly with the head of the sprocket bolt (see illustration 16.9). On L, N and R model, do not forget the wiring clip with the lower bolt. If removed, attach the speedometer cable to the drive housing. Install the gearchange linkage arm onto the end of the shaft, aligning the punch marks, and tighten the pinch bolt (see illustration 16.1a).
19 Locate the brake caliper bracket lug into the swingarm (see illustration) and attach the brake hose guide (see illustration 13.3).
20 Install the rear wheel (see Chapter 7).
21 Check and adjust the drive chain slack (see Chapter 1). Check the operation of the rear suspension before taking the machine on the road.

14 Swingarm – inspection and bearing renewal

Inspection

1 Remove the sleeve (J and K models) or collar (L, N, and R models) from the left-hand

13.15b . . . and tighten it as described, using the special tool, to the specified torque

13.16 Finally tighten the pivot bolt nut to the specified torque

13.19 Locate the lug on the bracket into the slot in the swingarm (arrowed)

Frame, suspension and final drive 6•23

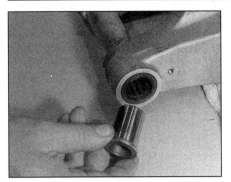

14.1 Remove the sleeve or collar

side of the swingarm **(see illustration)**. Lever out the seals from each side, taking care not to damage them.
2 Thoroughly clean all components, removing all traces of dirt, corrosion and grease.
3 Inspect all components closely, looking for obvious signs of wear such as heavy scoring, and cracks or distortion due to accident damage. Any damaged or worn component must be renewed.
4 Check the swingarm pivot bolt for straightness by rolling it on a flat surface such as a piece of plate glass (first wipe off all old grease and remove any corrosion using fine emery cloth). If the equipment is available, place the axle in V-blocks and measure the runout using a dial indicator. If the axle is bent or the runout excessive, renew it.

Bearing renewal

5 Remove the sleeve (J and K models) or collar (L, N, and R models) from the left-hand side of the swingarm **(see illustration 14.1)**. Lever out the seals from each side, taking care not to damage them, and remove the circlip from the right-hand side. Refer to *Tools and Workshop Tips (Section 5)* in the Reference section and remove the bearings, then clean them and inspect them for wear or damage. A needle bearing is fitted in the left-hand side, and two ball bearings are in the right-hand side. If the bearings do not run smoothly and freely or if there is excessive freeplay, they must be renewed. The needle bearing in the left-hand side of the swingarm should be renewed as a matter of course if it is removed.

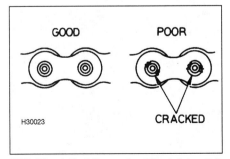

15.8 Check that the ends of the joining link are properly staked

6 Do not forget to install the bearing spacer between the bearings in the swingarm. When installing the new needle bearing, lubricate it with molybdenum disulphide grease and set it in the swingarm to a depth of exactly 4 mm. Check the condition of the dust seals and renew them if they are damaged or deteriorated.

15 Drive chain – removal, cleaning and installation

Removal

Note: *The original equipment drive chain fitted to these models has a staked-type master (joining) link which can be disassembled using either Honda service tool, Pt. No. 07HMH-MR10100 for J and K models, 07HMH-MR10101 for L and N models or 07HMH-MR10102 for R models, or one of several commercially-available drive chain cutting/staking tools. Such chains can be recognised by the master link side plate's identification marks (and usually its different colour), as well as by the staked ends of the link's two pins which look as if they have been deeply centre-punched, instead of peened over as with all the other pins.*

⚠ **Warning: NEVER install a drive chain which uses a clip-type master (split) link. Use ONLY the correct service tools to secure the staked-type of master link – if you do not have access to such tools, have the chain renewed by a dealer to be sure of having it securely installed.**

1 Locate the joining link in a suitable position to work on by rotating the back wheel.
2 Slacken the drive chain as described in Chapter 1.
3 Split the chain at the joining link using the chain cutter, following carefully the manufacturer's operating instructions (see also Section 8 in *Tools and Workshop Tips* in the Reference Section). Remove the chain from the bike, noting its routing through the swingarm.

Cleaning

4 Soak the chain in kerosene (paraffin) for approximately five or six minutes.
Caution: Don't use gasoline (petrol), solvent or other cleaning fluids which might damage its internal sealing properties. Don't use high-pressure water. Remove the chain, wipe it off, then blow dry it with compressed air immediately. The entire process shouldn't take longer than ten minutes – if it does, the O-rings in the chain rollers could be damaged.

Installation

⚠ **Warning: If you do not have access to a chain riveting tool, have the chain fitted by a dealer.**

5 Unscrew the gearchange lever linkage arm pinch bolt and remove the arm from the shaft, noting the alignment punch marks **(see illustration 16.1a)**. If no marks are visible, make your own before removing the arm so that it can be correctly aligned with the shaft on installation. If required, detach the speedometer cable from the front sprocket cover. Unscrew the bolts securing the sprocket cover and remove it, noting the guide plate **(see illustration 16.1b)**. Let the cover hang by the cable, if it wasn't detached. On L, N and R models, note the wiring clip secured by the lower bolt.
6 Install the drive chain through the swingarm sections and around the front sprocket, leaving the two ends in a convenient position to work on.
7 Refer to Section 8 in *Tools and Workshop Tips* in the Reference Section. Install the new joining link from the inside with the four O-rings correctly located between the link plates. Install the new side plate with its identification marks facing out. On R models, measure the amount that the joining link pins project from the side plate and check they are within the measurements specified at the beginning of the Chapter. Stake the new link using the drive chain cutting/staking tool, following carefully the instructions of both the chain manufacturer and the tool manufacturer. DO NOT re-use old joining link components.
8 After staking, check the joining link and staking for any signs of cracking **(see illustration)**. If there is any evidence of cracking, the joining link, O-rings and side plate must be renewed. On R models, measure the diameter of the staked ends in two directions and check that it is evenly staked and within the measurements specified at the beginning of the Chapter.
9 Install the sprocket cover with its guide plate, making sure the speedometer cable drive socket on the inside of the cover engages correctly with the head of the sprocket bolt. On L, N and R model, do not forget the wiring clip with the lower bolt. If removed, attach the speedometer cable to the drive housing. Install the gearchange linkage arm onto the end of the shaft, aligning the punch marks, and tighten the pinch bolt.
10 On completion, adjust and lubricate the chain following the procedures described in Chapter 1.
Caution: Use only the recommended lubricant.

16 Sprockets – check and renewal

Check

1 Unscrew the gearchange lever linkage arm pinch bolt and remove the arm from the shaft, noting the alignment punch marks **(see**

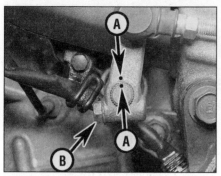

16.1a Note the alignment punch marks (A), then unscrew the bolt (B) and slide the arm off the shaft

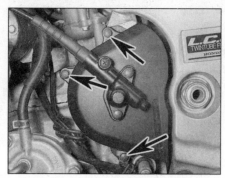

16.1b Unscrew the bolts (arrowed) and remove the cover

16.5 Unscrew the bolt (arrowed) . . .

illustration). If no marks are visible, make your own before removing the arm so that it can be correctly aligned with the shaft on installation. If required, detach the speedometer cable from the front sprocket cover. Unscrew the bolts securing the sprocket cover and remove it, noting the guide plate (see illustration). Let the cover hang by the cable, if it wasn't detached. On L, N and R models, note the wiring clip secured by the lower bolt.

2 Check the wear pattern on both sprockets (see Chapter 1, Section 1). If the sprocket teeth are worn excessively, renew the chain and both sprockets as a set. Whenever the sprockets are inspected, the drive chain should be inspected also (see Chapter 1). If you are renewing the chain, renew the sprockets as well.

3 Adjust and lubricate the chain following the procedures described in Chapter 1.

Caution: Use only the recommended lubricant.

Renewal

Front sprocket

4 Unscrew the gearchange lever linkage arm pinch bolt and remove the arm from the shaft, noting the alignment punch marks (see illustration 16.1a). If no marks are visible, make your own before removing the arm so that it can be correctly aligned with the shaft on installation. If required, detach the speedometer cable from the front sprocket cover. Unscrew the bolts securing the sprocket cover and remove it, noting how the guide plate locates (see illustration 16.1b). Let the cover hang by the cable, if it wasn't detached. On L, N and R models, note the wiring clip secured by the lower bolt.

5 Have an assistant apply the rear brake, then unscrew the sprocket bolt and remove the washer (see illustration).

6 Slide the sprocket and chain off the shaft and slip the sprocket out of the chain (see illustration). If there is not enough slack on the chain to remove the sprocket, refer to Chapter 1 and adjust the drive chain until it is fully slack.

7 Engage the new sprocket with the chain and slide it on the shaft (see illustration). Take up any slack in the chain.

8 Install the sprocket bolt with its washer and tighten it to the torque setting specified at the beginning of the Chapter, using the method employed on removal to prevent the sprocket from turning (see illustrations).

9 Locate the guide plate onto the sprocket cover, then install the cover, making sure the speedometer cable drive socket on the inside of the cover engages correctly with the head of the sprocket bolt (see illustration). On L, N and R models, do not forget the wiring clip with the lower bolt. If removed, attach the speedometer cable to the drive housing. Install the gearchange linkage arm onto the end of the shaft, aligning the punch marks, and tighten the pinch bolt (see illustration 16.1a).

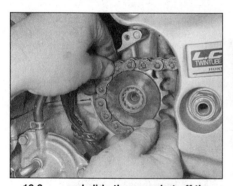

16.6 . . . and slide the sprocket off the shaft

16.7 Fit the new sprocket into the chain and slide it onto the shaft . . .

16.8a . . . then install the bolt and washer . . .

16.8b . . . and tighten it to the specified torque

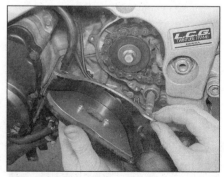

16.9 Locate the guide plate and install the cover

Frame, suspension and final drive 6•25

16.11 Lift out the sprocket coupling

16.12a Sprocket bolts (arrowed) – J and K models

16.12b Sprocket nuts (arrowed) – L, N and R models

10 On completion, adjust and lubricate the chain following the procedures described in Chapter 1.
Caution: Use only the recommended lubricant.

Rear sprocket

11 Remove the rear wheel (see Chapter 7). On J and K models, lift the sprocket coupling out of the wheel **(see illustration)**.
12 Unscrew the bolts (J and K models) or nuts (L, N and R models) securing the sprocket to the coupling and remove the sprocket, noting which way round it fits **(see illustrations)**.
13 On L, N and R models, check that the sprocket studs are secure in the coupling. If any are loose, remove them all and clean their threads, then apply a suitable non-permanent thread-locking compound and tighten them securely. To ease removal and tightening of the studs, thread two nuts onto the stud and use one as a locknut and the other to unscrew and tighten the studs.
14 Install the sprocket onto the coupling with the stamped tooth number facing out. Apply some engine oil to the threads of the bolts or nuts and tighten them to the torque setting specified at the beginning of the Chapter. On J and K models, fit the coupling into the wheel **(see illustration 16.11)**.
15 Install the rear wheel (see Chapter 7), then adjust and lubricate the chain following the procedures described in Chapter 1.

17 Sprocket coupling/rubber damper – check and renewal

1 Remove the rear wheel (see Chapter 7).
2 Lift the sprocket coupling out of the wheel **(see illustration)**. Check the coupling for cracks and damage, and renew it if necessary.
3 Remove the rubber dampers from the wheel and check them for cracks, hardening and general deterioration **(see illustrations)**; renew them if necessary.
4 Checking and renewal procedures for the coupling bearing are in Chapter 7.
5 Installation is the reverse of the removal procedure.

17.2 Lift out the sprocket coupling

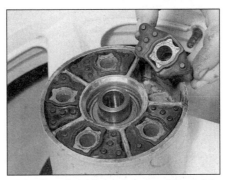

17.3a Damper segments – J and K models

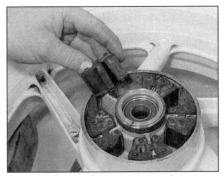

17.3b Damper segments – L, N and R models

Notes

Chapter 7
Brakes, wheels and tyres

Contents

Brake fluid level check . . . . .see *Daily (pre-ride) checks*	Rear brake disc – inspection, removal and installation 8
Brake light switches – check and replacement . . . . .see Chapter 9	Rear brake master cylinder – removal, overhaul
Brake pad wear check . . . . .see Chapter 1	and installation 9
Brake hoses, pipes and unions – inspection and renewal 10	Rear brake pads – renewal 6
Brake system bleeding 11	Rear wheel – removal and installation 15
Brake system check . . . . .see Chapter 1	Tyres – general information and fitting 17
Front brake calipers – removal, overhaul and installation 3	Tyres – pressure, tread depth and
Front brake disc – inspection, removal and installation 4	condition . . . . .see *Daily (pre-ride) checks*
Front brake master cylinder – removal, overhaul and installation . . . 5	Wheels – general check . . . . .see Chapter 1
Front brake pads – renewal 2	Wheel bearings – check . . . . .see Chapter 1
Front wheel – removal and installation 14	Wheel bearings – renewal 16
General information 1	Wheels – alignment check 13
Rear brake caliper – removal, overhaul and installation 7	Wheels – inspection and repair 12

Degrees of difficulty

Easy, suitable for novice with little experience	**Fairly easy,** suitable for beginner with some experience 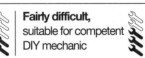	**Fairly difficult,** suitable for competent DIY mechanic	**Difficult,** suitable for experienced DIY mechanic	**Very difficult,** suitable for expert DIY or professional

Specifications

Note: *Models are identified by their production code letter – refer to 'Identification numbers' at the front of this manual for details.*

Brakes

Brake fluid type . . . . DOT 4
Front caliper bore ID
    New . . . . 25.400 to 25.450 mm
    Service limit . . . . 25.460 mm
Front caliper piston OD
    New . . . . 25.335 to 25.368 mm
    Service limit . . . . 25.330 mm
Rear caliper bore ID
    New . . . . 38.180 to 38.230 mm
    Service limit . . . . 38.240 mm
Rear caliper piston OD
    New . . . . 38.098 to 38.148 mm
    Service limit . . . . 38.090 mm
Front master cylinder bore ID
    New . . . . 12.700 to 12.743 mm
    Service limit . . . . 12.760 mm
Front master cylinder piston OD
    New . . . . 12.657 to 12.684 mm
    Service limit . . . . 12.650 mm
Rear master cylinder bore ID
    New . . . . 14.000 to 14.043 mm
    Service limit . . . . 14.055 mm

Brakes (continued)

Rear master cylinder piston OD
- New .... 13.957 to 13.984 mm
- Service limit .... 13.945 mm

Front disc minimum thickness
- Standard .... 4.0 mm
- Service limit .... 3.5 mm

Front disc maximum runout .... 0.3 mm

Rear disc minimum thickness
- Standard .... 5.0 mm
- Service limit .... 4.0 mm

Rear disc maximum runout .... 0.14 mm

Wheels

Maximum wheel runout (front and rear)
- Axial (side-to-side) .... 2.0 mm
- Radial (out-of-round) .... 2.0 mm

Maximum axle runout (front and rear) .... 0.2 mm

Tyres

Tyre pressures .... see *Daily (pre-ride)* checks

Tyre sizes
- J and K models
  - Front 120/60-R17 55H
  - Rear 150/60-R18 67H
- L, N and R models
  - Front 120/60-R17 55H
  - Rear 150/60-R17 66H

Torque settings

Note: *Where a specified setting is not given for a particular bolt, the general settings listed at the beginning apply. The dimension given applies to the diameter of the thread, not the head.*

5 mm bolt/nut	5 Nm
6 mm bolt/nut	10 Nm
8 mm bolt/nut	22 Nm
10 mm bolt/nut	35 Nm
12 mm bolt/nut	55 Nm
6 mm flange bolt with 8 mm head	9 Nm
6 mm flange bolt/nut with 10 mm head	12 Nm
8 mm flange bolt/nut	27 Nm
10 mm flange bolt/nut	40 Nm
Brake pad retaining pins	17 Nm
Brake pad retaining pin plugs	3 Nm
Front caliper slider pin	13 Nm
Front caliper bracket slider pin	23 Nm
Front brake caliper mounting bolts	27 Nm
Front master cylinder clamp bolts	12 Nm

Rear brake caliper mounting bolt
- J and K models .... 25 Nm
- L, N and R models .... 23 Nm

Rear caliper slider pin (L, N and R models) .... 28 Nm
Rear master cylinder mounting bolts .... 12 Nm

Brake hose banjo bolts
- J and K models .... 30 Nm
- L, N and R models .... 35 Nm

Front brake disc bolts
- J and K models .... 15 Nm
- L, N and R models .... 20 Nm

Rear brake disc bolts
- J and K models .... 40 Nm
- L, N and R models .... 43 Nm

Front axle bolt .... 60 Nm
Axle clamp bolts .... 22 Nm

Rear axle nut
- J and K models .... 90 Nm
- L, N and R models .... 95 Nm

Brakes, wheels and tyres 7•3

1 General information

All models covered in this manual are fitted with cast alloy wheels designed for tubeless tyres only. Both front and rear brakes are hydraulically operated disc brakes.

On all models, the front brakes are twin piston sliding calipers, and the rear brake is a single piston sliding caliper.

Caution: *Disc brake components rarely require disassembly. Do not disassemble components unless absolutely necessary. If a hydraulic brake line is loosened, the entire system must be disassembled, drained, cleaned and then properly filled and bled upon reassembly. Do not use solvents on internal brake components. Solvents will cause the seals to swell and distort. Use only clean brake fluid or denatured alcohol for cleaning. Use care when working with brake fluid as it can injure your eyes and it will damage painted surfaces and plastic parts.*

HAYNES HiNT *Models are identified by their production code letter – refer to 'Identification numbers' at the front of this manual for details.*

2 Front brake pads – renewal

Warning: *The dust created by the brake system may contain asbestos, which is harmful to your health. Never blow it out with compressed air and don't inhale any of it. An approved filtering mask should be worn when working on the brakes.*

1 Unscrew the pad retaining pin plug followed by the pad retaining pin and withdraw the pin, noting how it fits **(see illustration)**. Slide the pads out of the caliper **(see illustration)**.
2 Inspect the surface of each pad for

2.1a Remove the plug . . .

contamination and check that the friction material has not worn down level with or beyond the wear limit cutout in the pad edge **(see illustration)**. If either pad is worn down to, or beyond, the wear limit, is fouled with oil or grease, or heavily scored or damaged by dirt and debris, both pads must be renewed as a set. Note that it is not possible to degrease the friction material; if the pads are contaminated in any way they must be renewed. Also check that each pad has worn evenly at each end, and that each has the same amount of wear as the other. If uneven wear is noticed, one or more of the pistons are probably sticking in the caliper, in which case the calipers must be overhauled (see Section 3).
3 If the pads are in good condition clean them carefully, using a fine wire brush which is completely free of oil and grease to remove all traces of road dirt and corrosion. Using a pointed instrument, clean out the grooves in the friction material and dig out any embedded particles of foreign matter. Any areas of glazing may be removed using emery cloth.
4 Check the condition of the brake disc (see Section 4).
5 Remove all traces of corrosion from the pad pin. Inspect the pin for signs of damage and renew it if necessary.
6 If new pads are being installed, push the pistons as far back into the caliper as possible. A good way of doing this is to insert one of the old pads between the outside of the disc and the piston, then push the caliper against the pad and disc using hand pressure

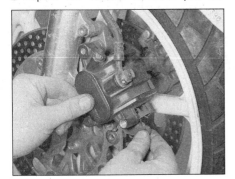

2.6 Push the caliper against the pad to create the extra room

2.8a Install the pads . . .

2.1b . . . then unscrew the pin and remove the pads

2.2 If the pads have worn to or beyond the wear limit cutout (arrowed) they must be renewed

(see illustration). Due to the increased friction material thickness of new pads, it may be necessary to remove the master cylinder reservoir cover and diaphragm and remove some fluid.
7 Smear the backs of the pads and the shank of the pad pin with copper-based grease, making sure that none gets on the front or sides of the pads.
8 Installation of the pads is the reverse of removal. Make sure the pad spring is correctly positioned in the caliper mouth and the pad plate is in place on the caliper bracket. Insert the pads into the caliper so that the friction material faces the disc, making sure they locate correctly against the pad spring and pad plate, then push up on the pads to align the holes and slide the pad retaining pin through **(see illustrations and 2.1b)**. Make

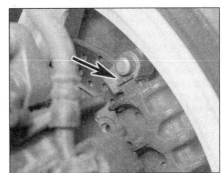

2.8b . . . making sure they locate correctly in the pad plate on the bracket (arrowed)

7•4 Brakes, wheels and tyres

3.2 Note the alignment of the hose before removing the banjo bolt (arrowed)

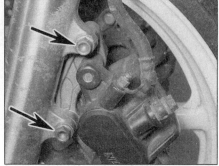

3.3a Unscrew the caliper mounting bolts (arrowed) . . .

3.3b . . . and slide the caliper off the disc

sure the pin passes through the hole in each pad. Tighten the pad retaining pin to the torque setting specified at the beginning of the Chapter. Install the pad pin plug and tighten it to the specified torque **(see illustration 2.1a)**.

9 Top up the master cylinder reservoir if necessary (see *Daily (pre-ride) checks*), and refit the reservoir cover and diaphragm.

10 Operate the brake lever several times to bring the pads into contact with the disc. Check the operation of the brake before riding the motorcycle.

3 Front brake calipers – removal, overhaul and installation

Warning: *If a caliper indicates the need for an overhaul (usually due to leaking fluid or sticky operation), all old brake fluid should be flushed from the system. Also, the dust created by the brake system may contain asbestos, which is harmful to your health. Never blow it out with compressed air and don't inhale any of it. An approved filtering mask should be worn when working on the brakes. Do not, under any circumstances, use petroleum-based solvents to clean brake parts. Use clean brake fluid, brake cleaner or denatured alcohol only.*

Removal

1 If the calipers are being overhauled, remove the brake pads (see Section 2). If the calipers are just being displaced or removed, the pads can be left in place.

2 If the calipers are just being displaced and not completely removed or overhauled, do not disconnect the brake hose. If the calipers are being overhauled, note the alignment of the brake hose on the caliper then unscrew the brake hose banjo bolt and separate the hose from the caliper **(see illustration)**. Plug the hose end or wrap a plastic bag tightly around it to minimise fluid loss and prevent dirt entering the system. Discard the sealing washers as new ones must be used on installation. **Note:** *If you are planning to overhaul the caliper and don't have a source of compressed air to blow out the pistons, just loosen the banjo bolt at this stage and retighten it lightly. The bike's hydraulic system can then be used to force the pistons out of the body once the pads have been removed. Disconnect the hose once the pistons have been sufficiently displaced.*

3 Unscrew the caliper mounting bolts, and slide the caliper off the disc **(see illustrations)**.

Overhaul

4 Slide the caliper off the caliper bracket, noting how it fits **(see illustration)**. Remove the pad plate on the caliper bracket and the pad spring inside the caliper mouth, noting how they fit. Clean the exterior of the caliper with denatured alcohol or brake system cleaner. Make sure all old grease is removed from the slider pins.

5 Displace the pistons as far as possible from the caliper body, either by pumping them out by operating the front brake lever, or by forcing them out using compressed air. If the compressed air method is used, place a wad of rag between the pistons and the caliper to act as a cushion, then use compressed air directed into the fluid inlet to force the pistons out of the body. Use only low pressure to ease the pistons out and make sure both pistons are displaced at the same time. If the air pressure is too high and the pistons are forced out, the caliper and/or pistons may be damaged. Mark each piston head and caliper bore with a felt marker to ensure that the pistons can be matched to their original bores on reassembly.

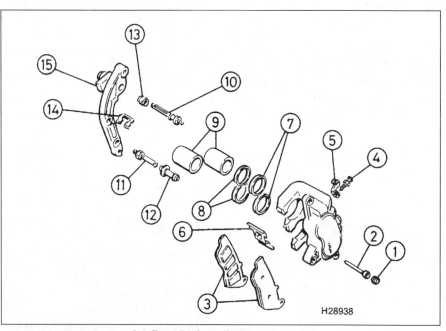

3.4 Front brake caliper components

1 Pad retaining pin plug
2 Pad retaining pin
3 Brake pads
4 Bleed valve
5 Bleed valve cap
6 Pad spring
7 Fluid seals
8 Dust seals
9 Pistons
10 Slider pin
11 Slider pin
12 Dust boot
13 Dust boot
14 Pad plate
15 Caliper bracket

3.6 Remove the dust seal with a plastic or wooden tool (a pencil works well) to avoid damage to the bore and seal groove

3.17 Tighten the caliper mounting bolts to the specified torque

 Warning: Never place your fingers in front of the pistons in an attempt to catch or protect them when applying compressed air, as serious injury could result.

6 Using a wooden or plastic tool, remove the dust seals from the caliper bores **(see illustration)**. Discard them as new ones must be used on installation. If a metal tool is being used, take great care not to damage the caliper bores.

7 Remove and discard the piston seals in the same way.

8 Clean the pistons and bores with denatured alcohol, clean brake fluid or brake system cleaner. If compressed air is available, use it to dry the parts thoroughly (make sure it's filtered and unlubricated).

Caution: Do not, under any circumstances, use a petroleum-based solvent to clean brake parts.

9 Inspect the caliper bores and pistons for signs of corrosion, nicks and burrs and loss of plating. If surface defects are present, the caliper assembly must be renewed. If the necessary measuring equipment is available, compare the dimensions of the pistons and bores to those given in the Specifications Section of this Chapter, renewing any component that is worn beyond the service limit. If the caliper is in bad shape the master cylinder should also be checked.

10 Clean off all traces of corrosion from the slider pins and their bores in the caliper and bracket. Renew the rubber dust boots if they are damaged or deteriorated. If a pin is loose, remove them both and clean the threads. Apply a suitable non-permanent thread locking compound and tighten them to the specified torque.

11 Lubricate the new piston seals with clean brake fluid and install them in their grooves in the caliper bores.

12 Lubricate the new dust seals with clean brake fluid and install them in their grooves in the caliper bores.

13 Lubricate the pistons with clean brake fluid and install them closed-end first into the caliper bores. Using your thumbs, push the pistons all the way in, making sure they enter the bore squarely.

14 If removed, fit the pads spring and pad plate, making sure they are correctly positioned as noted on removal **(see illustration 3.4)**. Apply a smear of copper or silicone based grease to the slider pins, then slide the caliper onto the bracket and check that it is able to move freely.

Installation

15 If the caliper has not been overhauled, slide the caliper off the caliper bracket, noting how it fits, and clean all old grease from the slider pins. Renew the rubber boots if they are damaged or deteriorated. Apply a smear of copper or silicone based grease to the slider pins, then slide the caliper onto the bracket and check that it is able to move freely.

16 Slide the caliper onto the brake disc, making sure the pads sit squarely each side of the disc (if they weren't removed) **(see illustration 3.3b)**.

17 Install the caliper mounting bolts, and tighten them to the torque setting specified at the beginning of the Chapter **(see illustration)**.

4.2 Set up a dial gauge with its probe contacting the brake disc, then rotate the wheel to check for runout

18 If removed, connect the brake hose to the caliper, using new sealing washers on each side of the fitting. Align the hose as noted on removal. Tighten the banjo bolt to the torque setting specified at the beginning of the Chapter. Top up the master cylinder reservoir with DOT 4 brake fluid (see *Daily (pre-ride) checks*) and bleed the hydraulic system as described in Section 11.

19 If removed, install the brake pads (see Section 2).

20 Check that there are no fluid leaks and thoroughly test the operation of the brake before riding the motorcycle.

4 Front brake discs – inspection, removal and installation

Inspection

1 Visually inspect the surface of the disc for score marks and other damage. Light scratches are normal after use and won't affect brake operation, but deep grooves and heavy score marks will reduce braking efficiency and accelerate pad wear. If a disc is badly grooved it must be machined or renewed.

2 To check disc runout, position the bike on an auxiliary stand and support it so that the wheel is raised off the ground. Mount a dial gauge on a fork leg, with the plunger on the gauge touching the surface of the disc about 10 mm from the outer edge **(see illustration)**. Rotate the wheel and watch the gauge needle, comparing the reading with the limit listed in the Specifications at the beginning of the Chapter. If the runout is greater than the service limit, check the wheel bearings for play (see Chapter 1). If the bearings are worn, renew them (see Section 16) and repeat this check. If the disc runout is still excessive, it will have to be renewed, although machining by an engineer may be possible.

3 The disc must not be machined or allowed

7•6 Brakes, wheels and tyres

4.3a The minimum disc thickness is marked on each disc

4.3b Using a micrometer to measure disc thickness

4.5 Each disc is secured by six bolts (arrowed)

to wear down to a thickness less than the service limit listed in this Chapter's Specifications and as marked on the disc itself **(see illustration)**. The thickness of the disc can be checked with a micrometer **(see illustration)**. If the thickness of the disc is less than the service limit, it must be renewed.

Removal

4 Remove the front wheel (see Section 14). **Caution: Do not lay the wheel down and allow it to rest on the disc – the disc could become warped. Set the wheel on wood blocks so the disc doesn't support the weight of the wheel.**

5 Mark the relationship of the disc to the wheel, so it can be installed in the same position. Unscrew the disc retaining bolts, loosening them a little at a time in a criss-cross pattern to avoid distorting the disc, then remove the disc from the wheel **(see illustration)**. On L, N and R models, discard the bolts as Honda specify new ones must be used. On J and K models, clean the threads of the bolts.

Installation

6 Install the disc on the wheel, making sure the marked side is on the outside. Align the previously applied matchmarks (if you're reinstalling the original disc).

7 Install the bolts, on J and K models using a suitable non-permanent thread-locking compound, and tighten them in a criss-cross pattern evenly and progressively to the torque setting specified at the beginning of the Chapter.

8 Clean off all grease from the brake disc(s) using acetone or brake system cleaner. If a new brake disc has been installed, remove any protective coating from its working surfaces.

9 Install the wheel (see Section 14).

10 Operate the brake lever several times to bring the pads into contact with the disc. Check the operation of the brake carefully before riding the bike.

5 Front brake master cylinder – removal, overhaul and installation

1 If the master cylinder is leaking fluid, or if the lever does not produce a firm feel when the brake is applied, and bleeding the brakes does not help (see Section 11), and the hydraulic hoses are all in good condition, then master cylinder overhaul is recommended **(see illustration)**.

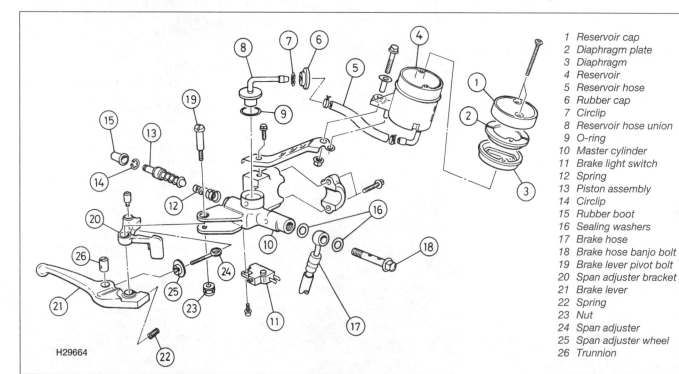

1 Reservoir cap
2 Diaphragm plate
3 Diaphragm
4 Reservoir
5 Reservoir hose
6 Rubber cap
7 Circlip
8 Reservoir hose union
9 O-ring
10 Master cylinder
11 Brake light switch
12 Spring
13 Piston assembly
14 Circlip
15 Rubber boot
16 Sealing washers
17 Brake hose
18 Brake hose banjo bolt
19 Brake lever pivot bolt
20 Span adjuster bracket
21 Brake lever
22 Spring
23 Nut
24 Span adjuster
25 Span adjuster wheel
26 Trunnion

5.1 Front brake master cylinder components

Brakes, wheels and tyres

5.3 Slacken the two reservoir cover screws

5.4 Disconnect the brake light switch wiring connectors (arrowed)

5.6 Note the alignment of the hose before removing the banjo bolt (arrowed)

2 Before disassembling the master cylinder, read through the entire procedure and make sure that you have obtain all new parts required. Also, you will need some new DOT 4 brake fluid, some clean rags and internal circlip pliers. **Note:** *If the master cylinder is just being displaced and not completely removed or overhauled, do not disconnect the brake hose or drain the reservoir.*

Caution: *Disassembly, overhaul and reassembly of the brake master cylinder must be done in a spotlessly clean work area to avoid contamination and possible failure of the brake hydraulic system components. To prevent damage to the paint from spilled brake fluid, always cover the fuel tank when working on the master cylinder.*

Removal

3 Loosen, but do not remove, the screws holding the reservoir cover in place **(see illustration)**. Access to the screws is restricted by the windshield. If a short or angled screwdriver is not available, remove the fairing to access the screws (see Chapter 8).

4 Disconnect the electrical connectors from the brake light switch **(see illustration)**.

5 Remove the front brake lever (see Chapter 6).

6 Unscrew the brake hose banjo bolt and separate the hose from the master cylinder, noting its alignment **(see illustration)**. Discard the sealing washers as they must be renewed. Wrap the end of the hose in a clean rag and suspend it in an upright position or bend it down carefully and place the open end in a clean container. The objective is to prevent excessive loss of brake fluid, fluid spills and system contamination.

7 Slacken the bolt securing the reservoir bracket to the master cylinder **(see illustration)**. Unscrew the master cylinder clamp bolts, then lift the master cylinder away from the handlebar **(see illustration)**.

8 Remove the reservoir cover retaining screws and lift off the cover, the diaphragm plate and the rubber diaphragm. Drain the brake fluid from the reservoir into a suitable container. Remove the reservoir bracket bolt, then release the clamp securing the reservoir hose to the union on the master cylinder and detach the hose. Wipe any remaining fluid out of the reservoir with a clean rag.

9 If required, remove the screw securing the brake light switch to the bottom of the master cylinder and remove the switch.

Overhaul

10 Carefully remove the dust boot from the end of the piston **(see illustration)**.

11 Using circlip pliers, remove the circlip and slide out the piston assembly and the spring, noting how they fit **(see illustration)**. Lay the parts out in the proper order to prevent confusion during reassembly **(see illustration)**.

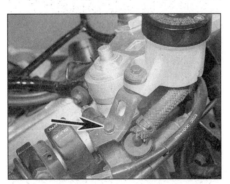

5.7a Slacken the bracket bolt (arrowed) . . .

5.7b . . . then unscrew the clamp bolts (arrowed) and remove the master cylinder assembly

5.10 Remove the rubber boot from the end of the master cylinder piston . . .

5.11a . . . then depress the piston and remove the circlip using a pair of internal circlip pliers

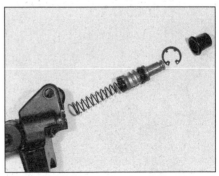

5.11b Lay out the internal parts as shown, even if new parts are being used, to avoid confusion on reassembly

7•8 Brakes, wheels and tyres

5.28 Make sure the diaphragm is correctly seated, then fit the plate and cover

12 Remove the fluid reservoir hose union rubber cap, then remove the circlip and detach the union from the master cylinder. Discard the O-ring as a new one must be used. Inspect the reservoir hose for cracks or splits and renew if necessary.
13 Clean all parts with clean brake fluid or denatured alcohol. If compressed air is available, use it to dry the parts thoroughly (make sure it's filtered and unlubricated).

Caution: Do not, under any circumstances, use a petroleum-based solvent to clean brake parts.

14 Check the master cylinder bore for corrosion, scratches, nicks and score marks. If the necessary measuring equipment is available, compare the dimensions of the piston and bore to those given in the Specifications Section of this Chapter. If damage or wear is evident, the master cylinder must be renewed. If the master cylinder is in poor condition, then the caliper(s) should be checked as well. Check that the fluid inlet and outlet ports in the master cylinder are clear.
15 The dust boot, circlip, piston, seal, primary cup and spring are included in the new piston/seal kit. Use all of the new parts, regardless of the apparent condition of the old ones. If the seal and cup are not already on the piston, fit them according to the layout of the old piston assembly.
16 Install the new spring in the master cylinder so that its tapered end faces the piston.

17 Lubricate the new piston, seal and cup with clean brake fluid. Install the assembly into the master cylinder, making sure it is the correct way round **(see illustration 5.11b)**. Make sure the lips on the cup do not turn inside out when they are slipped into the bore. Depress the piston and install the new circlip, making sure that it locates in the master cylinder groove **(see illustration 5.11a)**.
18 Install the new rubber dust boot, making sure the lip is seated correctly in the piston groove **(see illustration 5.10)**.
19 Fit a new O-ring onto the reservoir hose union, then press the union into the master cylinder and secure it with the circlip. Fit the rubber cap over the circlip.
20 Inspect the reservoir cover rubber diaphragm and renew it if it is damaged or deteriorated.

Installation

21 Install the brake light switch.
22 Attach the master cylinder to the handlebar and fit the clamp with its UP mark facing up, then tighten the bolts to the torque setting specified at the beginning of the Chapter **(see illustration 5.7b)**.
23 Connect the brake hose to the master cylinder, using new sealing washers on each side of the union, and aligning the hose as noted on removal **(see illustration 5.6)**. Tighten the banjo bolt to the torque setting specified at the beginning of this Chapter.
24 Install the brake lever (see Chapter 6).
25 Mount the reservoir onto the master cylinder and tighten its bolt securely **(see illustration 5.7a)**. Connect the reservoir hose to the union and secure it with the clamp.
26 Connect the brake light switch wiring **(see illustration 5.4)**.
27 Fill the fluid reservoir with new DOT 4 brake fluid as described in *Daily (pre-ride) checks*. Refer to Section 11 of this Chapter and bleed the air from the system.
28 Fit the rubber diaphragm, making sure it is correctly seated, the diaphragm plate and the cover onto the master cylinder reservoir **(see illustration)**.
29 Check the operation of the front brake before riding the motorcycle.

6 Rear brake pads – renewal

Warning: The dust created by the brake system may contain asbestos, which is harmful to your health. Never blow it out with compressed air and don't inhale any of it. An approved filtering mask should be worn when working on the brakes.

1 Unscrew the pad retaining pin plug and the pad retaining pin, then unscrew the caliper mounting bolt **(see illustrations)**.
2 Swing the caliper up at the back and slide out the pads **(see illustration)**. Where fitted, remove the shims from the back of the pads, noting how they fit.
3 Inspect the surface of each pad for contamination and check that the friction material has not worn level with or beyond the wear grooves in the pad face or the cutouts in the pad edge **(see illustration 2.2)**. If either pad is worn down to, or beyond, the wear limit, fouled with oil or grease, or heavily scored or damaged by dirt and debris, both pads must be renewed as a set. Note that it is not possible to degrease the friction material; if the pads are contaminated in any way they must be renewed.
4 If the pads are in good condition clean them carefully, using a fine wire brush which is completely free of oil and grease to remove all traces of road dirt and corrosion. Using a pointed instrument, clean out the grooves in the friction material and dig out any embedded particles of foreign matter. Any areas of glazing may be removed using emery cloth.
5 Check the condition of the brake disc (see Section 8).
6 Remove all traces of corrosion from the pad pin. Inspect the pin for signs of damage and renew it if necessary.
7 If new pads are being installed, push the pistons as far back into the caliper as possible. A good way of doing this is to insert one of the old pads between the outside of the disc and the piston, then push the caliper against the pad and disc using hand pressure

6.1a Remove the plug . . .

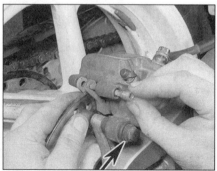

6.1b . . . and the pin, then unscrew the mounting bolt (arrowed)

6.2 Swing the caliper up and remove the pads

Brakes, wheels and tyres 7•9

6.7 Push the caliper against the pad to create the extra room

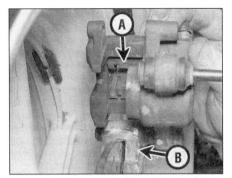

6.9a Make sure the pad spring (A) and pad plate (B) are correctly located . . .

6.9b . . . then install the pads

(see illustration). Due to the increased friction material thickness of new pads, it may be necessary to remove the master cylinder reservoir cover and diaphragm and remove some fluid.

8 Where removed, fit the shims onto the back of the pads, making sure the arrow points in the direction of normal disc rotation. Smear the backs of the pads and the shank of the pad pin with copper-based grease, making sure that none gets on the front or sides of the pads.

9 Installation of the pads is the reverse of removal. Make sure the pad spring is correctly positioned in the caliper and the pad plate is clipped to the caliper bracket **(see illustration)**. Insert the pads into the caliper so that the friction material faces the disc, making sure they locate correctly against the pad spring and engage the pad plate **(see illustration)**, then push up on the pads to align the holes and slide the pad retaining pin through **(see illustration 6.1b)**. Make sure the pin passes through the hole in each pad. Tighten the pad retaining pin finger tight.

10 Swing the caliper down and tighten the mounting bolt to the torque setting specified at the beginning of the Chapter **(see illustration)**. Tighten the pad retaining pin to the specified torque, then install the pad pin plug and tighten that **(see illustration 6.1a)**.

11 Top up the master cylinder reservoir if necessary (see *Daily (pre-ride) checks*), and refit the reservoir cover and diaphragm.

12 Operate the brake pedal several times to bring the pads into contact with the disc. Check the operation of the brake before riding the motorcycle.

7 Rear brake caliper – removal, overhaul and installation

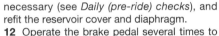

Warning: If a caliper indicates the need for an overhaul (usually due to leaking fluid or sticky operation), all old brake fluid should be flushed from the system. Also, the dust created by the brake system may contain asbestos, which is harmful to your health. Never blow it out with compressed air and don't inhale any of it. An approved filtering mask should be worn when working on the brakes. Do not, under any circumstances, use petroleum-based solvents to clean brake parts. Use clean brake fluid, brake cleaner or denatured alcohol only.

Removal

1 If the caliper is being overhauled, remove the brake pads (see Section 6). If the caliper is just being removed, the pads can be left in place.

2 If the caliper is just being displaced and not completely removed or overhauled, do not disconnect the brake hose. If the caliper is being overhauled, note the alignment of the

6.10 Tighten the mounting bolt to the specified torque

brake hose on the caliper then unscrew the brake hose banjo bolt and separate the hose from the caliper **(see illustration)**. Plug the hose end or wrap a plastic bag tightly around it to minimise fluid loss and prevent dirt entering the system. Discard the sealing washers as new ones must be used on installation. **Note:** *If you are planning to overhaul the caliper and don't have a source of compressed air to blow out the piston, just loosen the banjo bolt at this stage and retighten it lightly. The bike's hydraulic system can then be used to force the piston out of the body once the pads have been removed. Disconnect the hose once the piston has been sufficiently displaced.*

3 Unscrew the caliper mounting bolt, then swing the caliper up at the back and slide it sideways off of the bracket **(see illustrations)**.

7.2 Note the alignment of the hose, then unscrew the brake hose banjo bolt (arrowed)

7.3a Unscrew the mounting bolt . . .

7.3b . . . then swing the caliper up . . .

7•10 Brakes, wheels and tyres

7.3c ... and draw it off the bracket – L, N and R models type shown

Overhaul

4 Remove the pad spring in the caliper and the pad plate on the caliper bracket, noting how they fit (see illustration 6.9a). Clean the exterior of the caliper with denatured alcohol or brake system cleaner (see illustration). Make sure all old grease is removed from the slider pin and the mounting bolt shank. On J and K models, the slider pin is threaded into the caliper bracket, on L, N and R models it is in the caliper.

5 Displace the piston as far as possible from the caliper body, either by pumping it out by operating the brake pedal, or by forcing it out using compressed air. If the compressed air method is used, place a wad of rag between the piston and the caliper to act as a cushion, then use compressed air directed into the fluid inlet to force the piston out of the body. Use only low pressure to ease the piston out. If the air pressure is too high and the piston is forced out, the caliper and/or piston may be damaged.

 Warning: Never place your fingers in front of the piston in an attempt to catch or protect it when applying compressed air, as serious injury could result.

6 Using a wooden or plastic tool, remove the dust seal from the caliper bore (see illustration 3.6). Discard it as a new one must be used on installation. If a metal tool is being used, take great care not to damage the caliper bore.

7 Remove and discard the piston seal in the same way.

8 Clean the piston and bore with denatured alcohol, clean brake fluid or brake system cleaner. If compressed air is available, use it to dry the parts thoroughly (make sure it's filtered and unlubricated).

Caution: Do not, under any circumstances, use a petroleum-based solvent to clean brake parts.

9 Inspect the caliper bore and piston for signs of corrosion, nicks and burrs and loss of plating. If surface defects are present, the caliper assembly must be renewed. If the necessary measuring equipment is available, compare the dimensions of the piston and bore to those given in the Specifications Section of this Chapter, renewing any component that is worn beyond the service limit. If the caliper is in bad shape the master cylinder should also be checked.

10 Clean off all traces of corrosion from the slider pin and mounting bolt shank. Renew the rubber boots if they are damaged or deteriorated. If the pin is loose, remove it and clean the threads. Apply a suitable non-permanent thread-locking compound and tighten it to the specified torque, where given.

11 Lubricate the new piston seal with clean brake fluid and install it in its groove in the caliper bore.

12 Lubricate the new dust seal with clean brake fluid and install it in its groove in the caliper bore.

13 Lubricate the piston with clean brake fluid and install it closed-end first into the caliper bore. Using your thumbs, push the piston all the way in, making sure it enters the bore squarely.

14 If removed, fit the pad spring in the caliper and the pad plate on the bracket, making sure they are correctly positioned as noted on removal (see illustration 6.9a). Apply a smear of copper or silicone based grease to the slider pin, then slide the caliper onto the bracket and check that it is able to move freely (see illustration 7.3c).

Installation

15 If the caliper has not been overhauled, clean all old grease from the slider pin and the mounting bolt shank. Renew the rubber boots if they are damaged or deteriorated. Apply a smear of copper or silicone based grease to the slider pin and bolt shank, then slide the caliper onto the bracket (see illustration 7.3c). Check that it is able to move freely.

16 If removed, install the brake pads (see Section 6).

17 Swing the caliper down onto the brake disc, making sure the pads sit squarely on each side of the disc (see illustration 7.3b).

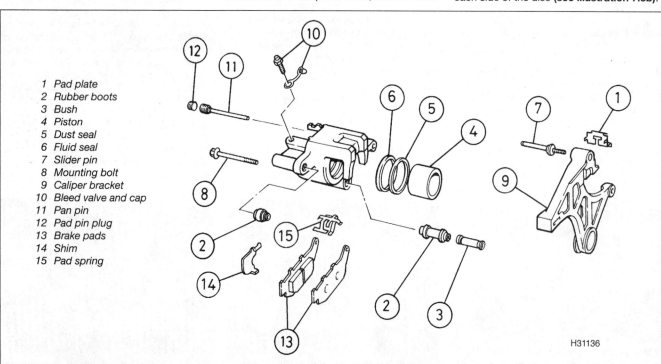

1 Pad plate
2 Rubber boots
3 Bush
4 Piston
5 Dust seal
6 Fluid seal
7 Slider pin
8 Mounting bolt
9 Caliper bracket
10 Bleed valve and cap
11 Pan pin
12 Pad pin plug
13 Brake pads
14 Shim
15 Pad spring

7.4 Rear brake caliper components – J and K type shown

Brakes, wheels and tyres 7•11

7.18a Install the bolt . . .

7.18b . . . and tighten it to the specified torque

18 Install the caliper mounting bolt, and tighten it to the torque setting specified at the beginning of the Chapter **(see illustrations)**.
19 If removed, connect the brake hose to the caliper, using new sealing washers on each side of the fitting. Align the hose as noted on removal **(see illustration 7.2)**. Tighten the banjo bolt to the torque setting specified at the beginning of the Chapter. Top up the master cylinder reservoir with DOT 4 brake fluid (see *Daily (pre-ride) checks*) and bleed the hydraulic system as described in Section 11.
20 Check for leaks and thoroughly test the operation of the rear brake before riding the motorcycle.

8 Rear brake disc – inspection, removal and installation

Inspection

1 Visually inspect the surface of the disc for score marks and other damage. Light scratches are normal after use and won't affect brake operation, but deep grooves and heavy score marks will reduce braking efficiency and accelerate pad wear. If a disc is badly grooved it must be machined or renewed.
2 To check disc runout, position the bike on an auxiliary stand and support it so that the wheel is raised off the ground. Mount a dial gauge on the swingarm, with the plunger on the gauge touching the surface of the disc about 10 mm from the outer edge **(see illustration 4.2)**. Rotate the wheel and watch the gauge needle, comparing the reading with the limit listed in the Specifications at the beginning of the Chapter. If the runout is greater than the service limit, check the wheel bearings for play (see Chapter 1). If the bearings are worn, renew them (see Section 16) and repeat this check. If the disc runout is still excessive, it will have to be renewed, although machining by an engineer may be possible.
3 The disc must not be machined or allowed to wear down to a thickness less than the service limit listed in this Chapter's Specifications and as marked on the disc itself **(see illustration)**. The thickness of the disc can be checked with a micrometer **(see illustration 4.3b)**. If the thickness of the disc is less than the service limit, it must be renewed.

Removal

4 Remove the rear wheel (see Section 15).
5 Mark the relationship of the disc to the hub, so it can be installed in the same position. Counter-hold the disc bolts and unscrew the nuts, loosening them a little at a time in a criss-cross pattern to avoid distorting the disc, then withdraw the bolts and remove the disc from the wheel **(see illustrations)**. On L, N and R models, discard the bolts as Honda specify new ones must be used. On J and K models, clean the threads of the bolts.

Installation

6 Install the disc on the hub, making sure the marked side is on the outside. Align the previously applied matchmarks (if you're reinstalling the original disc).
7 Install the bolts and tighten them in a criss-cross pattern evenly and progressively to the torque setting specified at the beginning of the Chapter. Note that on J and K models a suitable non-permanent thread locking compound must be applied to the bolt threads before installation, and on L, N and R models new bolts must be used.
8 Clean off all grease from the brake disc(s) using acetone or brake system cleaner. If a new brake disc has been installed, remove any protective coating from its working surfaces.
9 Install the rear wheel (see Section 15).
10 Operate the brake pedal several times to bring the pads into contact with the disc. Check the operation of the brake carefully before riding the bike.

9 Rear brake master cylinder – removal, overhaul and installation

1 If the master cylinder is leaking fluid, or if the lever does not produce a firm feel when the brake is applied, and bleeding the brake does not help (see Section 11), and the hydraulic hoses are all in good condition, then master cylinder overhaul is recommended.
2 Before disassembling the master cylinder, read through the entire procedure and make sure that you obtain all parts required. Also, you will need some new DOT 4 brake fluid, some clean rags and internal circlip pliers.
Note: *To prevent damage to the paint from spilled brake fluid, always cover the surrounding components when working on the master cylinder.*
Caution: *Disassembly, overhaul and reassembly of the brake master cylinder must be done in a spotlessly clean work area to avoid contamination and possible failure of the brake hydraulic system components.*

8.3 The minimum disc thickness is marked on each disc

8.5a The disc is secured by three bolts (arrowed) on J and K models . . .

8.5b . . . and by four bolts (arrowed) on L, N and R models

7•12 Brakes, wheels and tyres

9.4a Remove the cover screws

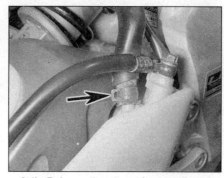

9.4b Release the clamp (arrowed) and detach the hose

9.5 Note the alignment of the hose then unscrew the banjo bolt (arrowed)

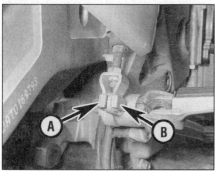

9.6 Remove the split pin (A) and withdraw the clevis pin (B)

Removal

3 Remove the seat cowling (see Chapter 8).
4 Remove the master cylinder fluid reservoir cover screws, but leave the cover in place **(see illustration)**. Unscrew the bolt securing the reservoir to the frame, then remove the reservoir cover and pour the fluid into a container. Release the clamp securing the reservoir hose to the union on the master cylinder and detach the hose **(see illustration)**.
5 Unscrew the brake hose banjo bolt and separate the brake hose from the master cylinder, noting its alignment **(see illustration)**. Discard the two sealing washers as they must be renewed. Wrap the end of the hose in a clean rag and suspend the hose in

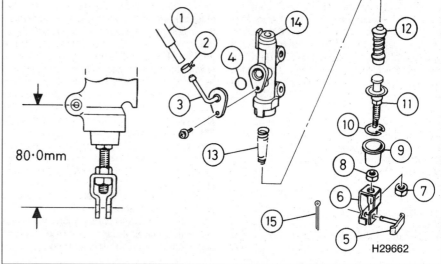

9.8a Rear master cylinder components

1 Reservoir hose
2 Reservoir hose clamp
3 Reservoir hose union
4 O-ring
5 Clevis pin
6 Clevis
7 Clevis base nut
8 Locknut
9 Rubber boot
10 Circlip
11 Pushrod
12 Piston/seal assembly
13 Spring
14 Master cylinder
15 Split pin

an upright position or bend it down carefully and place the open end in a clean container. The objective is to prevent excessive loss of brake fluid, fluid spills and system contamination.

6 Remove the split pin from the clevis pin securing the brake pedal to the master cylinder pushrod **(see illustration)**. Withdraw the clevis pin and separate the pedal from the pushrod. Discard the split pin as a new one must be used.
7 Unscrew the two bolts securing the heel plate and master cylinder to the footrest bracket and remove them, noting the collars fitted between the plate and cylinder on L, N and R models.

Overhaul

8 If required, mark the position of the clevis locknut on the pushrod, then slacken the locknut and thread the clevis and its base nut off the pushrod **(see illustrations)**.
9 Dislodge the rubber dust boot from the base of the master cylinder to reveal the pushrod retaining circlip **(see illustration)**.
10 Depress the pushrod and, using circlip

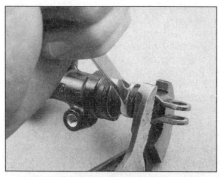

9.8b Hold the clevis and slacken the locknut

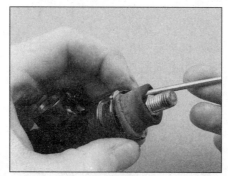

9.9 Remove the dust boot from the pushrod

Brakes, wheels and tyres 7•13

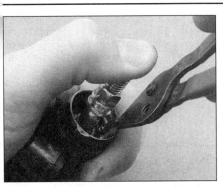

9.10 Depress the piston and remove the circlip from the cylinder

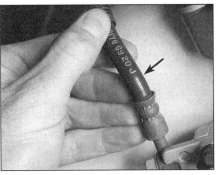

10.2 Flex the brake hoses and check for cracks, bulges and leaking fluid

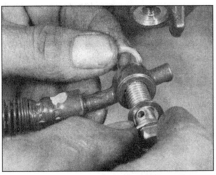

10.4 Remove the banjo bolt and separate the hose from the caliper; there is a sealing washer on each side of the fitting

pliers, remove the circlip **(see illustration)**. Slide out the piston assembly and spring. If they are difficult to remove, apply low pressure compressed air to the fluid outlet. Lay the parts out in the proper order to prevent confusion during reassembly.

11 Clean all of the parts with clean brake fluid or denatured alcohol.
Caution: Do not, under any circumstances, use a petroleum-based solvent to clean brake parts. If compressed air is available, use it to dry the parts thoroughly (make sure it's filtered and unlubricated).

12 Check the master cylinder bore for corrosion, scratches, nicks and score marks. If the necessary measuring equipment is available, compare the dimensions of the piston and bore to those given in the Specifications Section of this Chapter. If damage is evident, the master cylinder must be renewed. If the master cylinder is in poor condition, then the caliper should be checked as well.

13 If required, unscrew the fluid reservoir hose union screw and detach the elbow from the master cylinder. Discard the O-ring as a new one must be used. Inspect the reservoir hose for cracks or splits and renew it if necessary.

14 The dust boot, circlip, piston, seal, primary cup and spring are included in the piston/seal kit. Use all of the new parts, regardless of the apparent condition of the old ones. If the seal and cup are not already on the piston, fit them according to the layout of the old piston assembly.

15 Install the new spring in the master cylinder so that its tapered end faces the piston.

16 Lubricate the new piston, seal and cup with clean brake fluid. Install the assembly into the master cylinder, making sure it is the correct way round. Make sure the lips on the cup do not turn inside out when they are slipped into the bore.

17 Install and depress the pushrod, then fit the new circlip, making sure it is properly seated in the groove **(see illustration 9.10)**.

18 Install the new rubber dust boot, making sure the lip is seated properly in the groove **(see illustration 9.9)**.

19 If removed, fit a new O-ring to the fluid reservoir hose union, then install the union onto the master cylinder and secure it with its screw.

Installation

20 If removed, install the clevis locknut, the clevis and the base nut onto the master cylinder pushrod end. Position the clevis as noted on removal, then tighten the clevis locknut securely **(see illustration 9.8b)**. If in doubt, Honda recommend that the distance between the lower mounting bolt hole on the master cylinder and the pin hole on the clevis is 80 mm **(see illustration 9.8a)**. However the pedal height can be altered to suit individual tastes. This is best done after the assembly is installed, and is done by slackening the clevis locknut, then turning the pushrod itself using a spanner on the flats on the top of the rod, until the desired pedal height is obtained. Tighten the clevis locknut securely on completion.

21 Install the master cylinder and heel plate onto the footrest bracket, not forgetting the collars that fit between them on L, N and R models. Tighten the mounting bolts to the torque setting specified at the beginning of the Chapter.

22 Align the brake pedal with the master cylinder pushrod clevis, then slide in the clevis pin and secure it using a new split pin **(see illustration 9.6)**. Bend the split pin ends securely.

23 Connect the brake hose banjo bolt to the master cylinder, using a new sealing washer on each side of the banjo union. Ensure that the hose is positioned so that it butts against the lug and tighten the banjo bolt to the specified torque setting **(see illustration 9.5)**.

24 Secure the fluid reservoir to the frame with its retaining bolt. Ensure that the hose is correctly routed, then connect it to the union on the master cylinder and secure it with the clamp **(see illustration 9.4b)**. Check that the hose is secure and clamped at the reservoir end as well. If the clamps have weakened, use new ones.

25 Fill the fluid reservoir with new DOT 4 brake fluid (see *Daily (pre-ride) checks*) and bleed the system following the procedure in Section 11.

26 Install the seat cowling (see Chapter 8). Check the operation of the brake carefully before riding the motorcycle.

10 Brake hoses, pipes and unions – inspection and renewal

Inspection

1 Brake hose and pipe condition should be checked regularly and the hoses renewed at the specified interval (see Chapter 1).

2 Twist and flex the rubber hoses while looking for cracks, bulges and seeping fluid **(see illustration)**. Check extra carefully around the areas where the hoses connect with the banjo fittings, as these are common areas for hose failure.

3 Inspect the metal brake pipe linking the front caliper hoses and the banjo union fittings connected to all the brake hoses. If the fittings are rusted, scratched or cracked, renew them.

Renewal

4 The brake hoses have banjo union fittings on each end, and the brake pipe linking the two front caliper hoses has joint nuts **(see illustration)**. Cover the surrounding area with plenty of rags and unscrew the banjo bolt or joint nut at each end of the hose or pipe, noting its alignment. Free the hose or pipe from any clips or guides and remove it. Discard the sealing washers on the hose banjo unions.

5 Position the new hose or pipe, making sure it isn't twisted or otherwise strained, and either abut the hose union pipe against the lug on the component casting, or fit it into the slot between two lugs, where present. Otherwise align the hose or pipe as noted on removal. Install the hose banjo bolts using new sealing washers on both sides of the unions. Tighten the banjo bolts to the torque settings specified at the beginning of this Chapter. Make sure the hoses and pipes are correctly aligned and routed clear of all moving components.

6 Flush the old brake fluid from the system, refill with new DOT 4 brake fluid (see *Daily*

7•14 Brakes, wheels and tyres

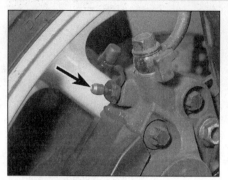

11.6a Brake caliper bleed valve

11.6b To bleed the brakes, you need a spanner, a short section of clear tubing, and a clear container half-filled with brake fluid

(pre-ride) checks) and bleed the air from the system (see Section 11). Check the operation of the brakes carefully before riding the motorcycle.

11 Brake system – bleeding

1 Bleeding the brakes is simply the process of removing all the air bubbles from the brake fluid reservoirs, the hoses and the brake calipers. Bleeding is necessary whenever a brake system hydraulic connection is loosened, when a component or hose is renewed, or when the master cylinder or caliper is overhauled. Leaks in the system may also allow air to enter, but leaking brake fluid will reveal their presence and warn you of the need for repair.
2 To bleed the brakes, you will need some new DOT 4 brake fluid, a length of clear vinyl or plastic tubing, a small container partially filled with clean brake fluid, some rags and a spanner to fit the brake caliper bleed valves.
3 Cover the fuel tank and other painted components to prevent damage in the event that brake fluid is spilled.
4 If bleeding the rear brake, remove the seat cowling (see Chapter 8) for access to the fluid reservoir.
5 Remove the reservoir cover, diaphragm plate and diaphragm and slowly pump the brake lever (front brake) or pedal (rear brake) a few times, until no air bubbles can be seen floating up from the holes in the bottom of the reservoir. Doing this bleeds the air from the master cylinder end of the line. Loosely refit the reservoir cover.
6 Pull the dust cap off the bleed valve **(see illustration)**. Attach one end of the clear vinyl or plastic tubing to the bleed valve and submerge the other end in the brake fluid in the container **(see illustration)**.
7 Remove the reservoir cover and check the fluid level. Do not allow the fluid level to drop below the lower mark during the bleeding process.
8 Carefully pump the brake lever or pedal three or four times and hold it in (front) or down (rear) while opening the caliper bleed valve. When the valve is opened, brake fluid will flow out of the caliper into the clear tubing and the lever will move toward the handlebar or the pedal will move down.
9 Retighten the bleed valve, then release the brake lever or pedal gradually. Repeat the process until no air bubbles are visible in the brake fluid leaving the caliper and the lever or pedal is firm when applied. On completion, disconnect the bleeding equipment, then tighten the bleed valve to the torque setting specified at the beginning of the chapter and install the dust cap.
10 Install the diaphragm and cover assembly, wipe up any spilled brake fluid and check the entire system for leaks.

> **HAYNES HINT** If it's not possible to produce a firm feel to the lever or pedal the fluid my be aerated. Let the brake fluid in the system stabilise for a few hours and then repeat the procedure when the tiny bubbles in the system have settled out.

12 Wheels – inspection and repair

1 In order to carry out a proper inspection of the wheels, it is necessary to support the bike

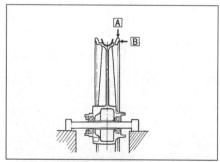

12.2 Check the wheel for radial (out-of-round) runout (A) and axial (side-to-side) runout (B)

upright so that the wheel being inspected is raised off the ground. Position the motorcycle on an auxiliary stand. Clean the wheels thoroughly to remove mud and dirt that may interfere with the inspection procedure or mask defects. Make a general check of the wheels (see Chapter 1) and tyres (see Daily (pre-ride) checks).
2 Attach a dial gauge to the fork slider or the swingarm and position its stem against the side of the rim **(see illustration)**. Spin the wheel slowly and check the axial (side-to-side) runout of the rim. In order to accurately check radial (out of round) runout with the dial gauge, the wheel would have to be removed from the machine, and the tyre from the wheel. With the axle clamped in a vice and the dial gauge positioned on the top of the rim, the wheel can be rotated to check the runout.
3 An easier, though slightly less accurate, method is to attach a stiff wire pointer to the fork slider or the swingarm and position the end of it a fraction of an inch from the wheel (where the wheel and tyre join). If the wheel is true, the distance from the pointer to the rim will be constant as the wheel is rotated. **Note:** If wheel runout is excessive, check the wheel bearings very carefully before renewing the wheel.
4 The wheels should also be visually inspected for cracks, flat spots on the rim and other damage. Look very closely for dents in the area where the tyre bead contacts the rim. Dents in this area may prevent complete sealing of the tyre against the rim, which leads to deflation of the tyre over a period of time. If damage is evident, or if runout in either direction is excessive, the wheel will have to be renewed. Never attempt to repair a damaged cast alloy wheel.

13 Wheels – alignment check

1 Misalignment of the wheels, which may be due to a cocked rear wheel or a bent frame or fork yokes, can cause strange and possibly serious handling problems. If the frame or yokes are at fault, repair by a frame specialist or renewal are the only alternatives.
2 To check the alignment you will need an assistant, a length of string or a perfectly straight piece of wood and a ruler. A plumb bob or other suitable weight will also be required.
3 In order to make a proper check of the wheels it is necessary to support the bike in an upright position, using an auxiliary stand. Measure the width of both tyres at their widest points. Subtract the smaller measurement from the larger measurement, then divide the difference by two. The result is the amount of offset that should exist between the front and rear tyres on both sides.
4 If a string is used, have your assistant hold

Brakes, wheels and tyres

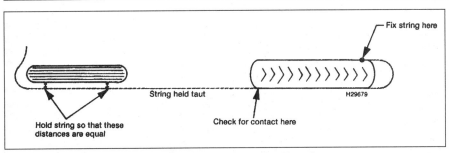

13.5 Wheel alignment check using string

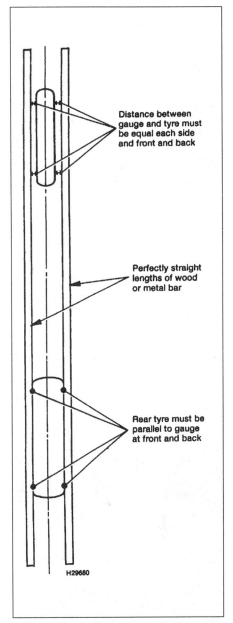

13.7 Wheel alignment check using a straight-edge

one end of it about halfway between the floor and the rear axle, touching the rear sidewall of the tyre.

5 Run the other end of the string forward and pull it tight so that it is roughly parallel to the floor **(see illustration)**. Slowly bring the string into contact with the front sidewall of the rear tyre, then turn the front wheel until it is parallel with the string. Measure the distance from the front tyre sidewall to the string.

6 Repeat the procedure on the other side of the motorcycle. The distance from the front tyre sidewall to the string should be equal on both sides.

7 As was previously pointed out, a perfectly straight length of wood or metal bar may be substituted for the string **(see illustration)**. The procedure is the same.

8 If the distance between the string and tyre is greater on one side, or if the rear wheel appears to be cocked, refer to Chapter 1, Section 1 and check that the chain adjuster markings coincide on each side of the swingarm.

9 If the front-to-back alignment is correct, the wheels still may be out of alignment vertically.

10 Using the plumb bob, or other suitable weight, and a length of string, check the rear wheel to make sure it is vertical. To do this, hold the string against the tyre upper sidewall and allow the weight to settle just off the floor. When the string touches both the upper and lower tyre sidewalls and is perfectly straight, the wheel is vertical. If it is not, place thin spacers under one leg of the stand.

11 Once the rear wheel is vertical, check the front wheel in the same manner. If both wheels are not perfectly vertical, the frame and/or major suspension components are bent.

14 Front wheel – removal and installation

Removal

1 Position the motorcycle on an auxiliary stand and support it under the crankcase so that the front wheel is off the ground. Always make sure the motorcycle is properly supported.

2 Remove the mounting bolts from one of the brake calipers and slide the caliper off the disc **(see illustration 3.3a and b)**. Support the caliper with a piece of wire or a bungee cord so that no strain is placed on its hydraulic hose. There is no need to disconnect the hose from the caliper, or to displace both calipers, unless required. **Note:** *Do not operate the front brake lever with the caliper(s) removed.*

3 Slacken the axle clamp bolts on the bottom of the right-hand fork, then unscrew the axle bolt from the right-hand end of the axle **(see illustration)**.

4 Slacken the axle clamp bolts on the bottom of the left-hand fork **(see illustration)**. Support the wheel, then withdraw the axle from the left-hand side and carefully lower the wheel. Use a screwdriver inserted through the holes in the end of the axle as a lever to aid removal **(see illustration)**.

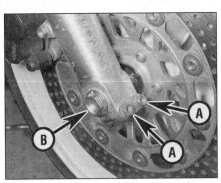

14.3 Slacken the axle clamp bolts (A), then remove the axle bolt (B)

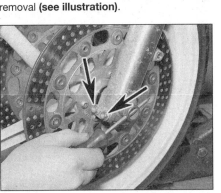

14.4a Slacken the axle clamp bolts (arrowed), then use a screwdriver or rod as a grip . . .

14.4b . . . and withdraw the axle

7•16 Brakes, wheels and tyres

14.5 Remove the spacers from the wheel

14.8 Note the directional arrow on the wheel

14.11a Install the axle bolt . . .

5 Remove the long wheel spacer from the right-hand side of the wheel and the short spacer from the left-hand side for safekeeping **(see illustration)**.
Caution: Don't lay the wheel down and allow it to rest on a disc – the disc could become warped. Set the wheel on wood blocks so the disc doesn't support the weight of the wheel.
6 Check the axle for straightness by rolling it on a flat surface such as a piece of plate glass (first wipe off all old grease and remove any corrosion using fine emery cloth). If the equipment is available, place the axle in V-blocks and measure the runout using a dial gauge. If the axle is bent or the runout exceeds the limit specified, renew it.
7 Refer to Section 16 if wheel bearing renewal is required.

14.11b . . . and tighten it to the specified torque

Installation

8 Apply a smear of grease to the inside of the wheel spacers, and also to the outside where they fit into the wheel. Fit the long spacer into the right-hand side of the wheel and the short spacer into the left-hand side **(see illustration 14.5)**. Each side of the wheel can be identified using the directional arrow cast into one of the spokes near the rim **(see illustration)**. The arrow denotes the normal direction of rotation of the wheel.
9 Manoeuvre the wheel into position, making sure the directional arrow is pointing in the normal direction of rotation. Apply a thin coat of grease to the axle.
10 Lift the wheel into place between the fork sliders, making sure the spacers remain in position. Slide the axle in from the left-hand side **(see illustration 14.4b)**.
11 Install the axle bolt and tighten it to the torque setting specified at the beginning of the Chapter **(see illustrations)**. Use a screwdriver inserted through the holes in the end of the axle to counter-hold it **(see illustration 14.4a)**.
12 Tighten the axle clamp bolts on the bottom of each fork to the specified torque setting **(see illustration)**.
13 Install the brake caliper, making sure the pads sit squarely on each side of the disc **(see illustration 3.3b)**. Tighten the caliper mounting bolts to the specified torque setting **(see illustration 3.17)**.
14 Apply the front brake a few times to bring

the pads back into contact with the discs. Move the motorcycle off its stand, apply the front brake and pump the front forks a few times to settle all components in position.
15 Check for correct operation of the front brake before riding the motorcycle.

15 Rear wheel – removal and installation

Removal

1 Position the motorcycle on an auxiliary stand so that the wheel is off the ground. If required, remove the lower fairing so that the stand can be fitted (see Chapter 8).
2 Unscrew the axle nut, and on L, N and R models remove the adjuster position marker from the end of the axle **(see illustrations)**.
3 Support the wheel then withdraw the axle from the left-hand side and lower the wheel to the ground **(see illustration 15.10a)**. On L, N and R models retrieve the other adjustment position marker. On J and K models, the markers will probably stay on the ends of the swingarm, but they can be removed if required. Note how the caliper bracket locates against the swingarm, and support it so that it will not fall off. If required, displace the brake caliper bracket from the swingarm, noting how it fits, and tie it to the top of the frame, making sure no strain is placed on the hose **(see illustration 15.7)**.

14.12 Tighten all the clamp bolts to the specified torque

15.2a Unscrew the nut (arrowed) . . .

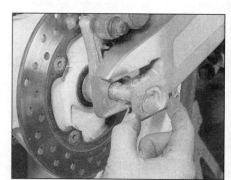

15.2b . . . and on L, N and R models, remove the position marker

Brakes, wheels and tyres 7•17

15.4 Disengage the chain and remove the wheel

15.6 Remove the collars

15.7 Locate the lug on the bracket into the slot in the swingarm (arrowed)

4 Disengage the chain from the sprocket and remove the wheel from the swingarm (see illustration). **Caution: Do not lay the wheel down and allow it to rest on the disc or the sprocket – they could become warped. Set the wheel on wood blocks so the disc or the sprocket doesn't support the weight of the wheel. Do not operate the brake pedal with the wheel removed.**

5 Check the axle for straightness by rolling it on a flat surface such as a piece of plate glass (if the axle is corroded, first remove the corrosion with fine emery cloth). If the equipment is available, place the axle in V-blocks and check the runout using a dial gauge. If the axle is bent or the runout exceeds the limit specified at the beginning of the Chapter, renew it.

6 Remove the collar from each side of the wheel for safekeeping (see illustration). Refer to Section 16 if wheel bearing renewal is required.

Installation

7 Apply a thin coat of grease to the lips of each grease seal, and also to the collars and the axle. Slide the right-hand adjustment position marker onto the swingarm on J and K models and onto the axle on L, N and R models, making sure it is the correct way round. If displaced, locate the brake caliper bracket into the swingarm (see illustration).

8 If removed, install the collars into the wheel (see illustration 15.6). Manoeuvre the wheel so that it is in between the ends of the swingarm and apply a thin coat of grease to the axle. Make sure the brake caliper bracket is still correctly positioned against the swingarm.

9 Engage the drive chain with the sprocket and lift the wheel into position (see illustration 15.4). Make sure the collars and caliper bracket remain correctly in place, and that the brake disc fits squarely into the caliper with the pads positioned correctly each side of the disc.

10 Install the axle from the left-hand side; on L, N and R models ensure that the left-hand adjustment marker is in position. As the axle is inserted, make sure that on J and K models it passes through the chain adjusters, and on all models that it passes through the caliper

bracket (see illustration). Check that everything is correctly aligned, then on L, N and R models fit the right-hand adjustment position marker (see illustration 15.2b). Fit the nut and tighten it to the specified torque setting, counter-holding the axle head on the other side of the wheel (see illustrations).

11 Check the chain slack as described in Chapter 1 and adjust if necessary.

12 Operate the brake pedal several times to bring the pads into contact with the disc. Check the operation of the rear brake carefully before riding the bike.

16 Wheel bearings – renewal

Front wheel bearings

Note: *Always renew the wheel bearings in pairs. Never renew the bearings individually. Avoid using a high pressure cleaner on the wheel bearing area.*

1 Remove the wheel (see Section 14).
2 Set the wheel on blocks so as not to allow the weight of the wheel to rest on either of the brake discs.
3 Prise out the seal on each side of the wheel using a flat-bladed screwdriver, taking care not to damage the rim of the hub (see illustration). Discard the seals as new ones should be used.

15.10a Slide in the axle from the left . . .

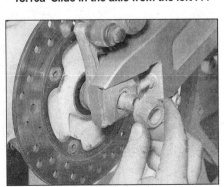

15.10b . . . then fit the position marker (L, N and R models) and the nut . . .

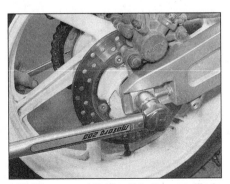

15.10c . . . and tighten it to the specified torque

16.3 Lever out the grease seals

7•18 Brakes, wheels and tyres

16.4a Use a drift to knock out the bearings

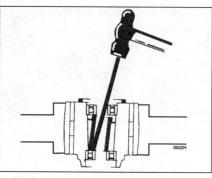

16.4b Locate the drift as shown when driving out the bearing

16.9 The bearing can be driven in using a suitable socket

4 Using a metal rod (preferably a brass drift punch) inserted through the centre of the left-hand bearing, tap evenly around the inner race of the right-hand bearing to drive it from the hub **(see illustrations)**. The bearing spacer will also come out.

5 Lay the wheel on its other side so that the left-hand bearing faces down. Drive the bearing out of the wheel using the same technique as above.

6 If the bearings are of the unsealed type or are only sealed on one side, clean them with a high flash-point solvent (one which won't leave any residue) and blow them dry with compressed air (don't let the bearings spin as you dry them). Apply a few drops of oil to the bearing. **Note:** *If the bearing is sealed on both sides don't attempt to clean it.*

7 Hold the outer race of the bearing and rotate the inner race – if the bearing doesn't turn smoothly, has rough spots or is noisy, renew it.

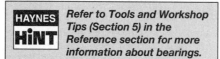

Haynes HiNT *Refer to Tools and Workshop Tips (Section 5) in the Reference section for more information about bearings.*

8 If the bearing is good and can be re-used, wash it in solvent once again and dry it, then pack the bearing with grease.

9 Thoroughly clean the hub area of the wheel. First install the left-hand side bearing into its recess in the hub, with the marked or sealed side facing outwards. Using the old bearing, a bearing driver or a socket large enough to contact the outer race of the bearing, drive it in until it's completely seated **(see illustration)**.

10 Turn the wheel over and install the bearing spacer. Drive the right-hand side bearing into place as described above.

11 Apply a smear of grease to the lips of the seals, then press them into the wheel. Gently drive them into place using a seal or bearing driver, a suitable socket or a flat piece of wood **(see illustration)**.

12 Clean off all grease from the brake discs using acetone or brake system cleaner then install the wheel (see Section 14).

Rear wheel bearings

13 Remove the rear wheel (see Section 15). Lift the sprocket coupling out of the wheel, noting how it fits **(see illustration)**.

14 Set the wheel on blocks so as not to allow the weight of the wheel to rest on the brake disc.

15 On J and K models, remove the collar from the left-hand side of the wheel **(see illustration)**. If not already done, also remove the collar from the right-hand side of the wheel **(see illustration)**. The collar could be a tight fit and may have to be driven out from the other side using a suitable long socket or piece of tubing **(see illustration)**.

16 On L, N and R models, if not already done, remove the collar from the right-hand side of the wheel **(see illustration 15.6)**.

17 Lever out the grease seal on the right-

16.11 The seal can be driven in using a flat piece of wood

16.13 Lift the sprocket coupling out of the wheel

16.15a On J and K models, remove the collar from the left-hand side . . .

16.15b . . . and if not already done, from the right-hand side (arrowed)

16.15c If the collar is tight, drive it out as shown

Brakes, wheels and tyres 7•19

16.17 Lever out the grease seal

16.22 Driving in the bearing using a socket

16.23 Renew the O-ring (arrowed) if necessary

hand side of the wheel using a flat-bladed screwdriver, taking care not to damage the rim of the hub **(see illustration)**. Discard the seal as a new one should be used.

18 Using a metal rod (preferably a brass drift punch) inserted through the centre of one bearing, tap evenly around the inner race of the other bearing to drive it from the hub **(see illustrations 16.4a and b)**. The bearing spacer will also come out.

19 Lay the wheel on its other side so that the remaining bearing faces down. Drive the bearing out of the wheel using the same technique as above.

20 Refer to Steps 6 to 8 above and check the bearings.

21 Thoroughly clean the hub area of the wheel. First install the right-hand bearing into its recess in the hub, with the marked or sealed side facing outwards. Using the old bearing, a bearing driver or a socket large enough to contact the outer race of the bearing, drive it in squarely until it's completely seated.

22 Turn the wheel over and install the bearing spacer. Drive the left-hand side bearing into place as described above **(see illustration)**. On J and K models, fit the collar into the left-hand bearing **(see illustration 16.15a)**.

23 On L, N and R models, check the condition of the hub O-ring and renew it if it is damaged or deteriorated **(see illustration)**.

24 Apply a smear of grease to the lips of the new grease seal, and press it into the right-hand side of the wheel, using a seal or bearing driver, a suitable socket or a flat piece of wood to drive it into place if necessary **(see illustration 16.11)**.

25 Clean off all grease from the brake disc using acetone or brake system cleaner. Fit the sprocket coupling assembly onto the wheel **(see illustration 16.13)**, then install the wheel (see Section 15).

Sprocket coupling bearing

26 Remove the rear wheel (see Section 15). Lift the sprocket coupling out of the wheel, noting how it fits **(see illustration 16.13)**.

27 On J and K models, if not already done, remove the collar from the outside of the sprocket coupling **(see illustration)**. The collar could be a tight fit and may have to be driven out from the inside using a suitable long socket or piece of tubing **(see illustration)**.

28 On L, N and R models, if not already done, remove the collar from the outside of the sprocket coupling **(see illustration)**. Also remove the spacer from the inside of the coupling bearing, noting which way round it fits. The spacer could be a tight fit and may have to be driven out using a suitable socket or piece of tubing **(see illustration)**.

29 Using a flat-bladed screwdriver, lever out the grease seal from the outside of the coupling **(see illustration)**.

30 Support the coupling on blocks of wood

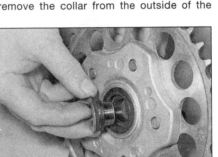

16.27a Remove the collar . . .

16.27b . . . driving it out from the inside if it is tight

16.28a Remove the collar from the outside

16.28b Drive out the spacer if it is tight

16.29 Lever out the grease seal

7•20 Brakes, wheels and tyres

16.30 Drive out the bearing from the inside

16.34 Renew the O-ring if necessary

and drive the bearing out from the inside using a bearing driver or socket **(see illustration)**.
31 Refer to Steps 6 to 8 above and check the bearings.
32 Thoroughly clean the bearing recess then install the bearing into the coupling, with the marked or sealed side facing out. Using the old bearing, a bearing driver or a socket large enough to contact the outer race of the bearing, drive it in until it is completely seated.
33 On L, N and R models, install the spacer into the inside of the coupling.
34 On J and K models, check the condition of the coupling O-ring and renew it if it is damaged or deteriorated **(see illustration)**.
35 Apply a smear of grease to the lips of the new seal, and press it into the coupling, using a seal or bearing driver, a suitable socket or a flat piece of wood to drive it into place if necessary **(see illustration 16.11)**. Insert the collar into the outside of the sprocket coupling.
36 Check the sprocket coupling/rubber damper (see Chapter 6).
37 Clean off all grease from the brake disc using acetone or brake system cleaner. Fit the sprocket coupling into the wheel **(see illustration 16.13)**, then install the wheel (see Section 15).

17 Tyres – general information and fitting

General information
1 The wheels fitted to all models are designed to take tubeless tyres only. Tyre sizes are given in the Specifications at the beginning of this Chapter.
2 Refer to *Daily (pre-ride) checks* at the beginning of this manual for tyre maintenance.

Fitting new tyres
3 When selecting new tyres, ensure that front and rear tyre types are compatible, the correct size and correct speed rating; if necessary seek advice from a tyre fitting specialist **(see illustration)**.
4 It is recommended that tyres are fitted by a motorcycle tyre specialist rather than attempted in the home workshop. This is particularly relevant in the case of tubeless tyres because the force required to break the seal between the wheel rim and tyre bead is substantial, and is usually beyond the capabilities of an individual working with normal tyre levers. Additionally, the specialist will be able to balance the wheels after tyre fitting.
5 Note that punctured tubeless tyres can in some cases be repaired, but such repairs must be carried out by a tyre fitting specialist.

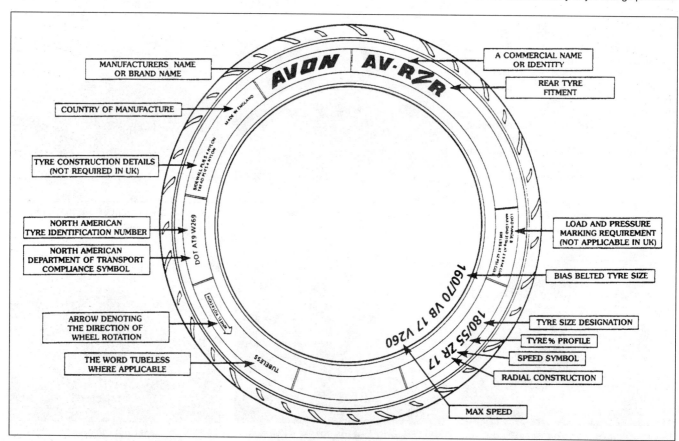

17.3 Common tyre sidewall markings

Chapter 8
Bodywork

Contents

Fairing panels – removal and installation 5
Front mudguard – removal and installation 6
General information 1
Seats – removal and installation 2
Seat cowling – removal and installation 3
Rear view mirrors – removal and installation 4

Degrees of difficulty

| Easy, suitable for novice with little experience | Fairly easy, suitable for beginner with some experience | Fairly difficult, suitable for competent DIY mechanic | Difficult, suitable for experienced DIY mechanic | Very difficult, suitable for expert DIY or professional |

1 General information

This Chapter covers the procedures necessary to remove and install the body parts. Since many service and repair operations on these motorcycles require the removal of the body parts, the procedures are grouped here and referred to from other Chapters.

In the case of damage to the body parts, it is usually necessary to remove the broken component and renew it (or used) one. The material that the body panels are composed of doesn't lend itself to conventional repair techniques. Note that there are however some companies that specialise in 'plastic welding' and there are a number of bodywork repair kits now available for motorcycles.

When attempting to remove any body panel, first study it closely, noting any fasteners and associated fittings, to be sure of returning everything to its correct place on installation. In some cases the aid of an assistant will be required when removing panels, to help avoid the risk of damage to paintwork. Once the evident fasteners have been removed, try to withdraw the panel as described but DO NOT FORCE IT – if it will not release, check that all fasteners have been removed and try again. Where a panel engages another by means of tabs, be careful not to break the tab or its mating slot or to damage the paintwork. Remember that a few moments of patience at this stage will save you a lot of money in replacing broken fairing panels!

When installing a body panel, first study it closely, noting any fasteners and associated fittings removed with it, to be sure of returning everything to its correct place. Check that all fasteners are in good condition, including all trim nuts or clips and damping/rubber mounts; any of these must be renewed if faulty before the panel is reassembled. Check also that all mounting brackets are straight and repair or renew them if necessary before attempting to install the panel. Where assistance was required to remove a panel, make sure your assistant is on hand to install it.

Tighten the fasteners securely, but be careful not to overtighten any of them or the panel may break (not always immediately) due to the uneven stress. Where quick-release fasteners are fitted, turn them 90° anti-clockwise to release them, and 90° clockwise to secure them.

 Models are identified by their production code letter – refer to 'Identification numbers' at the front of this manual for details.

8•2 Bodywork

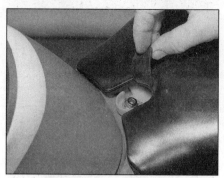

2.1a Lift the flap and unscrew the bolt . . .

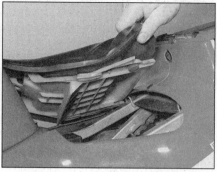

2.1b . . . and remove the seat

2.2a Insert the key and turn the lock . . .

2.2b . . . and remove the seat

2.3a Insert the key and turn the lock to raise the seat up

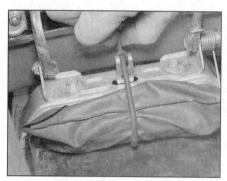

2.3b Remove the toolkit . . .

2 Seats – removal and installation

Removal

1 To remove the rider's seat, lift up the flap at the front of the seat and unscrew the bolt **(see illustration)**. Remove the seat, noting how the tabs at the back locate **(see illustration)**.

2 To remove the passenger seat on J and K models, insert the ignition key into the seat lock located behind the rider's seat, and turn it clockwise to unlock the passenger's seat **(see illustration)**. Remove the passenger's seat **(see illustration)**.

3 To open the passenger seat on L, N and R models, insert the ignition key into the seat lock located behind the seat, then turn it clockwise and raise the seat **(see illustration)**. To remove the seat, release the strap and remove the toolkit, then unscrew the two bolts securing the bracket assembly and remove the seat, noting how the hinge spring locates **(see illustrations)**.

Installation

4 Installation is the reverse of removal. Make sure the tabs at the rear of the rider's seat, and the tab at the rear of the passenger seat on J and K models locate correctly into the brackets **(see illustrations 2.1b and 2.2b)**. On L, N and R models, make sure the hinge spring end locates correctly **(see illustration 2.3c)**. Push down on the passenger seat to engage the latch.

3 Seat cowling – removal and installation

Removal

1 Remove the seats (see Section 2).

2 On J and K models, remove the two bolts securing each side of the seat cowling, the bolt securing the central retaining bracket, and the screw on each underside at the back **(see illustrations)**. Carefully pull each side of

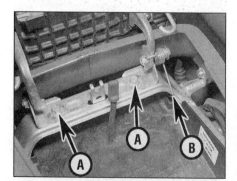

2.3c . . . then unscrew the two bolts (A) and remove the seat, noting how the spring end (B) locates

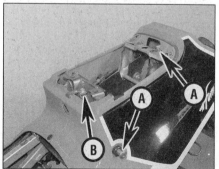

3.2a Remove the two bolts (A) on each side, the central bolt (B) . . .

3.2b . . . and the two screws (arrowed)

Bodywork 8•3

3.2c Pull each front side away to release the lugs from the grommets (arrowed) . . .

3.2d . . . and remove the cowling

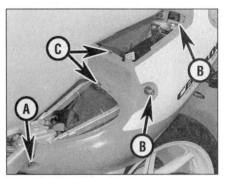

3.3a Remove the screw (A), the two bolts (B), the two clips (C) . . .

the front of the cowling out away from the frame to free the two lugs on each side from their rubber grommets **(see illustration)**. Carefully draw the cowling rearwards and off the bike, taking care not to bend the front sides excessively **(see illustration)**.

3 On L, N and R models, remove the screw and the two bolts securing each side of the seat cowling, the two clips joining the halves in the middle, and the two screws securing the underside at the back **(see illustrations)**. Carefully draw the cowling rearwards and off the bike, taking care not to bend the front sides excessively **(see illustration)**.

Installation

4 Installation is the reverse of removal. On L, N and R models, make sure the arm on the lock barrel fits against the release mechanism lever.

 HAYNES HiNT *Note that a small amount of lubricant (liquid soap or similar) applied to the mounting rubber grommets of the seat cowling will assist the lugs to engage without the need for undue pressure.*

3.3b . . . and the two screws (arrowed) . . .

3.3c . . . and remove the cowling

4 Rear view mirrors – removal and installation

Removal

1 Unscrew the two nuts securing each mirror and remove the mirror along with its rubber insulator pads **(see illustrations)**.

Installation

2 Installation is the reverse of removal.

5 Fairing panels – removal and installation

Fairing – J and K models

Note: *The windshield is riveted to the fairing and should not be removed unless it is being renewed. Windshields are available separately from the fairing. A rivet gun is required to secure a new windshield to the fairing.*

4.1a Unscrew the nuts (arrowed) . . .

4.1b . . . and remove the mirror

8•4 Bodywork

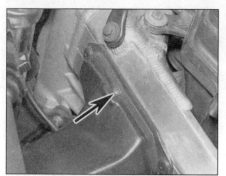

5.2a Remove the screw (arrowed) and detach the duct . . .

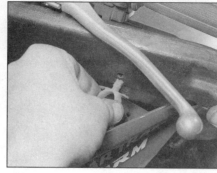

5.2b . . . then release the wiring clip from its underside

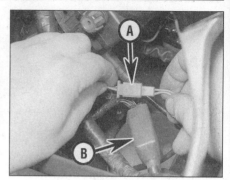

5.3 Disconnect the turn signal wiring connector (A) on each side and fusebox wiring connector (B)

Removal

1 Remove the fairing side panels (see below).
2 To improve access to the turn signal and fusebox wiring connectors (though they can be reached from the underside of the fairing), remove the screw securing the left-hand air duct to the frame **(see illustration)**. Lift the duct to access the wiring clip underneath, then press the ends of the clip together and withdraw it from the duct **(see illustration)**. Remove the duct. If required, also remove the right-hand duct – it doesn't have a wiring clip.
3 Disconnect the turn signal and fusebox wiring connectors **(see illustration)**. Remove the rear view mirrors (see Section 4).

4 Remove the screws securing the sides of the fairing and the bolt securing the front of the fairing to the fairing bracket **(see illustrations)**. Carefully draw the fairing forward and off the bike **(see illustration)**.
5 Installation is the reverse of removal. Make sure the wiring connectors are correctly and securely connected. If not removed, make sure the fairing locates correctly onto the air ducts.

Fairing – L, N and R models

Removal

6 Remove the fairing side panels (see below).

7 Release the trim clip securing the back of the wiring connector bracket, and release the front from the locating pin **(see illustration)**. Also lift the fusebox off its holder.
8 Disconnect the turn signal wiring connectors on each side **(see illustration)**.
9 Remove the rear view mirrors (see Section 4).
10 Remove the trim clips securing the underside of the fairing at the front **(see illustration)**.
11 Unscrew the bolt securing the front of the fairing to the fairing bracket **(see illustration)**. Carefully draw the fairing forward and off the

5.4a Remove the screw (arrowed) on each side . . .

5.4b . . . and the bolt at the front . . .

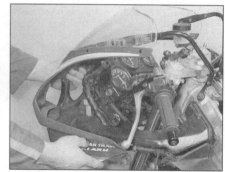

5.4c . . . and remove the fairing

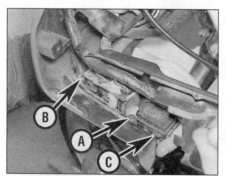

5.7 Release the trim clip (A) and free the wiring connector bracket from the pin (B). Also displace the fusebox (C)

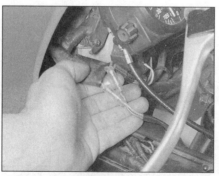

5.8 Disconnect the turn signal wiring connectors

5.10 Release the two trim clips (arrowed) . . .

Bodywork 8•5

5.11a ... then unscrew the bolt (arrowed) ...

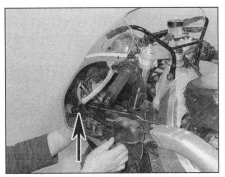

5.11b ... and remove the fairing, noting how the trim panels locate at the front (arrowed) ...

5.11c ... and in the middle

bike, noting how the front trim panels locate (see illustrations).

Installation

12 Installation is the reverse of removal. Make sure the wiring connectors are correctly and securely connected. Make sure the tabs on the front and in the middle of the front trim panels locate into the slots in the fairing (see illustrations 5.11b and c).

Fairing side panels – all models

Removal

13 On J and K models, turn the three quick-release fasteners along the bottom edge of the panel 1/4 turn anti-clockwise, then

carefully draw the bottom edge out and down to release the tabs along the top edge from the slots in the fairing (see illustration).

14 On L, N and R models, release the two trim clips securing the top of the panel and turn the three quick-release fasteners 1/4 turn anti-clockwise (see illustration). Lift the top section off the locating pin in the middle (see illustration), then carefully draw the panel away to release the tabs along the front and top from the slots in the fairing.

Installation

15 Installation is the reverse of removal. Make sure the tabs locate correctly into the slots in the fairing.

Lower fairing – all models

Removal

16 Remove the fairing side panels (see above).

17 Remove the three screws securing each side, then carefully lower the fairing and manoeuvre it from under the bike (see illustrations).

Installation

18 Installation is the reverse of removal.

Air ducts (J and K models) and trim panels (L, N and R models)

Removal

19 On J and K models, an air duct runs between the front of the fairing and the frame on each side. Remove the screw securing the duct to the frame and remove the duct, noting how it locates onto the fairing at the front (see illustration 5.2a). When removing the left-hand duct, lift it to access the wiring clip underneath, then press the ends of the clip together and withdraw it from the duct (see illustration 5.2b).

20 On L, N and R models, to access the front trim panels, remove the fairing (see above), then remove the trim clip securing the back of the panel and remove the panel, noting how it

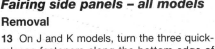

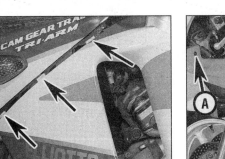

5.13 Note how the tabs (arrowed) locate in the fairing

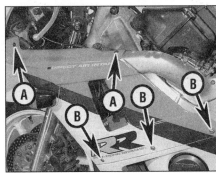

5.14a Release the trim clips (A) and the fasteners (B) ...

5.14b ... and lift the top off the locating pin (arrowed)

5.17a Remove the screws (arrowed) on each side ...

5.17b ... and remove the lower fairing

8•6 Bodywork

fits **(see illustration)**. To remove the rear trim panels, remove the two bolts at the back and remove the panel, noting how it fits **(see illustration)**.

Installation

21 Installation is the reverse of removal.

6 Front mudguard – removal and installation

Removal

1 Unscrew the four bolts securing the mudguard to the holder on each fork slider and remove the mudguard, noting how it fits **(see illustrations)**.

Installation

2 Installation is the reverse of removal.

5.20a Remove the trim clip (arrowed) to release the front trim panel

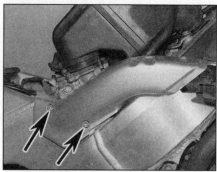

5.20b Unscrew the two bolts (arrowed) to release the rear trim panel

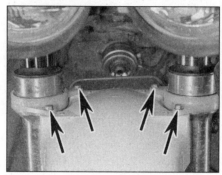

6.1a Unscrew the four bolts (arrowed) . . .

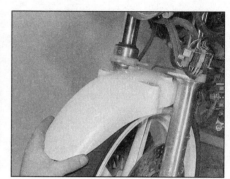

6.1b . . . and remove the mudguard

Chapter 9
Electrical system

Contents

Alternator – removal, inspection and installation	32
Battery – charging	4
Battery – removal, installation, inspection and maintenance	3
Brake light switches – check and replacement	14
Brake/tail light bulbs – renewal	9
Charging system – leakage and output test	31
Charging system testing – general information and precautions	30
Clutch switch – check and replacement	24
Diode – check and replacement	25
Electrical system – fault finding	2
Fuel pump – check, removal and installation	see Chapter 4
Fuses – check and renewal	5
General information	1
Handlebar switches – check	20
Handlebar switches – removal and installation	21
Headlight aim – check and adjustment	see Chapter 1
Headlight bulbs and sidelight bulbs – renewal	7
Headlight assembly – removal and installation	8
Horn – check and replacement	26
Ignition (main) switch – check, removal and installation	19
Ignition system components	see Chapter 5
Instrument and warning light bulbs – renewal	17
Instrument cluster and speedometer cable – removal and installation	15
Instruments – check and replacement	16
Lighting system – check	6
Neutral switch – check, removal and installation	22
Oil pressure switch – check and replacement	18
Regulator/rectifier – check and replacement	33
Sidestand switch – check and replacement	23
Starter motor – disassembly, inspection and reassembly	29
Starter motor – removal and installation	28
Starter relay – check and replacement	27
Tail light assembly – removal and installation	10
Turn signal bulbs – renewal	12
Turn signal assemblies – removal and installation	13
Turn signal circuit – check	11

Degrees of difficulty

| Easy, suitable for novice with little experience | Fairly easy, suitable for beginner with some experience | Fairly difficult, suitable for competent DIY mechanic | Difficult, suitable for experienced DIY mechanic | Very difficult, suitable for expert DIY or professional |

Specifications

Note: *Models are identified by their production code letter – refer to 'Identification numbers' at the front of this manual for details.*

Battery
Capacity
- J and K models 12V, 8Ah
- L, N and R models 12V, 6Ah

Voltage
- Fully charged .... 12.8 to 13.0V
- Discharged .... below 12.3V

Charging rate
- Normal
  - J and K models .... 0.9A for 5 to 10 hrs
  - L, N and R models .... 0.7A for 5 to 10 hrs
- Quick .... 3.0A for 1 hr

Charging system
Current leakage .... 1mA (max)

Regulated voltage output
- J and K models .... 14.0 to 15.0V @ 3000 rpm
- L, N and R models .... 14.0 to 16.0V @ 3000 rpm

Regulated current output .... 0.5A @ 3000 rpm
Stator coil resistance .... 0.1 to 0.5 ohms

Starter motor
Brush length
- Standard .... 12.0 to 13.0 mm
- Service limit (min) .... 6.5 mm

Fuses
Main 30A
Headlight 15A
Tail light, signal 10A
Ignition 10A
Fan 10A

Bulbs
Headlights
  UK spec .... 60/55W H4 halogen
  Japan spec .... 60/35W halogen
Sidelights (J, K, L and N models)
  UK spec 5.0W
  Japan spec 1.7W
Brake/tail lights
  UK spec 21/5W
  Japan spec 18/5W
Turn signal lights
  UK spec 21W
  Japan spec
    J and K models .... 18W (front), 15W (rear)
    L and N models .... 15W (front), 15W (rear)
    R models .... 18/5W (front), 15W (rear)
Instrument and warning lights .... 3.4W or 1.7W

Torque settings
Note: *Where a specified setting is not given for a particular bolt, the general settings listed at the beginning apply. The dimension given applies to the diameter of the thread, not the head.*

5 mm bolt/nut .... 5 Nm
6 mm bolt/nut .... 10 Nm
8 mm bolt/nut .... 22 Nm
10 mm bolt/nut .... 35 Nm
12 mm bolt/nut .... 55 Nm
6 mm flange bolt with 8 mm head .... 9 Nm
6 mm flange bolt/nut with 10 mm head .... 12 Nm
8 mm flange bolt/nut .... 27 Nm
10 mm flange bolt/nut .... 40 Nm
Footrest bracket bolts .... 27 Nm
Oil pressure switch .... 12 Nm
Ignition (main) switch bolts .... 25 Nm
Neutral switch .... 12 Nm
Sidestand switch bolt (L, N and R models) .... 10 Nm
Alternator stator bolts .... 10 Nm
Alternator rotor bolt
  J and K models .... 85 Nm
  L, N and R models .... 95 Nm
Alternator/clutch cover bolts .... 12 Nm

1 General information

All models have a 12-volt electrical system charged by a three-phase alternator with a separate regulator/rectifier.

The regulator maintains the charging system output within the specified range to prevent overcharging, and the rectifier converts the ac (alternating current) output of the alternator to dc (direct current) to power the lights and other components and to charge the battery. The alternator rotor is mounted on the right-hand end of the crankshaft.

The starter motor is mounted behind the cylinders. The starting system includes the motor, the battery, the relay and the various wires and switches. If the engine kill switch in the RUN position and the ignition (main) switch is ON, the starter relay allows the starter motor to operate only if the transmission is in neutral (neutral switch on) or, if the transmission is in gear, if the clutch lever is pulled into the handlebar and, on L, N and R models, the sidestand is up.

Note: *Keep in mind that electrical parts, once purchased, cannot be returned. To avoid unnecessary expense, make very sure the faulty component has been positively identified before buying a new part.*

 Models are identified by their production code letter – refer to 'Identification numbers' at the front of this manual for details.

2 Electrical system – fault finding

⚠️ **Warning:** *To prevent the risk of short circuits, the ignition (main) switch must always be OFF and the battery negative (-ve) terminal should be disconnected before any of the bike's other electrical components are disturbed. Don't forget to reconnect the terminal securely once work is finished or if battery power is needed for circuit testing.*

1 A typical electrical circuit consists of an electrical component, the switches, relays, etc. related to that component and the wiring and connectors that hook the component to both the battery and the frame. To aid in locating a problem in any electrical circuit, refer to the wiring diagrams at the end of this Chapter.

2 Before tackling any troublesome electrical circuit, first study the wiring diagram (see end of Chapter) thoroughly to get a complete picture of what makes up that individual circuit. Trouble spots, for instance, can often be narrowed down by noting if other components related to that circuit are operating properly or not. If several components or circuits fail at one time, chances are the fault lies in the fuse or earth (ground) connection, as several circuits often are routed through the same fuse and earth (ground) connections.

3 Electrical problems often stem from simple causes, such as loose or corroded connections or a blown fuse. Prior to any electrical fault finding, always visually check the condition of the fuse, wires and connections in the problem circuit. Intermittent failures can be especially frustrating, since you can't always duplicate the failure when it's convenient to test. In such situations, a good practice is to clean all connections in the affected circuit, whether or not they appear to be good. All of the connections and wires should also be wiggled to check for looseness which can cause intermittent failure.

4 If testing instruments are going to be utilised, use the wiring diagram to plan where you will make the necessary connections in order to accurately pinpoint the trouble spot.

5 The basic tools needed for electrical fault finding include a battery and bulb test circuit, a continuity tester, a test light, and a jumper wire. A multimeter capable of reading volts, ohms and amps is also very useful as an alternative to the above, and is necessary for performing more extensive tests and checks.

> **HAYNES HiNT** *Refer to Fault Finding Equipment in the Reference section for details of how to use electrical test equipment.*

3 Battery – removal, installation, inspection and maintenance

Caution: Be extremely careful when handling or working around the battery. The electrolyte is very caustic and an explosive gas (hydrogen) is given off when the battery is charging.

Removal and installation

1 Remove the rider's seat (see Chapter 8).
2 Release the battery strap, and on J and K models remove the battery cover **(see illustrations)**.
3 Unscrew the negative (-ve) terminal bolt first and disconnect the lead from the battery **(see illustration)**. Lift up the red insulating cover to access the positive (+ve) terminal, then unscrew the bolt and disconnect the lead. Lift the battery from the bike **(see illustration)**.
4 On installation, clean the battery terminals and lead ends with a wire brush or knife and emery paper. Reconnect the leads, connecting the positive (+ve) terminal first.

> **HAYNES HiNT** *Battery corrosion can be kept to a minimum by applying a layer of petroleum jelly to the terminals after the cables have been connected.*

5 Install the seat (see Chapter 8).

Inspection and maintenance

6 The battery fitted to the models covered in this manual is of the maintenance free (sealed) type, it therefore does not require topping up. However, the following checks should still be regularly performed.
7 Check the battery terminals and leads for tightness and corrosion. If corrosion is evident, unscrew the terminal screws and disconnect the leads from the battery, disconnecting the negative (-ve) terminal first, and clean the terminals and lead ends with a wire brush or knife and emery paper.

3.2a On J and K models, release the strap . . .

3.2b . . . and remove the cover

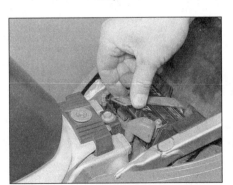
3.2c On L, N and R models, release the strap

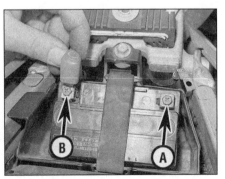
3.3a Disconnect the negative (-ve) terminal (A) first, then the positive (+ve) terminal (B) . . .

3.3b . . . and remove the battery

9•4 Electrical system

4.2 If the charger doesn't have ammeter built in, connect one in series as shown. DO NOT connect the ammeter between the battery terminals or it will be ruined

Reconnect the leads, connecting the negative (-ve) terminal last, and apply a thin coat of petroleum jelly to the connections to slow further corrosion.

8 The battery case should be kept clean to prevent current leakage, which can discharge the battery over a period of time (especially when it sits unused). Wash the outside of the case with a solution of baking soda and water. Rinse the battery thoroughly, then dry it.

9 Look for cracks in the case and renew the battery if any are found. If acid has been spilled on the frame or battery box, neutralise it with a baking soda and water solution, dry it thoroughly, then touch up any damaged paint.

10 If the motorcycle sits unused for long periods of time, disconnect the cables from the battery terminals, negative (-ve) terminal first. Refer to Section 4 and charge the battery once every month to six weeks.

11 The condition of the battery can be assessed by measuring the voltage present at the battery terminals. Connect the voltmeter positive (+ve) probe to the battery positive (+ve) terminal, and the negative (-ve) probe to the battery negative (-ve) terminal. When fully charged there should be more than 12.8 volts present. If the voltage falls to 12.3 volts the battery must be removed, disconnecting the negative (-ve) terminal first, and recharged as described below in Section 4.

4 Battery – charging

Caution: Be extremely careful when handling or working around the battery. The electrolyte is very caustic and an explosive gas (hydrogen) is given off when the battery is charging.

1 Remove the battery (see Section 3). Connect the charger to the battery, making sure that the positive (+ve) lead on the charger is connected to the positive (+ve) terminal on the battery, and the negative (-ve) lead is connected to the negative (-ve) terminal.

2 Honda recommend that the battery is charged at a maximum rate of 0.9 amps (J and K models) or 0.7 amps (L, N and R models) for 5 to 10 hours. Exceeding this figure can cause the battery to overheat, buckling the plates and rendering it useless. Few owners will have access to an expensive current controlled charger, so if a normal domestic charger is used check that after a possible initial peak, the charge rate falls to a safe level **(see illustration)**. If the battery becomes hot during charging **stop**. Further charging will cause damage. **Note:** *In emergencies the battery can be charged at a higher rate of around 3.0 amps for a period of 1 hour. However, this is not recommended and the low amp charge is by far the safer method of charging the battery.*

3 If the recharged battery discharges rapidly if left disconnected it is likely that an internal short caused by physical damage or sulphation has occurred. A new battery will be required. A sound item will tend to lose its charge at about 1% per day.

4 Install the battery (see Section 3).

5 If the motorcycle sits unused for long periods of time, charge the battery once every month to six weeks and leave it disconnected.

5 Fuses – check and renewal

1 The electrical system is protected by fuses of different ratings. All except the main fuse are housed in the fusebox, which is located in the left-hand side of the fairing. The main fuse is integral with the starter relay, which is under the seat cowling on the right-hand side of the bike on J and K models, and on the left on L, N and R models.

2 To access the fuses in the fusebox, unclip the fusebox lid **(see illustration)**. On J and K models, remove the left-hand air duct for access (see Chapter 8). On L, N and R models, access is quite restricted – it is easier to lift the fusebox off its clip and displace it upwards **(see illustrations)**. To access the main fuse, remove the seat cowling (see Chapter 8) and disconnect the starter relay wiring connector **(see illustrations)**.

5.2a Fusebox – J and K models

5.2b On L, N and R models, lift the box off its clip . . .

5.2c . . . for best access to the fuses

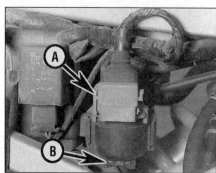

5.2d Disconnect the wiring connector (A) to access the main fuse. Note the spare fuse (B) – J and K models

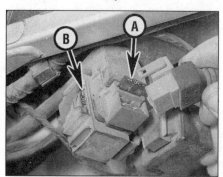

5.2e Main fuse (A), spare fuse (B) – L, N and R models

3 The fuses can be removed and checked visually. If you can't pull the fuse out with your fingertips, use a pair of suitable pliers **(see illustration)**. A blown fuse is easily identified by a break in the element **(see illustration)**. Each fuse is clearly marked with its rating and must only be replaced by a fuse of the correct rating. A spare fuse of each rating is housed in the fusebox, and a spare main fuse is housed in the bottom of the starter relay **(see illustrations 5.2d and e)**. If a spare fuse is used, always renew it so that a spare of each rating is carried on the bike at all times.

 Warning: Never put in a fuse of a higher rating or bridge the terminals with any other substitute, however temporary it may be. Serious damage may be done to the circuit, or a fire may start.

4 If a fuse blows, be sure to check the wiring circuit very carefully for evidence of a short-circuit. Look for bare wires and chafed, melted or burned insulation. If the fuse is renewed before the cause is located, the new fuse will blow immediately.

5 Occasionally a fuse will blow or cause an open-circuit for no obvious reason. Corrosion of the fuse ends and fusebox terminals may occur and cause poor fuse contact. If this happens, remove the corrosion with a wire brush or emery paper, then spray the fuse end and terminals with electrical contact cleaner.

6 Lighting system – check

1 The battery provides power for operation of the headlight, tail light, brake light and instrument cluster lights. If none of the lights operate, always check battery voltage before proceeding. Low battery voltage indicates either a faulty battery or a defective charging system. Refer to Section 3 for battery checks and Sections 30 and 31 for charging system tests. Also, check the condition of the fuses.

Headlight

2 If the headlight (HI or LO beam) fails to work, first check the fuse with the key ON (see

5.3a Use a pair of pliers to remove a fuse

Section 5), and then the bulbs (see Section 7). If they are both good, use jumper wires to connect the bulb directly to the battery terminals. If the beams come on, the problem lies in the wiring, the relay (HI beam only), or one of the switches in the circuit. Refer to Section 20 for the switch testing procedures, and also the wiring diagrams at the end of this Chapter.

3 If the HI beam relay is suspected of being faulty, the best way to determine this is to substitute it with another relay. Remove the fairing to access the relay – it is mounted above the headlight assembly **(see illustrations)**. If the beam in question then works, the faulty relay must be renewed. If another relay is not available, switch the ignition ON and the light switch ON, then flick the HI/LO beam switch to HI and listen for a click in the relay. If the relay clicks, disconnect the wiring connector and check for voltage at the black/red terminal of the relay **(see illustration)**. If voltage is present, and the wiring between the relay and the headlight is good, renew the relay. If no voltage is present, check the wiring between the relay and the fusebox (see wiring diagrams at the end of the Chapter). If the relay does not click, check for voltage at the blue/white (J, K, L, N models) or blue wire (R model) terminal of the relay. If voltage is present, renew the relay. If no voltage is present, check the wiring between the relays and the switches (see wiring diagrams at the end of the Chapter).

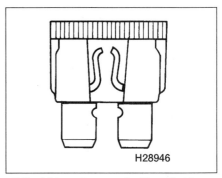

5.3b A blown fuse can be identified by a break in its element

Tail light

4 If the tail light fails to work, check the bulbs and the bulb terminals first, then the fuse, then check for battery voltage at the brown (J, K, L, N models) or brown/white wire (R model) terminal on the supply side of the tail light wiring connector. If voltage is present, check the earth (ground) circuit for an open or poor connection.

5 If no voltage is indicated, check the wiring between the tail light and the ignition switch, then check the switch. Also check the lighting switch.

Brake light

6 If the brake light fails to work, check the bulbs and the bulb terminals first, then the fuse, then check for battery voltage at the green/yellow terminal on the supply side of the tail light wiring connector, with the brake lever pulled in or the pedal depressed. If voltage is present, check the earth (ground) circuit for an open or poor connection.

7 If no voltage is indicated, check the brake light switches, then the wiring between the tail light and the switches.

8 See Section 14 for brake switch checks and Section 9 for tail/brake light bulb renewal.

Instrument and warning lights

9 See Section 17 for instrument and warning light bulb renewal.

Turn signal lights

10 See Section 11 for the turn signal circuit check.

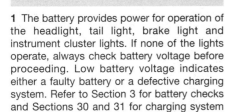

6.3a Headlight relay (arrowed) – J and K models

6.3b Headlight relay (arrowed) – L, N and R models

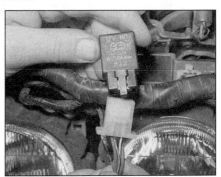

6.3c Displace the relay, then remove the rubber cover and disconnect the wiring connector

9•6 Electrical system

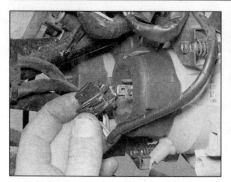

7.1a Disconnect the wiring connector . . .

7.1b . . . and remove the dust cover

7.2a Release the clip . . .

7 Headlight bulbs and sidelight bulbs – renewal

Note: *The headlight bulbs are of the quartz-halogen type. Do not touch the bulb glass as skin acids will shorten the bulb's service life. If the bulb is accidentally touched, it should be wiped carefully when cold with a rag soaked in methylated spirit and dried before fitting.*

 Warning: Allow the bulbs time to cool before removing them if the headlight has just been on.

Headlight

1 Disconnect the relevant wiring connector from the back of the headlight assembly and remove the rubber dust cover, noting how it fits **(see illustrations)**.
2 Release the bulb retaining clip, noting how it fits, then remove the bulb **(see illustrations)**.
3 Fit the new bulb, bearing in mind the information in the **Note** above. Make sure the tabs on the bulb fit correctly in the slots in the bulb housing, and secure it in position with the retaining clip.
4 Install the dust cover, making sure it is correctly seated and with the 'TOP' mark at the top, and connect the wiring connector.
5 Check the operation of the headlight.

 HAYNES HiNT *Always use a paper towel or dry cloth when handling new bulbs to prevent injury if the bulb should break and to increase bulb life.*

Sidelight

6 Pull the bulbholder out of its socket in the base of the headlight, then carefully pull the bulb out of the holder **(see illustrations)**.
7 Install the new bulb in the bulbholder, then install the bulbholder by pressing it in. Make sure the rubber cover is correctly seated.
8 Check the operation of the sidelight.

8 Headlight assembly – removal and installation

J and K models

Removal

1 Remove the fairing (see Chapter 8).
2 Remove the instrument cluster (see Section 15).
3 The headlights are mounted individually. Disconnect the headlight and sidelight wiring connectors **(see illustrations)**. Release the wiring from any clips or ties on the headlight.
4 Each headlight is secured by three bolts – one at the top and two at the bottom. Unscrew the bolts and remove the headlight, noting how it fits **(see illustrations)**.
5 To separate the beam unit from its holder, first mark each unit according to its side and note the positions of the beam adjusting screws. Check the current aim setting by

7.2b . . . and remove the bulb

7.6a Remove the bulbholder . . .

7.6b . . . and pull out the bulb

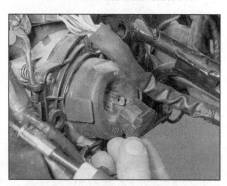

8.3a Disconnect the headlight wiring connector . . .

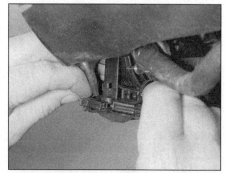

8.3b . . . and the sidelight wiring connector

Electrical system 9•7

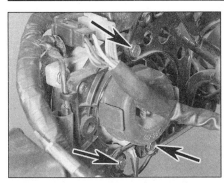

8.4a Unscrew the bolts (arrowed) . . .

8.4b . . . and remove the headlight

8.5 Remove the pivot screw (arrowed)

measuring the amount of thread projecting from the holder on each beam adjusting screw. Remove the beam unit pivot screw and open the pivot cover **(see illustration)**. Also remove the two headlight beam adjusting screws. Lift the beam unit off the holder, noting how the pivot locates, and remove the adjuster springs. Remove the captive nuts from the beam unit.

Installation

6 Installation is the reverse of removal. Make sure all the wiring is correctly connected and secured. Check the operation of the headlights and sidelights. Check the headlight aim (see Chapter 1).

L, N and R models

Removal

Note: *R models are not equipped with sidelight bulbs.*

7 Remove the fairing (see Chapter 8).
8 To remove an individual headlight, disconnect the headlight and sidelight wiring connectors **(see illustrations)**. Release the wiring from any clips or ties on the headlight. Carefully pull the top inner corner of the headlight away from the mounting bracket to release the pivot ball from its socket **(see illustration)**. Now slide the headlight sideways out of the bracket, pushing in on the adjuster springs to help release them from their slots in the bracket if necessary.
9 To remove the complete headlight assembly, disconnect the headlight and sidelight wiring connectors for both headlights **(see illustrations 8.8a and b)**. Release the

8.8a Disconnect the headlight wiring connector . . .

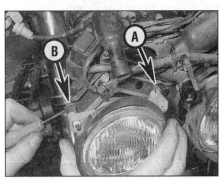

8.8c Pull the unit off the pivot (A) and slide it out to release the adjuster (B) from the bracket

wiring from any clips or ties on the headlight. Unscrew the three bolts securing the headlight bracket and remove the headlight assembly, noting how it fits **(see illustration)**.

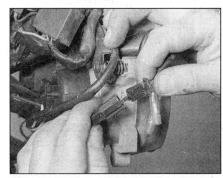

8.8b . . . and the sidelight wiring connector

8.9 Unscrew the three bolts (arrowed) and remove the assembly

Installation

10 Installation is the reverse of removal. Make sure all the wiring is correctly connected and secured. Check the operation of the headlights and sidelights. Check the headlight aim (see Chapter 1).

9 Brake/tail light bulbs – renewal

1 Remove (J and K models) or raise (L, N and R models) the passenger seat (see Chapter 8).
2 Turn the bulbholder anti-clockwise and withdraw it from the tail light **(see illustration)**.
3 Push the bulb into the holder and twist it anti-clockwise to remove it **(see illustration)**.

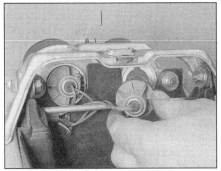

9.2 Release the bulbholder . . .

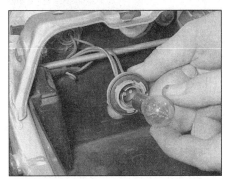

9.3 . . . and remove the bulb

9•8 Electrical system

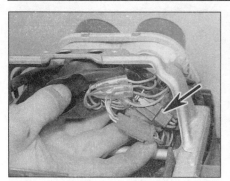

10.2a Tail light wiring connector (arrowed) – J and K models

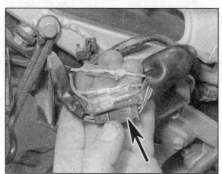

10.2b Tail light wiring connector (arrowed) – L, N and R models

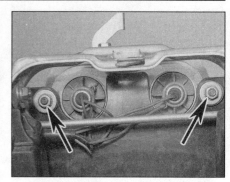

10.3 Unscrew the nuts (arrowed) and remove the tail light

Check the socket terminals for corrosion and clean them if necessary. Line up the pins of the new bulb with the slots in the socket, then push the bulb in and turn it clockwise until it locks into place. **Note:** *The pins on the bulb are offset so it can only be installed one way. It is a good idea to use a paper towel or dry cloth when handling the new bulb to prevent injury if the bulb should break and to increase bulb life.*
4 Install the bulbholder into the tail light and turn it clockwise to secure it.
5 Install the seat (see Chapter 8).

10 Tail light assembly – removal and installation

Removal

1 Remove the seat cowling (see Chapter 8).
2 Either draw back the rubber boot (in front of the tail light on J and K models, and on the right-hand side of the sub-frame on L, N and R models), and disconnect the tail light assembly wiring connector **(see illustrations)**, or turn the bulbholders anti-clockwise and withdraw them **(see illustration 9.2)**.
3 Unscrew the two nuts securing the tail light assembly and carefully remove it **(see illustration)**. If required, turn the bulbholders anti-clockwise and withdraw them from the tail light.

Installation

4 Installation is the reverse of removal. Check the operation of the tail light and the brake light.

11 Turn signal circuit – check

1 The battery provides power for operation of the turn signal lights, so if they do not operate, always check the battery voltage first. Low battery voltage indicates either a faulty battery or a defective charging system. Refer to Section 3 for battery checks and Sections 30 and 31 for charging system tests. Also, check the tail/signal fuse (see Section 5) and the switch (see Section 20).
2 Most turn signal problems are the result of a burned out bulb or corroded socket. This is especially true when the turn signals function properly in one direction, but fail to flash in the other direction. Check the bulbs and the sockets (see Section 12).
3 If the bulbs and sockets are good, remove the seat cowling (see Chapter 8) for access to the relay, which is mounted on the right-hand side on J and K models and the left-hand side on L, N and R models **(see illustrations)**. Disconnect the wiring connector and check for power at the turn signal relay black/brown wire (J, K, L, N models) or brown/wire wire (R model) with the ignition ON **(see illustration)**. Turn the ignition OFF when the check is complete.
4 If no power was present at the relay, check the wiring from the relay to the ignition (main) switch for continuity.
5 If power was present at the relay, using the appropriate wiring diagram at the end of this Chapter, check the wiring between the relay, turn signal switch and turn signal lights for continuity. If the wiring and switch are sound, renew the relay.

12 Turn signal bulbs – renewal

J and K models
Front

1 Remove either the air duct or the fairing side panel for the best access to the turn signal (see Chapter 8).
2 Turn the bulbholder anti-clockwise and withdraw it from the lens **(see illustration 13.2)**.
3 Push the bulb into the holder and twist it anti-clockwise to remove it. Check the socket terminals for corrosion and clean them if necessary. Line up the pins of the new bulb with the slots in the socket, then push the bulb in and turn it clockwise until it locks into place.
4 Fit the bulbholder back into the lens and turn it clockwise, making sure it is securely held.

Rear

5 Remove the screw securing the turn signal lens and remove the lens, noting how it fits

11.3a Turn signal relay (arrowed) – J and K models

11.3b Turn signal relay (arrowed) – L, N and R models

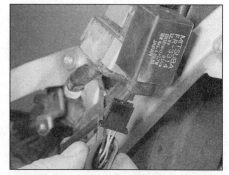

11.3c Disconnect the wiring connector and test as described

Electrical system

12.5 Remove the screw (arrowed) and detach the lens

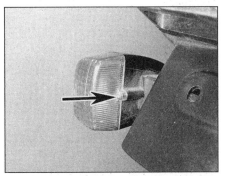

12.8 Remove the screw (arrowed) and detach the lens . . .

12.9 . . . then release the bulbholder . . .

(see illustration). Remove the rubber gasket if it is free, and discard it if it is damaged or deteriorated.
6 Push the bulb into the holder and twist it anti-clockwise to remove it. Check the socket terminals for corrosion and clean them if necessary. Line up the pins of the new bulb with the slots in the socket, then push the bulb in and turn it clockwise until it locks into place.
7 Fit the lens onto the holder. Use a new rubber gasket if required, and make sure it is properly seated and not pinched by the lens.

L, N and R models

8 Remove the screw securing the turn signal lens and remove the lens, noting how it fits **(see illustration)**.
9 Turn the bulbholder anti-clockwise and withdraw it from the lens **(see illustration)**.
10 Push the bulb into the holder and twist it anti-clockwise to remove it **(see illustration)**. Check the socket terminals for corrosion and clean them if necessary. Line up the pins of the new bulb with the slots in the socket, then push the bulb in and turn it clockwise until it locks into place. The front turn signals on the R model double as running lights and use dual filament bulbs; the pins on the bulb are offset so it can only be installed one way.
11 Fit the bulbholder back into the lens and turn it clockwise, making sure it is securely held.
12 Fit the lens onto the holder. Locate the clip on the side of the lens behind the tab in the housing **(see illustration)**.

13 Turn signal assemblies – removal and installation

Removal

1 Disconnect the turn signal wiring connectors. On the front turn signals, they are on the inside of the fairing (on J and K models, you can either disconnect the connector or

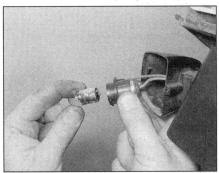

12.10 . . . and remove the bulb

remove the bulbholder) **(see illustrations)**. On the rear turn signals, remove the seat (J and K models) or the seat cowling (L, N and R

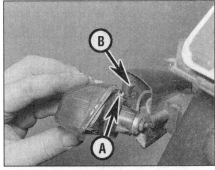

12.12 Locate the clip (A) behind the tab (B)

models) (see Chapter 8) – the connectors are inside a rubber boot which is secured by a clip **(see illustrations)**.

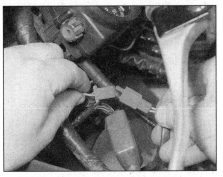

13.1a Front turn signal wiring connector – J and K models

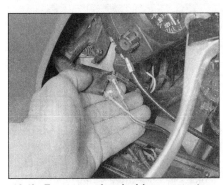

13.1b Front turn signal wiring connectors – L, N and R models

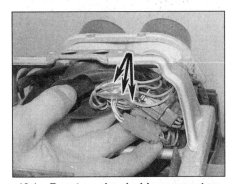

13.1c Rear turn signal wiring connectors (arrowed) – J and K models

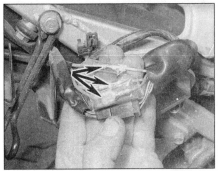

13.1d Rear turn signal wiring connectors (arrowed) – L, N and R models

9•10 Electrical system

13.2 Turn signal screw (arrowed)

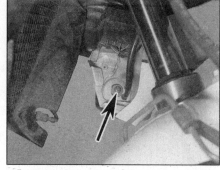

13.4a Front turn signal screw (arrowed)

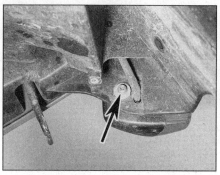

13.4b Rear turn signal screw (arrowed)

2 On J and K models, to remove a front turn signal, remove the screw securing the signal to the inside of the fairing and remove it, noting how it fits **(see illustration)**.

3 On J and K models, to remove a rear signal, either remove the screw securing the turn signal to its bracket, or remove the screw securing the bracket to the inside of the mudguard **(see illustration 13.4b)**. Remove the turn signal, taking care not to snag the wiring as you pull it through.

4 On L, N and R models, remove the screw securing the assembly to either the inside of the fairing or rear mudguard **(see illustrations)**. Remove the assembly, noting the arrangement of the mounting base and collar, and taking care not to snag the wiring as you pull it through.

Installation

5 Installation is the reverse of removal. Check the operation of the turn signals.

14 Brake light switches – check and replacement

Circuit check

1 Before checking any electrical circuit, check the bulb (see Section 9) and fuse (see Section 5).
2 Using a multimeter or test light connected to a good earth (ground), with the ignition ON check for voltage at one of the terminals on the brake light switch wiring connector(s) – there will be battery voltage on one terminal and zero on the other with the lever/pedal at rest **(see illustration 14.5 or 14.8)**. If there's no voltage present at either, check the wiring between the switch and the tail/signal fuse (see the *wiring diagrams* at the end of this Chapter).
3 If voltage is available, touch the probe of the test light to the other terminal of the switch, then pull the brake lever in or depress the brake pedal. If no reading is obtained or the test light doesn't light up, renew the switch.
4 If a reading is obtained or the test light does light up, yet the bulbs still do not come on, check the wiring between the switch and the brake light bulbs (see the *wiring diagrams* at the end of this Chapter).

Switch replacement

Front brake lever switch

5 The switch is mounted on the underside of the brake master cylinder. Disconnect the wiring connectors from the switch **(see illustration)**.
6 Remove the single screw securing the switch to the bottom of the master cylinder and remove the switch **(see illustration)**.
7 Installation is the reverse of removal. The switch isn't adjustable.

Rear brake pedal switch

8 The switch is mounted on the inside of the right-hand footrest bracket **(see illustration)**. Remove the seat cowling for access to the connector (see Chapter 8). Trace the wiring from the switch and disconnect it at the connector **(see illustrations)**.

14.5 Disconnect the wiring connectors (arrowed) . . .

14.6 . . . then remove the screw (arrowed) securing the switch

14.8a Rear brake light switch (arrowed)

14.8b Switch wiring connector (arrowed) – J and K models

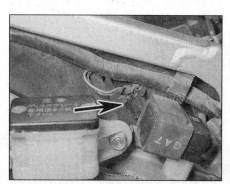

14.8c Switch wiring connector (arrowed) – L, N and R models

Electrical system

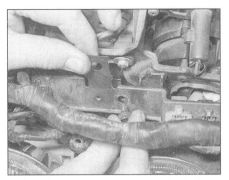

15.4 Remove the bolt and displace the bracket . . .

15.5a . . . then unscrew the bolts (arrowed) . . .

15.5b . . . and remove the cluster

9 Detach the lower end of the switch spring from the brake pedal, then unscrew and remove the switch. If access to the switch is too restricted, unscrew the two bolts securing the rider's right-hand footrest bracket, then swing the whole footrest/rear brake master cylinder assembly out, making sure no strain is placed on the brake and reservoir hoses.

10 Installation is the reverse of removal. If removed, tighten the footrest bracket bolts to the torque setting specified at the beginning of the Chapter. Make sure the brake light is activated just before the rear brake pedal takes effect. If adjustment is necessary, hold the switch and turn the adjusting ring on the switch body until the brake light is activated when required.

15 Instrument cluster and speedometer cable – removal and installation

15.8 Disconnect the wiring connectors . . .

15.9a . . . then unscrew the bolts (arrowed) . . .

illustration 6.3a). Unscrew the bolt securing the wiring connector bracket and draw it forward to expose the instrument cluster bolts (see illustration).
5 Unscrew the two bolts securing the instrument cluster and carefully remove it (see illustration).

Removal – L, N and R models
6 Remove the fairing (see Chapter 8).
7 Unscrew the knurled ring securing the speedometer cable to the back of the speedometer and detach the cable (see illustration 15.12).
8 Pull back the rubber boot above the headlight and disconnect the two black wiring connectors (see illustration).
9 Unscrew the two bolts securing the instrument cluster and carefully remove it (see illustration). Note how the two pegs locate in

the rubber grommets in the bracket (see illustration).

Installation – all models
10 Installation is the reverse of removal. Make sure that the speedometer cable and wiring connectors are correctly routed and secured.

Speedometer cable

Removal
11 Remove the fairing (see Chapter 8).
12 Unscrew the knurled ring securing the speedometer cable to the back of the speedometer and detach the cable (see illustration).
13 Remove the screw securing the lower end of the cable to the drive housing on the front sprocket cover and detach the cable (see illustration).

Instrument cluster

Removal – J and K models
1 Remove the fairing (see Chapter 8).
2 Unscrew the knurled ring securing the speedometer cable to the back of the speedometer and detach the cable (see illustration 15.12).
3 Disconnect the two black wiring connectors in the middle of the bracket above the headlights.
4 Displace the headlight relay (see

15.9b . . . and remove the cluster, noting how it locates (arrowed)

15.12 Unscrew the ring (arrowed) and detach the cable

15.13 Remove the screw (arrowed) and detach the cable

9•12 Electrical system

15.16 Fit the upper end into the speedometer and tighten the ring

15.17 Fit the lower end into the drive housing and tighten the screw

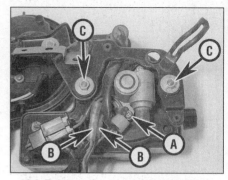

16.3 Remove the screw (A), disconnect the wiring connectors (B), then unscrew the nuts (C)

14 Withdraw the cable, releasing it from its guides, and remove it from the bike, noting its correct routing.

Installation

15 Route the cable up through its guides to the back of the instrument cluster.
16 Connect the cable upper end to the speedometer and tighten the retaining ring securely **(see illustration)**.
17 Connect the cable lower end to the drive housing and tighten the retaining screw securely **(see illustration)**.
18 Check that the cable doesn't restrict steering movement or interfere with any other components, then install the fairing (see Chapter 8).

16 Instruments – check and replacement

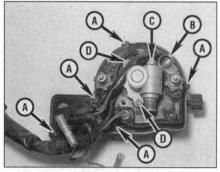

16.5 Remove the cover screws (A), the blanking cap (B), the screw (C) and the gearbox screws (D)

16.11 Remove the screw from the centre of the knob

3 Remove the screw securing the wiring clamp and disconnect the two bullet connectors, then unscrew the two nuts securing the speedometer/warning light assembly to the bracket and remove the assembly **(see illustration)**.
4 Using a very small Phillips screwdriver, remove the screw in the centre of the odometer trip knob and remove the knob **(see illustration 16.11)**.
5 Remove the five screws securing the speedometer/warning light assembly front cover and remove the cover **(see illustration)**.
6 Remove the blanking cap from the entry hole for the speedometer wiring **(see illustration 16.5)**.

7 Remove the remaining screw securing the speedometer to the casing **(see illustration 16.5)**.
8 Remove the two screws securing the speedometer gearbox and lift off the box **(see illustration 16.5)**. Carefully withdraw the speedometer from the front, feeding the wiring and its connector through the hole as you do.
9 Installation is the reverse of removal.

Replacement – L, N and R models

10 Remove the instrument cluster (see Section 15).
11 Using a very small Phillips screwdriver, remove the screw in the centre of the odometer trip knob and remove the knob **(see illustration)**.
12 Remove the screws securing the instrument cluster front cover, noting the positions of the wiring clips, and remove the cover **(see illustration)**.
13 Remove the two screws securing the speedometer gearbox and lift off the box **(see illustration)**.
14 Remove the blanking cap from the entry hole for the speedometer wiring **(see illustration 16.13)**. Disconnect the speedometer wiring connector **(see illustration 16.13)**.
15 Remove the two screws securing the speedometer to the casing **(see illustration 16.13)**. Carefully withdraw the speedometer from the front, feeding the wiring and its connector through the hole as you do.
16 Installation is the reverse of removal.

Speedometer

Check

1 Special instruments are required to properly check the operation of this meter. If it is believed to be faulty, take the speedometer to a dealer for assessment.

Replacement – J and K models

2 Remove the instrument cluster (see Section 15).

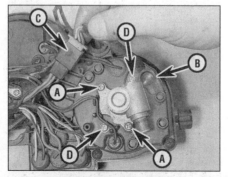

16.12 Remove the screws (arrowed) and the front cover

16.13 Gearbox screws (A), blanking cap (B), wiring connector (C), speedometer screws (D)

Electrical system 9•13

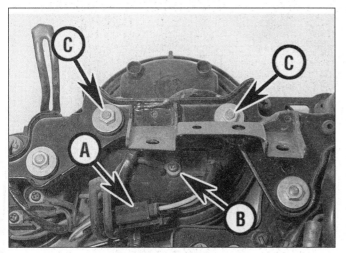

16.19 Disconnect the wiring connector (A), then remove the screw (B) and the nuts (C)

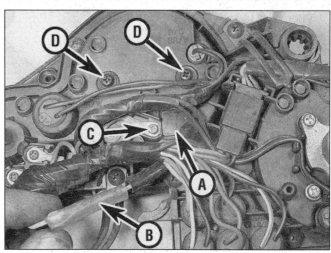

16.26 Remove the blanking cap (A), disconnect the wiring connector (B), remove the screw (C) and detach the wire, then remove the tachometer screws (D)

Tachometer

Check

17 Special instruments are required to properly check the operation of this meter. If it is believed to be faulty, take the tachometer to a dealer or automotive electrician for assessment.

Replacement – J and K models

18 Remove the instrument cluster (see Section 15).
19 Disconnect the tachometer wiring connector, and remove the screw securing the earth wire (see illustration).
20 Remove the two nuts and their washers securing the tachometer to the bracket (see illustration 16.19).
21 Carefully pull the tachometer light bulbholders out of the back of the tachometer, and withdraw the tachometer from the front.
22 Installation is the reverse of removal. Make sure the wires are correctly and securely connected.

Replacement – L, N and R models

23 Remove the instrument cluster (see Section 15).
24 Using a very small Phillips screwdriver, remove the screw in the centre of the odometer trip knob and remove the knob (see illustration 16.11).
25 Remove the screws securing the instrument cluster front cover, noting the positions of the wiring clips, and remove the cover (see illustration 16.12).
26 Remove the blanking cap from the entry hole for the tachometer wiring (see illustration). Disconnect the wiring connector. Also remove the screw securing the earth wire.
27 Remove the two screws securing the tachometer to the casing (see illustration 16.26). Carefully withdraw the tachometer

from the front, feeding the wiring and its connector through the hole as you do.
28 Installation is the reverse of removal.

Coolant temperature gauge

Check

29 See Chapter 3.

Replacement – J and K models

30 Remove the instrument cluster (see Section 15).
31 Remove the screw securing each wiring connector, making a note of which fits where (see illustration).
32 Carefully pull the light bulbholder out of the back of the temperature gauge (see illustration 16.31).
33 Remove the two nuts and their washers securing the temperature gauge to the bracket and carefully withdraw the gauge from the front (see illustration 16.31).
34 Installation is the reverse of removal. Make sure the wires are correctly and securely connected.

Replacement – L, N and R models

35 Remove the instrument cluster (see Section 15).

36 Using a very small Phillips screwdriver, remove the screw in the centre of the odometer trip knob and remove the knob (see illustrations 16.11).
37 Remove the screws securing the instrument cluster front cover, noting the positions of the wiring clips, and remove the cover (see illustration 16.12).
38 Remove the screw securing each wiring connector, making a note of which fits where, then carefully withdraw the temperature gauge from the front (see illustration).
39 Installation is the reverse of removal. Make sure the wires are correctly and securely connected.

17 Instrument and warning light bulbs – renewal

1 Remove the fairing (see Chapter 8).
2 Some of the bulbs are accessible with the instrument cluster in place, but access to others is quite restricted. If it is too restricted for the bulb you are changing, displace the instrument cluster (see Section 15) to improve

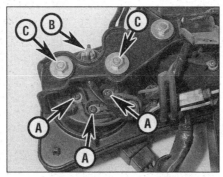

16.31 Remove the screws (A) and detach the wires, then pull out the bulbholder (B) and unscrew the nuts (C)

16.38 Remove the screws (arrowed) and detach the wires

9•14 Electrical system

17.3a Pull out the bulbholder...

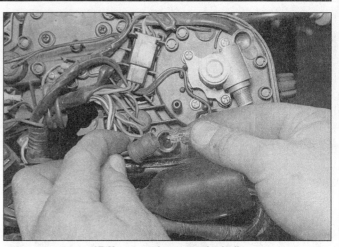

17.3b ...and remove the bulb

access – there is no need to disconnect any of the wiring connectors, though it may be necessary to disconnect the speedometer cable, depending on which bulb is being renewed.

3 Gently pull the bulbholder out of the back of the instrument casing, then pull the bulb out of the bulbholder **(see illustrations)**.

4 If the socket contacts are dirty or corroded, scrape them clean and spray with electrical contact cleaner before a new bulb is installed.

5 Carefully push the new bulb into the holder, then push the holder into the casing.

6 Install the fairing (see Chapter 8).

18 Oil pressure switch – check, removal and installation

Check

1 The oil pressure warning light should come on when the ignition (main) switch is turned ON and extinguish a few seconds after the engine is started. If the oil pressure warning light comes on whilst the engine is running, stop the engine immediately and carry out an oil level check (see *Daily (pre-ride) checks*), and if the level is correct, an oil pressure check (see Chapter 1).

2 If the oil pressure warning light does not come on when the ignition is turned on, check the bulb (see Section 17) and tail/signal fuse (see Section 5).

3 The oil pressure switch is screwed into the crankcase behind the cylinders and is accessed by removing the fuel tank and the carburettors (see Chapter 4). Pull the rubber cover off the switch and remove the screw securing the wiring connector **(see illustrations)**. With the ignition switched ON, earth (ground) the wire on the crankcase and check that the warning light comes on. If the light comes on, the switch is defective and must be renewed.

4 If the light still does not come on, check for voltage at the wire terminal. If there is no voltage present, check the wire between the switch, the instrument cluster and tail/signal fuse for continuity (see the *wiring diagrams* at the end of this Chapter).

5 If the warning light comes on whilst the engine is running, yet the oil pressure is satisfactory, remove the wire from the oil pressure switch. With the wire detached and the ignition switched ON the light should be out. If it is illuminated, the wire between the switch and instrument cluster must be earthed (grounded) at some point. If the wiring is good, the switch must be assumed faulty and renewed.

Removal

6 Remove the fuel tank and the carburettors (see Chapter 4).

7 Pull the rubber cover off the switch and remove the screw securing the wiring connector **(see illustrations 18.3a and b)**.

8 Unscrew the oil pressure switch and withdraw it from the crankcase.

Installation

9 Apply a suitable sealant to the upper

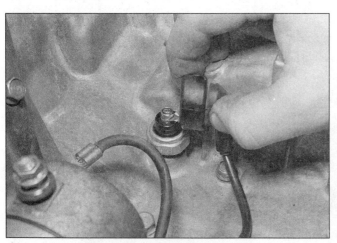

18.3a Pull back the rubber cover...

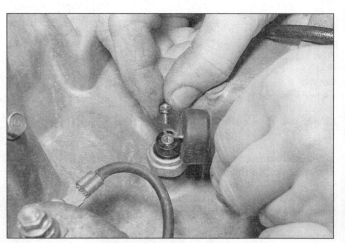

18.3b ...then remove the terminal screw and detach the wiring

Electrical system 9•15

19.1 Ignition switch wiring connector (arrowed) – L, N and R models

portion of the switch threads near the switch body, leaving the bottom 3 to 4 mm of thread clean. Install the switch in the crankcase and tighten it to the torque setting specified at the beginning of the Chapter.

10 Attach the wiring connector and secure it with the screw, then fit the rubber cover **(see illustrations 18.3b and a)**.

11 Install the carburettors and fuel tank (see Chapter 4). Run the engine and check that the switch operates correctly.

19 Ignition (main) switch – check, removal and installation

 Warning: To prevent the risk of short circuits, remove the rider's seat (see Chapter 8) and disconnect the battery negative (-ve) lead before making any ignition (main) switch checks.

Check

1 Remove the fairing (see Chapter 8). Trace the ignition (main) switch wiring back from the base of the switch and disconnect it at the black connector. On J and K models, the connector is on the right-hand end of the bracket above the headlights. On L, N and R models, the connector is in the bracket on the left-hand side of the bike **(see illustration)**.

2 Using an ohmmeter or a continuity tester, check the continuity of the connector terminal pairs (see the *wiring diagrams* at the end of this Chapter). Insert the key in the switch. Continuity should exist between the terminals connected by a solid line on the diagram when the switch is in the indicated position.

3 If the switch fails any of the tests, renew it.

Removal

4 Remove the fairing (see Chapter 8). Trace the ignition (main) switch wiring back from the base of the switch and disconnect it at the black connector. On J and K models, the connector is on the right-hand end of the bracket above the headlights. On L, N and R models, the connector is in the bracket on the left-hand side of the bike **(see illustration 19.1)**. Draw the wiring through to the switch, releasing it from any clips or ties and noting its routing.

5 Two Torx bolts mount the ignition switch to the underside of the top yoke **(see illustration)**. Ease of access to these bolts depends on the tools available. It may be necessary to either displace or remove the instrument cluster (see Section 15), or to displace the top yoke (see Chapter 6).

6 Remove the bolts and withdraw the switch from the top yoke.

Installation

7 Installation is the reverse of removal. Tighten the Torx bolts to the torque setting specified at the beginning of the Chapter. Make sure the wiring is securely connected and correctly routed.

20 Handlebar switches – check

1 Generally speaking, the switches are reliable and trouble-free. Most troubles, when they do occur, are caused by dirty or corroded contacts, but wear and breakage of internal parts is a possibility that should not be overlooked. If breakage does occur, the entire switch and related wiring harness will have to be renewed as individual parts are not available.

2 The switches can be checked for continuity using an ohmmeter or a continuity test light. Always disconnect the battery negative (-ve) lead, which will prevent the possibility of a short circuit, before making the checks.

3 Remove the fairing (see Chapter 8). Trace the wiring harness of the switch in question back to its connector(s) and disconnect it/them. On J and K models, the right-hand switch connector is the red one in the bracket above the headlight, and the left-hand switch connectors (one black, one white) are on the left-hand end of the bracket. On L, N and R models, the connectors are in the bracket on the left-hand side of the bike **(see illustration)**.

4 Check for continuity between the terminals of the switch harness with the switch in the various positions (ie switch off – no continuity, switch on – continuity) – see the *wiring diagrams* at the end of this Chapter.

5 If the continuity check indicates a problem exists, refer to Section 21, remove the switch and spray the switch contacts with electrical contact cleaner. If they are accessible, the contacts can be scraped clean with a knife or polished with crocus cloth. If switch components are damaged or broken, it will be obvious when the switch is disassembled.

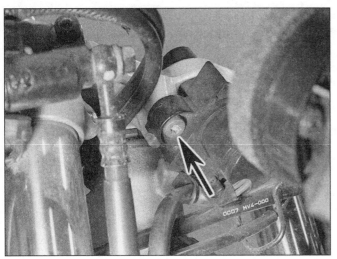

19.5 Access the bolts from the underside of the switch (arrowed)

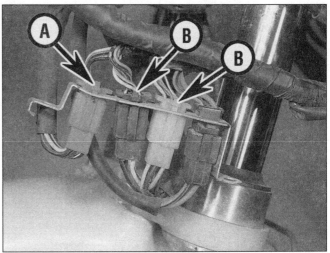

20.3 Right-hand switch wiring connector (A), left-hand switch wiring connectors (B) – L, N and R models

21.3a Right-hand switch screws (arrowed)

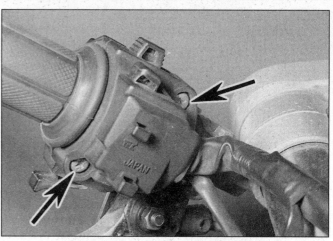

21.3b Left-hand switch screws (arrowed)

21 Handlebar switches – removal and installation

Removal

1 If the switch is to be removed from the bike, rather than just displaced from the handlebar, remove the fairing (see Chapter 8), then trace the wiring harness of the switch in question back to its connector(s) and disconnect it/them. On J and K models, the right-hand switch connector is the red one in the bracket above the headlight, and the left-hand switch connectors (one black, one white) are on the left-hand end of the bracket. On L, N and R models, the connectors are in the bracket on the left-hand side of the bike **(see illustration 20.3)**. Work back along the harness, freeing it from all the relevant clips and ties, noting its correct routing.

2 Disconnect the two wires from the front brake light switch (if removing the right-hand switch) or the clutch switch (if removing the left-hand switch) **(see illustration 14.5 or 24.2)**.

3 Unscrew the two handlebar switch screws and free the switch from the handlebar by separating the halves **(see illustrations)**.

Installation

4 Installation is the reverse of removal. Make sure the locating pin in the lower half of the switch locates in the hole in the underside of the handlebar.

22 Neutral switch – check, removal and installation

Check

1 Before checking the electrical circuit, check the bulb (see Section 17) and fuse (see Section 5).

2 The switch is located in the left-hand side of the engine below the front sprocket cover. On J and K models remove the lower fairing (see Chapter 8). On L, N and R models the fairing does not restrict access, but removing it gives more room. Make sure the transmission is in neutral. Detach the wiring connector from the switch **(see illustration overleaf)**.

3 With the connector disconnected and the ignition switched ON, the neutral light should be out. If not, the wire between the connector and instrument cluster must be earthed (grounded) at some point.

4 Check for continuity between the switch terminal and the crankcase. With the transmission in neutral, there should be continuity. With the transmission in gear, there should be no continuity. If the tests prove otherwise, then the switch is faulty.

5 If the continuity tests prove the switch is good, check for voltage at the wire terminal using a test light. If there's no voltage present, check the wire between the switch, instrument cluster and fusebox (see the *wiring diagrams* at the end of this Chapter).

6 On L, N and R models note that a diode is located in the power supply from the instrument cluster to the switch (see Section 25). Trace the light green wire from the switch up to the diode. Disconnect the diode from the wiring harness. Using an ohmmeter or continuity tester connect its probes between the light green and light green/red wire terminals on the diode, then reverse the probes. Continuity should only exist in one direction (as indicated on the body of the diode). If the diode shows the same condition in each direction it should be renewed.

Removal

7 The switch is located in the left-hand side of the engine below the front sprocket cover. On J and K models remove the lower fairing (see Chapter 8). On L, N and R models the fairing does not restrict access, but removing it gives more room.

8 Detach the wiring connector from the switch **(see illustration 22.2)**.

9 Unscrew the switch and withdraw it from the transmission casing.

Installation

10 Apply a smear of sealant to the threads of the switch, taking care not to cover the contact point.

11 Install the switch and tighten it to the torque setting specified at the beginning of the Chapter.

12 Check the operation of the neutral light.

13 Install the lower fairing (see Chapter 8).

23 Sidestand switch (L, N and R models) – check and replacement

Check

1 The sidestand switch is mounted on the back of the sidestand. The switch is part of the safety circuit which prevents or stops the engine running if the transmission is in gear whilst the sidestand is down, and prevents the engine from starting if the transmission is in gear unless the sidestand is up, and unless the clutch is pulled in. Before checking the electrical circuit, check the warning bulb (see Section 17) and the tail/signal fuse (see Section 5).

2 Remove the fuel tank (see Chapter 4). Trace the wiring back from the switch to its connector and disconnect it **(see illustration)**. Place the bike on an auxiliary stand or have an assistant hold it upright whilst the checks are made.

3 Check the operation of the switch using an ohmmeter or continuity test light. Connect the

Electrical system 9•17

22.2 Disconnect the wiring connector from the neutral switch

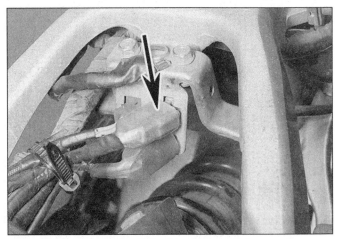

23.2 Sidestand switch wiring connector (arrowed)

meter to the green/white and green wires on the switch side of the connector. With the sidestand up there should be continuity (zero resistance) between the terminals, and with the stand down there should be no continuity (infinite resistance). Now connect the meter to the yellow/black and green wires on the switch side of the connector. With the sidestand down there should be continuity (zero resistance) between the terminals, and with the stand up there should be no continuity (infinite resistance).

4 If the switch does not perform as expected, it is defective and must be renewed.

5 If the switch is good, check the wiring between the various components in the starter safety circuit (see the *wiring diagrams* at the end of this book).

Replacement

6 The sidestand switch is mounted on the back of the sidestand (see illustration). Remove the lower fairing (see Chapter 8) and the fuel tank (see Chapter 4). Trace the wiring back from the switch to its connector and disconnect it (see illustration 23.2). Work back along the switch wiring, freeing it from any relevant retaining clips and ties, noting its correct routing.

7 Unscrew the switch bolt and remove the switch from the stand, noting how it fits.

8 Fit the new switch onto the sidestand, making sure the pin locates in the hole in the sidestand, and the lug for the spring on the stand bracket locates into the cutout in the switch body (see illustration 23.6). Tighten the bolt to the torque setting specified at the beginning of the Chapter.

9 Make sure the wiring is correctly routed up to the connector and retained by all the necessary clips and ties.

10 Reconnect the wiring connector and check the operation of the sidestand switch.

11 Install the fuel tank (see Chapter 4) and the lower fairing (see Chapter 8).

24 Clutch switch – check and replacement

Check

1 The clutch switch is housed in the clutch lever bracket. On J and K models the clutch switch prevents the engine from being started whilst in gear unless the clutch lever is pulled in. On L, N and R models the clutch switch prevents the engine from being started unless the clutch lever is pulled in and the sidestand is up.

2 To check the switch, disconnect the wiring connectors from the switch (see illustration). Connect the probes of an ohmmeter or a continuity test light to the two switch terminals. With the clutch lever pulled in, continuity should be indicated. With the clutch lever out, no continuity (infinite resistance) should be indicated. Reconnect the wiring connectors.

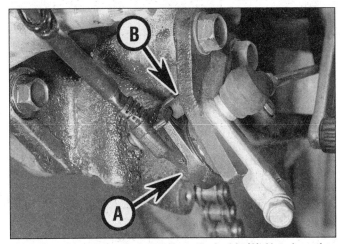

23.6 The sidestand switch bolt is on the inside (A). Note how the lug locates (B)

24.2 Clutch switch wiring connectors

9•18 Electrical system

25.2 Diode (arrowed) – J and K models

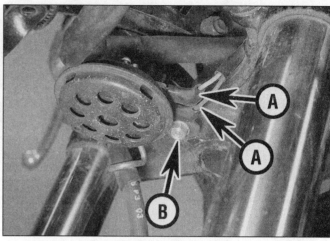

26.5 Disconnect the wiring connectors (A), then unscrew the bolt (B)

3 If the switch is good, check for voltage at one of the terminals on the clutch switch wiring connectors with the ignition ON – there will be battery voltage on one terminal and zero on the other. If voltage is indicated, check the other components in the starter circuit as described in the relevant sections of this Chapter. If no voltage is indicated, or if all components are good, check the wiring between the various components (see the *wiring diagrams* at the end of this book).

Replacement

4 Remove the clutch lever (see Chapter 6).
5 Disconnect the wiring connectors from the clutch switch **(see illustration 24.2)**. Using a small screwdriver, push the switch from the connector end and withdraw it from inside the bracket.
6 Installation is the reverse of removal. Make sure the ridge on the top of the switch locates in the cutout in the lever bracket, and push the switch fully home.

25 Diode – check and replacement

Check

1 Remove the fuel tank (see Chapter 4).
2 The diode is a small block that plugs into a connector in the main wiring harness **(see illustration)**. The diode is part of the safety circuit which prevents or stops the engine running if the transmission is in gear whilst the sidestand is down, and prevents the engine from starting if the transmission is in gear unless the sidestand is up and the clutch lever is pulled in.
3 Disconnect the diode from the harness **(see illustration 25.2)**.
4 On J and K models (two-pin diode), using an ohmmeter or continuity tester, connect its probes between the two terminals on the diode, then reverse the probes. The diode should only show continuity in one direction (as indicated by the symbol on the body of the diode). If the diode shows continuity or no continuity in both directions it should be renewed.
5 On L, N and R models (three-pin diode), using an ohmmeter or continuity tester, connect one probe to one of the outer terminals of the diode and the other probe to the middle terminal of the diode (the terminal which connects to the light green wire in the wire harness). Now reverse the probes. The diode should only show continuity in one direction (as indicated by the symbol on the body of the diode). If the diode shows continuity or no continuity in both directions it should be renewed. Repeat the tests between the other outer terminal and the middle terminal. The same results should be achieved. If it doesn't behave as stated, renew the diode.
6 If the diode is good, check the other components in the starter circuit as described in the relevant sections of this Chapter. If all components are good, check the wiring between the various components (see the *wiring diagrams* at the end of this book).

Replacement

7 Remove the fuel tank (see Chapter 4).
8 The diode is a small block that plugs into a connector in the main wiring harness **(see illustration 25.2)**. Disconnect the diode from the harness and connect the new one.

26 Horn – check and replacement

Check

1 The horn is mounted on the underside of the bottom yoke. If required, remove the fairing for improved access (see Chapter 8).
2 Disconnect the wiring connectors from the horn **(see illustration 26.5)**. Using two jumper wires, apply battery voltage directly to the terminals on the horn. If the horn sounds, check the switch (see Section 21) and the wiring between the switch and the horn (see the *wiring diagrams* at the end of this Chapter).
3 If the horn doesn't sound, renew it.

Replacement

4 The horn is mounted on the underside of the bottom yoke. If required, remove the fairing for improved access (see Chapter 8).
5 Disconnect the wiring connectors from the horn, then unscrew the bolt securing the horn and remove it from the bike **(see illustration)**.
6 Install the horn and securely tighten the bolt. Connect the wiring connectors to the horn.

27 Starter relay – check and replacement

Check

1 If the starter circuit is faulty, first check the main fuse and ignition/starter fuse (see Section 5).
2 Remove the seat cowling (see Chapter 8). The starter relay is located behind the seat cowling on the right-hand side of the bike on J and K models, and on the left on L, N and R models **(see illustration)**.
3 Disconnect the relay wiring connector to provide access to the rear terminals, then lift the rubber terminal cover and unscrew the bolt securing the starter motor lead **(see illustrations)**; position the lead well away from the relay terminal. Reconnect the wiring connector. With the ignition switch ON, the engine kill switch in the RUN position, the transmission in neutral and the clutch pulled

Electrical system

27.2 Starter relay – J and K models shown

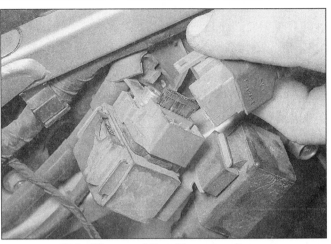

27.3a Disconnect the wiring connector . . .

in, press the starter switch. The relay should be heard to click.

4 If the relay doesn't click, switch off the ignition and remove the relay as described below; test it as follows.

5 Set a multimeter to the ohms x 1 scale and connect it across the relay's starter motor and battery lead terminals **(see illustration 27.3b)**. Using a fully-charged 12 volt battery and two insulated jumper wires, connect the positive (+ve) terminal of the battery to the yellow/red wire terminal of the relay, and the negative (-ve) terminal to the green/red wire terminal of the relay. At this point the relay should be heard to click and the multimeter read 0 ohms (continuity). If this is the case the relay is proved good. If the relay does not click when battery voltage is applied and indicates no continuity (infinite resistance) across its terminals, it is faulty and must be renewed.

6 If the relay is good, check for battery voltage between the yellow/red wire and the green/red wire when the starter button is pressed. Check the other components in the starter circuit as described in the relevant sections of this Chapter. If all components are good, check the wiring between the various components (see the *wiring diagrams* at the end of this book).

Replacement

7 Remove the seat cowling (see Chapter 8). The starter relay is located behind the seat cowling on the right-hand side of the bike on J and K models, and on the left on L, N and R models **(see illustration 27.2)**.

8 Disconnect the battery terminals, remembering to disconnect the negative (-ve) terminal first.

9 Disconnect the relay wiring connector, then unscrew the two bolts securing the starter motor and battery leads to the relay and detach the leads **(see illustrations 27.3a and b)**. Remove the relay with its rubber sleeve from its mounting lug on the frame.

10 Installation is the reverse of removal. Make sure the terminal nuts are securely tightened. Connect the negative (-ve) lead last when reconnecting the battery.

28 Starter motor – removal and installation

Removal

1 Remove the rider's seat (see Chapter 8). Disconnect the battery negative (-ve) lead. The starter motor is mounted on the crankcase behind the cylinder block.

2 On J and K models, remove the carburettors (see Chapter 4). On L, N and R models, either remove the carburettors or displace the fuel pump according to preference and the tools available for accessing the starter motor bolts (see Chapter 4).

3 Peel back the rubber terminal cover, then remove the nut securing the starter lead to the motor and detach the lead **(see illustration)**.

4 Unscrew the two bolts securing the starter motor to the crankcase, noting the earth lead attached to the front bolt **(see illustration)**.

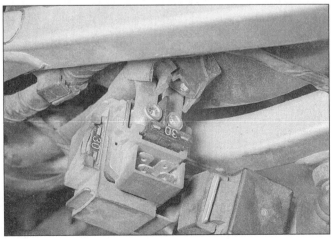

27.3b . . . then trace the lead from the starter motor and detach it from the relay by removing its bolt

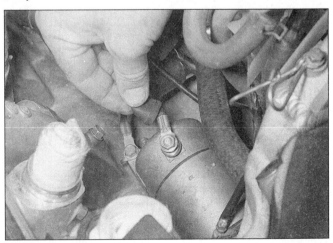

28.3 Pull back the cover, unscrew the nut and detach the lead

9•20 Electrical system

28.4 Unscrew the two bolts (arrowed), noting the lead

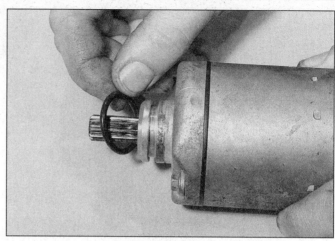

28.6 Fit a new O-ring . . .

28.7a . . . then install the starter motor

28.7b Do not forget to secure the earth lead with the front bolt

Slide the starter motor out from the crankcase and remove it from the machine **(see illustration 28.7a)**.
5 Remove the O-ring on the end of the starter motor and discard it as a new one must be used **(see illustration 28.6)**.

Installation

6 Install a new O-ring on the end of the starter motor and ensure it is seated in its groove **(see illustration)**. Apply a smear of engine oil to the O-ring to aid installation.
7 Manoeuvre the motor into position and slide it into the crankcase **(see illustration)**. Ensure that the starter motor teeth mesh correctly with those of the starter idle/reduction gear. Install the mounting bolts, not forgetting to fit the earth lead with the front bolt, and tighten them securely **(see illustration)**.

8 Connect the starter lead to the motor and secure it with the nut **(see illustration 28.3)**. Make sure the rubber cover is correctly seated over the terminal.
9 Connect the battery negative (-ve) lead and install the carburettors or fuel pump and the rider's seat.

29 Starter motor – disassembly, inspection and reassembly

Disassembly

1 Remove the starter motor (see Section 28).
2 Note the alignment marks between the main housing and the front and rear covers, or make your own if they aren't clear **(see illustration)**.
3 Unscrew the two long bolts, noting how the

29.2 Note the alignment marks between the housing and the covers (arrowed)

Electrical system 9•21

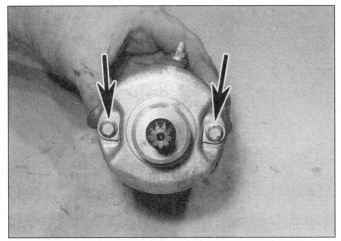

29.3 Unscrew and remove the two bolts (arrowed) . . .

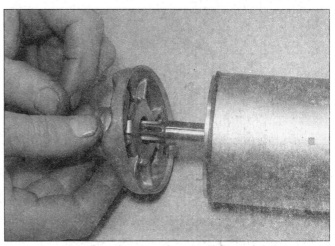

29.4a . . . then remove the front cover . . .

29.4b . . . the shims . . .

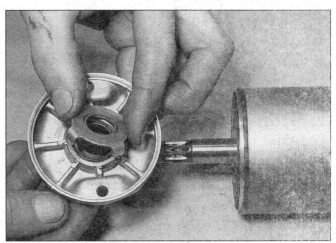

29.4c . . . and the tabbed washer

washers locate, and withdraw them from the starter motor **(see illustration)**. Discard their O-rings as new ones must be used.

4 Wrap some insulating tape around the teeth on the end of the starter motor shaft – this will protect the oil seal from damage as the front cover is removed. Remove the front cover from the motor **(see illustration)**. Remove the cover O-ring from the main housing and discard it as a new one must be used. Remove the shims from the front end of the armature shaft or the inside of the front cover, noting their correct fitted locations **(see illustration)**. Also remove the tabbed thrust washer from the front cover **(see illustration)**.

5 Remove the rear cover and brushplate assembly from the motor **(see illustration)**. Remove the cover O-ring from the main housing and discard it as a new one must be used. Remove the shims from the rear end of the armature shaft or from inside the rear

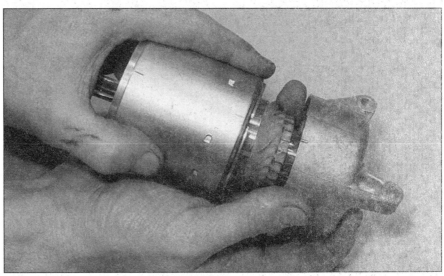

29.5 Remove the rear cover

9•22 Electrical system

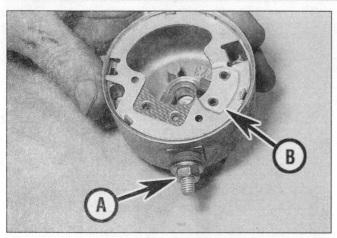

29.7 Unscrew the nut (A) and remove the washers, then remove the brushplate assembly (B)

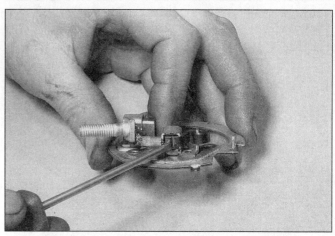

29.8 Lift the brush springs and withdraw the brushes

cover after the brushplate assembly has been removed.
6 Withdraw the armature from the main housing.
7 Noting the correct fitted location of each component, unscrew the terminal nut and remove it along with its washer, the insulating washers and O-ring **(see illustration)**. Withdraw the terminal bolt and brushplate assembly from the rear cover.
8 Lift the brush springs and slide the brushes out from their holders **(see illustration)**.

Inspection

9 The parts of the starter motor that are most likely to require attention are the brushes **(see illustration)**. Measure the length of the brushes and compare the results to the brush length listed in this Chapter's Specifications **(see illustration)**. If any of the brushes are worn beyond the service limit, renew the brush assembly. If the brushes are not worn excessively, nor cracked, chipped, or otherwise damaged, they may be re-used.

10 Inspect the commutator bars on the armature for scoring, scratches and discoloration. The commutator can be cleaned and polished with crocus cloth, but do not use sandpaper or emery paper. After cleaning, wipe away any residue with a cloth soaked in electrical system cleaner or denatured alcohol.
11 Using an ohmmeter or a continuity test light, check for continuity between the commutator bars **(see illustration)**. Continuity should exist between each bar and all of the others. Also, check for continuity between the commutator bars and the armature shaft **(see illustration)**. There should be no continuity (infinite resistance) between the commutator and the shaft. If the checks indicate otherwise, the armature is defective.
12 Check for continuity between each brush and the terminal bolt. There should be continuity (zero resistance). Check for continuity between the terminal bolt and the housing (when assembled). There should be no continuity (infinite resistance).
13 Check the front end of the armature shaft for worn, cracked, chipped and broken teeth. If the shaft is damaged or worn, renew the armature.
14 Inspect the end covers for signs of cracks or wear. Inspect the magnets in the main housing and the housing itself for cracks.
15 Inspect the insulating washers and front cover oil seal for signs of damage and renew them if necessary.

Reassembly

16 Slide the brushes back into position in their holders and place the brush spring ends onto the brushes **(see illustration)**.
17 Ensure that the inner rubber insulator is in place on the terminal bolt, then insert the bolt through the rear cover and fit the brushplate assembly in the rear cover, making sure its tab is correctly located in the slot in the cover **(see illustrations)**. Fit the O-ring and the

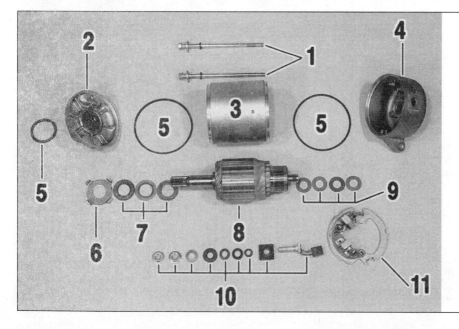

29.9a Starter motor components

1 Long bolts
2 Front cover
3 Main housing
4 Rear cover
5 O-rings
6 Tabbed washer
7 Front shims
8 Armature
9 Rear shims
10 Terminal bolt assembly
11 Brushplate assembly

Electrical system 9•23

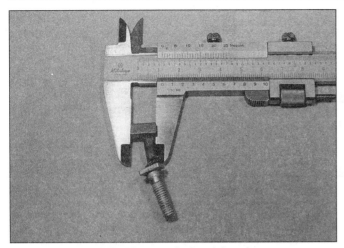

29.9b Measure the brush length

29.11a Continuity should exist between the commutator bars

29.11b There should be no continuity between the commutator bars and the armature shaft

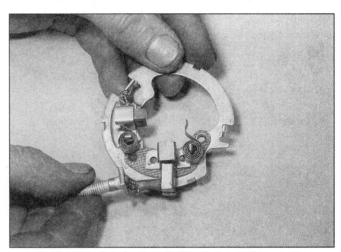

29.16 Fit the brushes into the brushplate and locate the springs . . .

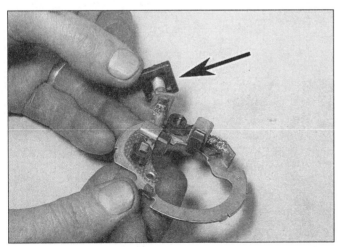

29.17a . . . then fit the inner insulator (arrowed) onto the terminal bolt

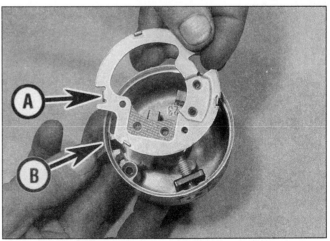

29.17b Fit the brushplate assembly into the rear cover, aligning the tab (A) with the slot (B) . . .

9•24 Electrical system

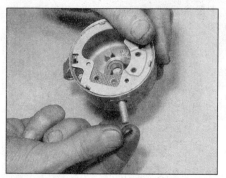

29.17c ... then fit the O-ring ...

29.17d ... the insulating washers ...

29.17e ... the plain washer and the nut

insulating washers over the terminal, then fit the standard washer and the nut **(see illustrations)**.

18 Slide the shims onto the rear end of the armature shaft, then lubricate the shaft with a drop of oil **(see illustration)**. Insert the armature into the rear cover, locating the brushes on the commutator bars as you do, taking care not to damage them **(see illustration)**. Check that each brush is securely pressed against the commutator by its spring and is free to move easily in its holder.

19 Fit a new O-ring onto the main housing, then fit the housing over the armature and onto the rear cover, aligning the marks made on removal **(see illustrations)**.

20 Apply a smear of grease to the lips of the front cover oil seal and fit a new O-ring onto

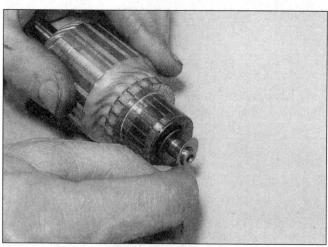

29.18a Fit the shims ...

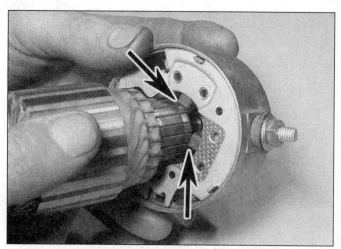

29.18b ... then insert the armature into the rear cover, making sure the brush ends (arrowed) locate correctly

29.19a Fit a new O-ring ...

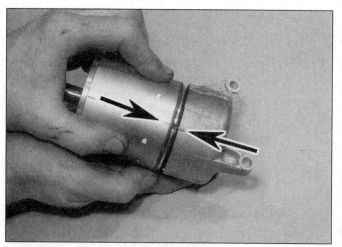

29.19b ... then install the main housing, aligning the marks (arrowed)

29.20a Fit a new O-ring onto the main housing . . .

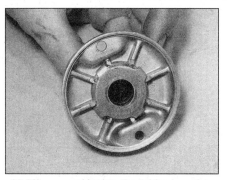

29.20b . . . and locate the tabbed washer in the front cover

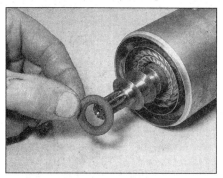

29.20c Fit the shims onto the shaft . . .

the front of the main housing **(see illustration)**. Fit the tabbed washer onto the cover, making sure the tabs locate correctly **(see illustration)**. Slide the shims onto the front of the armature shaft, then install the cover, aligning the marks made on removal **(see illustrations)**. Remove the protective tape from the shaft end.

21 Slide a new O-ring onto each of the long bolts. Check the marks made on removal are correctly aligned, then install the long bolts and tighten them securely, making sure the flat edge on each washer is against the raised section on the front cover **(see illustrations)**.

22 Install the starter motor (see Section 28).

30 Charging system testing – general information and precautions

1 If the performance of the charging system is suspect, the system as a whole should be checked first, followed by testing of the individual components. **Note:** *Before beginning the checks, make sure the battery is fully charged and that all system connections are clean and tight.*

2 Checking the output of the charging system and the performance of the various components within the charging system requires the use of a multimeter (with voltage, current and resistance checking facilities).

3 When making the checks, follow the procedures carefully to prevent incorrect connections or short circuits, as irreparable damage to electrical system components may result if short circuits occur.

4 If a multimeter is not available, the job of checking the charging system should be left to a dealer or automotive electrician.

31 Charging system – leakage and output test

1 If the charging system of the machine is thought to be faulty, remove the seat cowling (see Chapter 8) and perform the following checks.

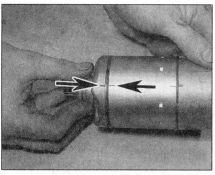

29.20d . . . then install the cover, aligning the marks (arrowed)

Leakage test

Caution: Always connect an ammeter in series, never in parallel with the battery, otherwise it will be damaged. Do not turn the ignition ON or operate the starter motor when the ammeter is connected – a sudden surge in current will blow the meter's fuse.

2 Turn the ignition switch OFF and disconnect the lead from the battery negative (-ve) terminal.

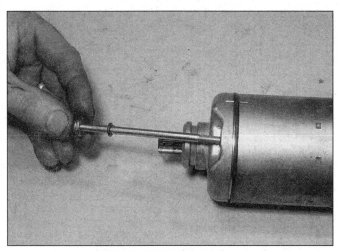

29.21a Install the long bolts . . .

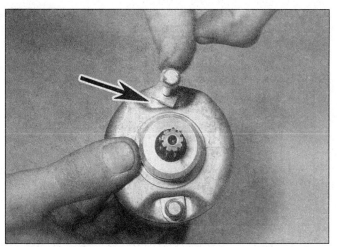

29.21b . . . locating the flat of each washer against the raised edge (arrowed)

9•26 Electrical system

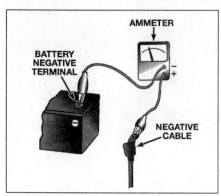

31.3 Checking the charging system leakage rate – connect the meter as shown

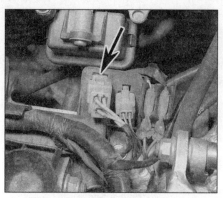

32.2a Alternator wiring connector (arrowed) – J and K models

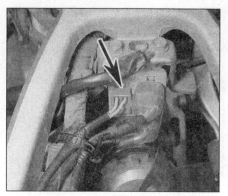

32.2b Alternator wiring connector (arrowed) – L, N and R models

3 Set the multimeter to the Amps function and connect its negative (-ve) probe to the battery negative (-ve) terminal, and positive (+ve) probe to the disconnected negative (-ve) lead **(see illustration)**. Always set the meter to a high amps range initially and then bring it down to the mA (milli Amps) range; if there is a high current flow in the circuit it may blow the meter's fuse.

4 If the current leakage indicated exceeds the amount specified at the beginning of the Chapter, there is probably a short circuit in the wiring. Disconnect the meter and reconnect the negative (-ve) lead to the battery, tightening it securely.

5 If leakage is indicated, use the wiring diagrams at the end of this book to systematically disconnect individual electrical components and repeat the test until the source is identified.

Output test

6 Start the engine and warm it up to normal operating temperature, then stop the engine. Remove the seat cowling (see Chapter 8).

7 Start the engine and allow it to idle. Connect a multimeter set to the 0 – 20 volts DC scale (voltmeter) across the terminals of the battery (positive (+ve) lead to battery positive (+ve) terminal, negative (-ve) lead to battery negative (-ve) terminal). Slowly increase the engine speed to 3000 rpm and note the reading obtained, then stop the engine and turn the ignition OFF; do not allow the engine to overheat. The regulated voltage should be as specified at the beginning of the Chapter. If the voltage is outside these limits, check the alternator and the regulator (see Sections 32 and 33).

8 To check the current output, disconnect the starter relay wiring connector **(see illustration 27.3a)** and remove the main fuse, then reconnect the connector. Connect a multimeter set to the 0 – 20 amps DC scale (ammeter) between the terminals of the main fuse (positive (+ve) lead to the left-hand terminal, negative (-ve) lead to right-hand terminal), inserting the probes into the base of the fuse sockets. Start the engine and allow it to idle, then slowly increase the engine speed to 3000 rpm and note the reading obtained. The regulated current should be as specified at the beginning of the Chapter. Stop the engine and turn the ignition OFF. If the current is outside these limits, check the alternator and the regulator (see Sections 32 and 33).

 Clues to a faulty regulator are constantly blowing bulbs, with brightness varying considerably with engine speed, and battery overheating.

32 Alternator – check, removal and installation

Check

1 Remove the fuel tank (see Chapter 4).

2 Trace the wiring back from the top of the alternator/clutch cover on the right-hand side of the engine and disconnect it at the white (J and K models) or red (L, N and R models) connector containing the three yellow wires **(see illustrations)**.

3 Using a multimeter set to the ohms x 1 (ohmmeter) scale measure the resistance between each of the yellow wires on the alternator side of the connector, taking a total of three readings, then check for continuity between each terminal and ground (earth). If the stator coil windings are in good condition the three readings should be within the range shown in the Specifications at the start of this Chapter and there should be no continuity (infinite resistance) between any of the terminals and earth. If not, the alternator stator coil assembly is at fault and should be renewed. **Note:** *Before condemning the stator coils, check the fault is not due to damaged wiring between the connector and coils.*

Removal

4 Remove the lower fairing (see Chapter 8) and the fuel tank (see Chapter 4). Trace the alternator wiring from the alternator/clutch cover and disconnect it at the white (J and K models) or red (L, N and R models) connector containing the three yellow wires **(see illustrations 32.2a and b)**. Free the wiring from any clips or guides and feed it through to the alternator/clutch cover.

5 Drain the engine oil (see Chapter 1).

6 Unscrew the two bolts securing the clutch cable bracket to the alternator/clutch cover on the right-hand side of the engine, then free the cable end from the release lever **(see illustrations)**. Position the cable clear of the engine.

7 Unscrew the remaining alternator/clutch

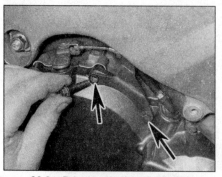

32.6a Remove the bracket bolts (arrowed) . . .

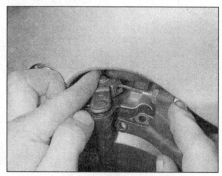

32.6b . . . and free the cable end from the lever

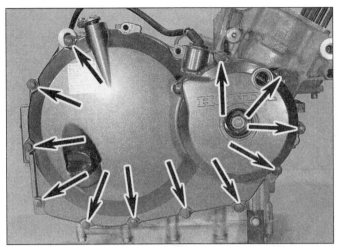
32.7 Unscrew the bolts (arrowed) and remove the cover

32.8 Unscrew the rotor bolt using a strap as shown, if available

cover bolts and remove the cover, being prepared to catch any residue oil **(see illustration)**. Discard the gasket as a new one must be used. Remove the dowels from either the cover or the crankcase if they are loose. Note the pushrod fitted in the bottom of the shaft in the cover and remove it for safekeeping if required **(see illustration 32.14)**. It is possible that it is stuck to the release bearing in the centre of the clutch release plate.

8 To remove the rotor bolt it is necessary to stop the rotor from turning. If a rotor holding strap or tool is not available, and the engine is in the frame, place the transmission in gear and have an assistant apply the rear brake, then unscrew the bolt **(see illustration)**.

9 To remove the rotor from the shaft it is necessary to use a rotor puller (the universal Honda tool pt. no. 07733-0020001 is ideal). Thread the rotor puller into the centre of the rotor and turn it until the rotor is displaced from the shaft **(see illustration)**. Remove the Woodruff key from its slot in the end of the crankshaft for safekeeping if it is loose **(see illustration 32.12a)**.

10 To remove the stator from the cover, unscrew the bolts securing the stator, and the bolt securing the wiring clamp, then remove the assembly, noting how the rubber wiring grommet fits **(see illustration)**.

Installation

11 Install the stator onto the cover, aligning the rubber wiring grommet with the groove **(see illustration 32.10)**. Apply a suitable non-permanent thread-locking compound to the stator bolt threads, then install the bolts and tighten them to the torque setting specified at the beginning of the Chapter. Apply a suitable sealant to the wiring grommet, then install it into the cut-out in the cover. Secure the wiring with its clamp.

12 Clean the tapered end of the crankshaft and the corresponding mating surface on the inside of the rotor with a suitable solvent. If

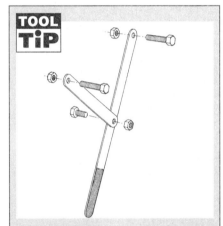
A rotor holding tool can easily be made using two strips of steel bolted together in the middle, and with a bolt through each end which locate into the recessed bores in the rotor.

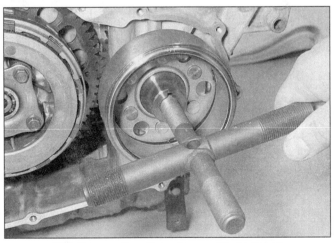
32.9 Remove the rotor using a puller as shown

32.10 Unscrew the stator bolts (A) and the wiring clamp bolt (B)

9•28 Electrical system

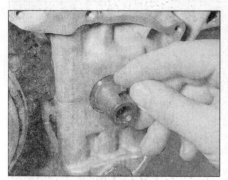

32.12a Fit the Woodruff key into its slot . . .

32.12b . . . then fit the rotor

32.13a Install the bolt and washer . . .

32.13b . . . and tighten the bolt to the specified torque

32.14 Fit the clutch pushrod if removed

32.15a Fit a new gasket, locating it onto the dowels (arrowed) . . .

removed, fit the Woodruff key into its slot in the crankshaft **(see illustration)**. Make sure that no metal objects have attached themselves to the magnet on the inside of the rotor, then install the rotor onto the shaft, aligning the slot in the rotor with the Woodruff key **(see illustration)**.

13 Install the rotor bolt and tighten it to the torque setting specified at the beginning of the Chapter, using the method employed on removal to prevent the rotor from turning **(see illustrations)**.

14 If removed, apply some grease to the clutch release pushrod, then turn the shaft in the cover until the pushrod seat is aligned and insert the pushrod **(see illustration)**.

15 If removed, insert the dowels in the crankcase. Install the alternator/clutch cover using a new gasket, making sure it locates correctly onto the dowels, then install all the cover bolts with the exception of the two which secure the cable bracket **(see illustrations)**. Tighten the bolts evenly in a criss-cross sequence to the specified torque setting.

16 Fit the clutch cable end into the release lever and position the bracket on the cover, then fit the two bolts and tighten them to the specified torque **(see illustrations 32.6b and a)**.

17 Feed the alternator wiring back to its connector, making sure it is correctly routed, and reconnect it **(see illustrations 32.2a and b)**. Install the fuel tank (see Chapter 4).

18 Refill the engine with oil (see Chapter 1).

19 Install the lower fairing (see Chapter 8).

33 Regulator/rectifier – check and replacement

Check

1 Remove the seat cowling (see Chapter 8). The regulator/rectifier is mounted on the left-hand side of the rear sub-frame on J and K models, and on the right on L, N and R models **(see illustrations)**. On J and K models, disconnect the wiring connector. On L, N and R models, unscrew the two mounting bolts, then disconnect the wiring connector.

2 Connect the meter positive (+ve) probe to the red/white terminal and the negative (-ve) probe to the green terminal on the wiring

32.15b . . . then install the cover

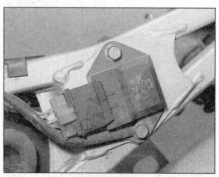

33.1a Regulator/rectifier – J and K models

33.1b Regulator/rectifier – L, N and R models

connector and check for voltage with the ignition switched ON. Full battery voltage should be present. Switch the ignition switch OFF.

3 Switch the multimeter to the resistance (ohms) scale. Check for continuity between the green terminal of the wiring connector and earth on the frame. There should be continuity.

4 Check the resistance between any two yellow terminals of the wiring connector. A resistance reading of 0.1 to 0.5 ohms should be obtained between any two Yellow terminals of the wiring connector.

5 If the above checks do not provide the expected results check the wiring between the battery, regulator/rectifier and alternator (see the *wiring diagrams* at the end of this book).

6 If the wiring checks out, the regulator/rectifier unit is probably faulty. To check the unit, remove it from the bike (see below) and use a multimeter set to the appropriate resistance scale to check the resistance between the various terminals on the regulator/rectifier **(see illustrations)**. If the readings do not compare closely with those shown in the accompanying table the regulator/rectifier unit can be considered faulty. **Note:** *The use of certain multimeters could lead to false readings being obtained, as could a low battery in the meter and contact between the meter probes and your fingers. If the above check shows the regulator/rectifier unit to be faulty, take the unit to a dealer for confirmation of its condition before renewing it.*

Replacement

7 Remove the seat cowling (see Chapter 8).
8 The regulator/rectifier is mounted on the left-hand side of the rear sub-frame on J and K models, and on the right on L, N and R models **(see illustration 33.1a or b)**.
9 Unscrew the two bolts securing the regulator/rectifier and remove it, on J and K models noting the lead secured by the bottom bolt.
10 Connect the wiring connector. Install the new unit and tighten its bolts securely, not forgetting the lead with the bottom bolt on J and K models.
11 Install the seat cowling (see Chapter 8).

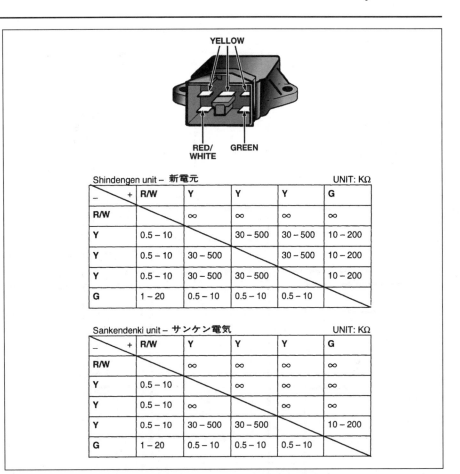

33.6a Test connections and date – J and K models
The table applicable depends on the manufacturer of the regulator/rectifier
∞ = *infinite resistance (no continuity)*

Shindengen unit – 新電元   UNIT: KΩ

− \ +	R/W	Y	Y	Y	G
R/W		∞	∞	∞	∞
Y	0.5 – 10		30 – 500	30 – 500	10 – 200
Y	0.5 – 10	30 – 500		30 – 500	10 – 200
Y	0.5 – 10	30 – 500	30 – 500		10 – 200
G	1 – 20	0.5 – 10	0.5 – 10	0.5 – 10	

Sankendenki unit – サンケン電気   UNIT: KΩ

− \ +	R/W	Y	Y	Y	G
R/W		∞	∞	∞	∞
Y	0.5 – 10		∞	∞	∞
Y	0.5 – 10	∞		∞	∞
Y	0.5 – 10	30 – 500	30 – 500		10 – 200
G	1 – 20	0.5 – 10	0.5 – 10	0.5 – 10	

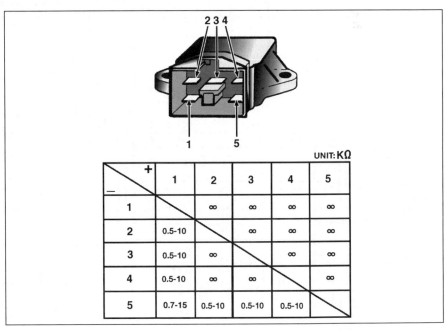

33.6b Test connections and data – L, N and R models
∞ = *infinite resistance (no continuity)*

− \ +	1	2	3	4	5
1		∞	∞	∞	∞
2	0.5-10		∞	∞	∞
3	0.5-10	∞		∞	∞
4	0.5-10	∞	∞		∞
5	0.7-15	0.5-10	0.5-10	0.5-10	

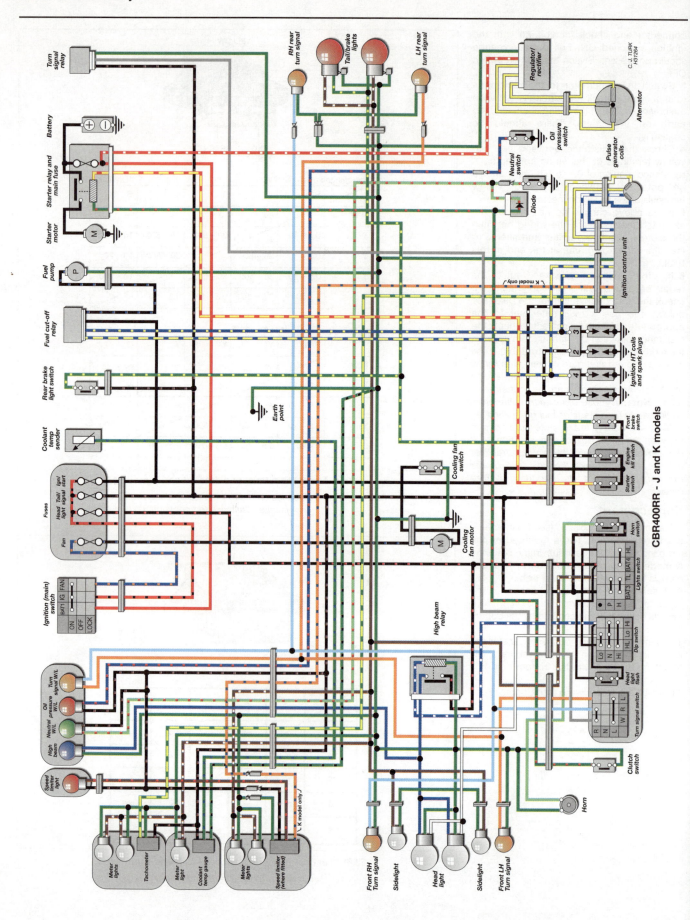

Electrical system 9•31

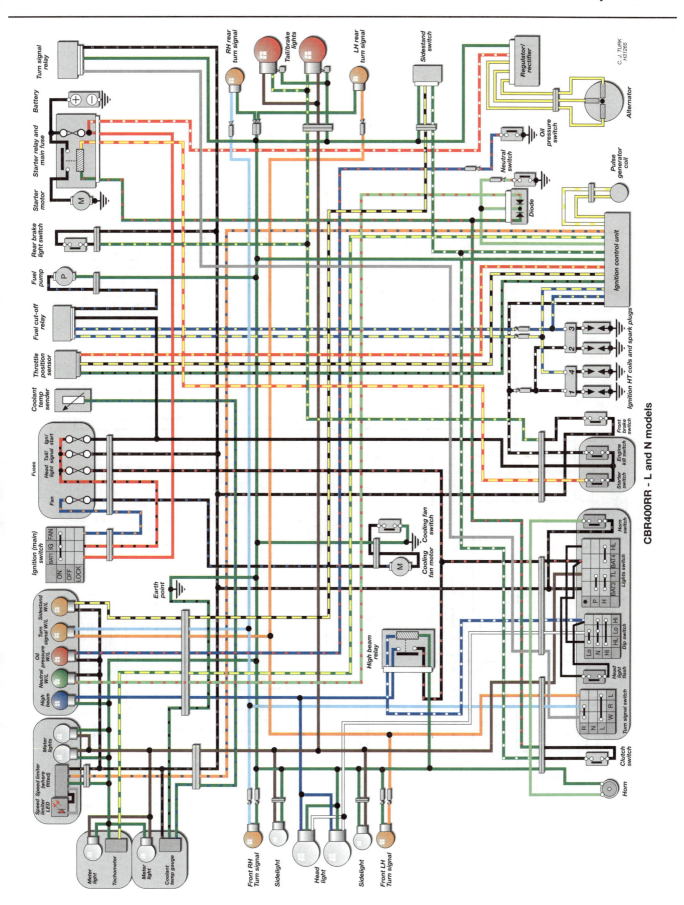

CBR400RR - L and N models

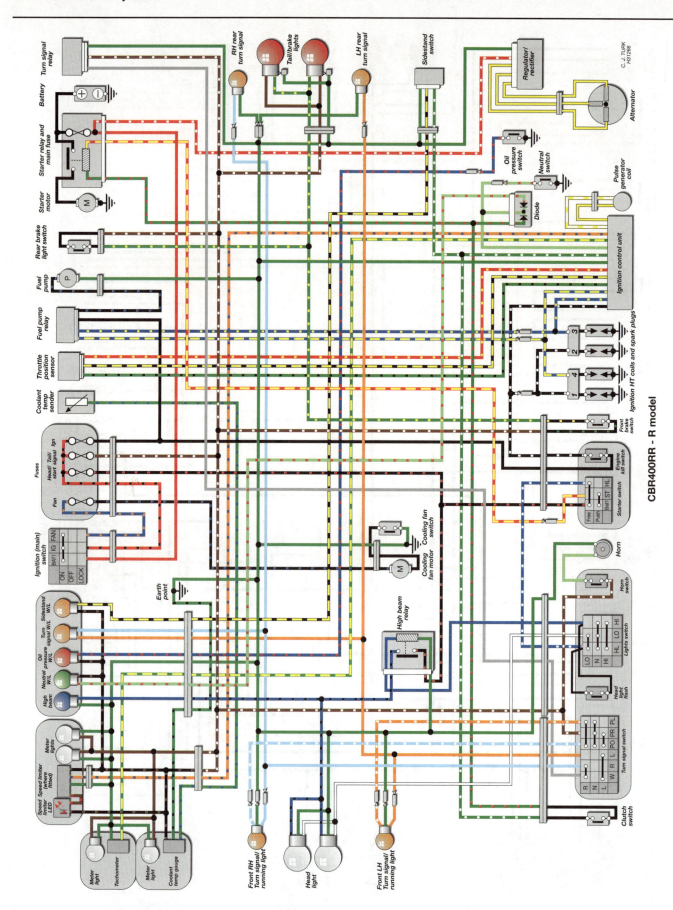

Reference

Dimensions and Weights .REF•1	Storage . . . . .REF•26
Tools and Workshop Tips . . . . .REF•2	Fault Finding . . . . .REF•28
Conversion Factors . . . . .REF•20	Fault Finding Equipment . . . . .REF•36
Motorcycle Chemicals and Lubricants . . . . .REF•21	Technical Terms Explained . . . . .REF•40
MOT Test Checks . . . . .REF•22	Index . . . . .REF•44

Dimensions and Weights

J model
Wheelbase	1370 mm
Overall length	2020 mm
Overall width	690 mm
Overall height	1110 mm
Ground clearance	120 mm
Weight (dry)	179 kg

K model
Wheelbase	1380 mm
Overall length	2202 mm
Overall width	675 mm
Overall height	1110 mm
Ground clearance	120 mm
Weight (dry)	182 kg

L, N and R models
Wheelbase	1365 mm
Overall length	1990 mm
Overall width	670 mm
Overall height	1080 mm
Ground clearance	125 mm
Weight (dry)	180 kg

REF•2 Tools and Workshop Tips

Buying tools

A toolkit is a fundamental requirement for servicing and repairing a motorcycle. Although there will be an initial expense in building up enough tools for servicing, this will soon be offset by the savings made by doing the job yourself. As experience and confidence grow, additional tools can be added to enable the repair and overhaul of the motorcycle. Many of the specialist tools are expensive and not often used so it may be preferable to hire them, or for a group of friends or motorcycle club to join in the purchase.

As a rule, it is better to buy more expensive, good quality tools. Cheaper tools are likely to wear out faster and need to be renewed more often, nullifying the original saving.

Warning: To avoid the risk of a poor quality tool breaking in use, causing injury or damage to the component being worked on, always aim to purchase tools which meet the relevant national safety standards.

The following lists of tools do not represent the manufacturer's service tools, but serve as a guide to help the owner decide which tools are needed for this level of work. In addition, items such as an electric drill, hacksaw, files, soldering iron and a workbench equipped with a vice, may be needed. Although not classed as tools, a selection of bolts, screws, nuts, washers and pieces of tubing always come in useful.

For more information about tools, refer to the Haynes *Motorcycle Workshop Practice TechBook* (Bk. No. 3470).

Manufacturer's service tools

Inevitably certain tasks require the use of a service tool. Where possible an alternative tool or method of approach is recommended, but sometimes there is no option if personal injury or damage to the component is to be avoided. Where required, service tools are referred to in the relevant procedure.

Service tools can usually only be purchased from a motorcycle dealer and are identified by a part number. Some of the commonly-used tools, such as rotor pullers, are available in aftermarket form from mail-order motorcycle tool and accessory suppliers.

Maintenance and minor repair tools

1. Set of flat-bladed screwdrivers
2. Set of Phillips head screwdrivers
3. Combination open-end and ring spanners
4. Socket set (3/8 inch or 1/2 inch drive)
5. Set of Allen keys or bits
6. Set of Torx keys or bits
7. Pliers, cutters and self-locking grips (Mole grips)
8. Adjustable spanners
9. C-spanners
10. Tread depth gauge and tyre pressure gauge
11. Cable oiler clamp
12. Feeler gauges
13. Spark plug gap measuring tool
14. Spark plug spanner or deep plug sockets
15. Wire brush and emery paper
16. Calibrated syringe, measuring vessel and funnel
17. Oil filter adapters
18. Oil drainer can or tray
19. Pump type oil can
20. Grease gun
21. Straight-edge and steel rule
22. Continuity tester
23. Battery charger
24. Hydrometer (for battery specific gravity check)
25. Anti-freeze tester (for liquid-cooled engines)

Tools and Workshop Tips   REF•3

Repair and overhaul tools

1 Torque wrench (small and mid-ranges)
2 Conventional, plastic or soft-faced hammers
3 Impact driver set
4 Vernier gauge
5 Circlip pliers (internal and external, or combination)
6 Set of cold chisels and punches
7 Selection of pullers
8 Breaker bars
9 Chain breaking/riveting tool set
10 Wire stripper and crimper tool
11 Multimeter (measures amps, volts and ohms)
12 Stroboscope (for dynamic timing checks)
13 Hose clamp (wingnut type shown)
14 Clutch holding tool
15 One-man brake/clutch bleeder kit

Specialist tools

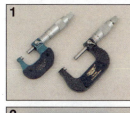   
 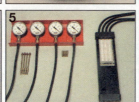  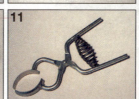 
 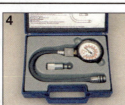   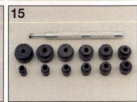

1 Micrometers (external type)
2 Telescoping gauges
3 Dial gauge
4 Cylinder compression gauge
5 Vacuum gauges (left) or manometer (right)
6 Oil pressure gauge
7 Plastigauge kit
8 Valve spring compressor (4-stroke engines)
9 Piston pin drawbolt tool
10 Piston ring removal and installation tool
11 Piston ring clamp
12 Cylinder bore hone (stone type shown)
13 Stud extractor
14 Screw extractor set
15 Bearing driver set

REF•4 Tools and Workshop Tips

1 Workshop equipment and facilities

The workbench

● Work is made much easier by raising the bike up on a ramp - components are much more accessible if raised to waist level. The hydraulic or pneumatic types seen in the dealer's workshop are a sound investment if you undertake a lot of repairs or overhauls **(see illustration 1.1)**.

1.1 Hydraulic motorcycle ramp

● If raised off ground level, the bike must be supported on the ramp to avoid it falling. Most ramps incorporate a front wheel locating clamp which can be adjusted to suit different diameter wheels. When tightening the clamp, take care not to mark the wheel rim or damage the tyre - use wood blocks on each side to prevent this.
● Secure the bike to the ramp using tie-downs **(see illustration 1.2)**. If the bike has only a sidestand, and hence leans at a dangerous angle when raised, support the bike on an auxiliary stand.

1.2 Tie-downs are used around the passenger footrests to secure the bike

● Auxiliary (paddock) stands are widely available from mail order companies or motorcycle dealers and attach either to the wheel axle or swingarm pivot **(see illustration 1.3)**. If the motorcycle has a centrestand, you can support it under the crankcase to prevent it toppling whilst either wheel is removed **(see illustration 1.4)**.

1.3 This auxiliary stand attaches to the swingarm pivot

1.4 Always use a block of wood between the engine and jack head when supporting the engine in this way

Fumes and fire

● Refer to the Safety first! page at the beginning of the manual for full details. Make sure your workshop is equipped with a fire extinguisher suitable for fuel-related fires (Class B fire - flammable liquids) - it is not sufficient to have a water-filled extinguisher.
● Always ensure adequate ventilation is available. Unless an exhaust gas extraction system is available for use, ensure that the engine is run outside of the workshop.
● If working on the fuel system, make sure the workshop is ventilated to avoid a build-up of fumes. This applies equally to fume build-up when charging a battery. Do not smoke or allow anyone else to smoke in the workshop.

Fluids

● If you need to drain fuel from the tank, store it in an approved container marked as suitable for the storage of petrol (gasoline) **(see illustration 1.5)**. Do not store fuel in glass jars or bottles.

1.5 Use an approved can only for storing petrol (gasoline)

● Use proprietary engine degreasers or solvents which have a high flash-point, such as paraffin (kerosene), for cleaning off oil, grease and dirt - never use petrol (gasoline) for cleaning. Wear rubber gloves when handling solvent and engine degreaser. The fumes from certain solvents can be dangerous - always work in a well-ventilated area.

Dust, eye and hand protection

● Protect your lungs from inhalation of dust particles by wearing a filtering mask over the nose and mouth. Many frictional materials still contain asbestos which is dangerous to your health. Protect your eyes from spouts of liquid and sprung components by wearing a pair of protective goggles **(see illustration 1.6)**.

1.6 A fire extinguisher, goggles, mask and protective gloves should be at hand in the workshop

● Protect your hands from contact with solvents, fuel and oils by wearing rubber gloves. Alternatively apply a barrier cream to your hands before starting work. If handling hot components or fluids, wear suitable gloves to protect your hands from scalding and burns.

What to do with old fluids

● Old cleaning solvent, fuel, coolant and oils should not be poured down domestic drains or onto the ground. Package the fluid up in old oil containers, label it accordingly, and take it to a garage or disposal facility. Contact your local authority for location of such sites or ring the oil care hotline.

Note: It is antisocial and illegal to dump oil down the drain. To find the location of your local oil recycling bank, call this number free.

In the USA, note that any oil supplier must accept used oil for recycling.

Tools and Workshop Tips

2 Fasteners - screws, bolts and nuts

Fastener types and applications

Bolts and screws

- Fastener head types are either of hexagonal, Torx or splined design, with internal and external versions of each type **(see illustrations 2.1 and 2.2)**; splined head fasteners are not in common use on motorcycles. The conventional slotted or Phillips head design is used for certain screws. Bolt or screw length is always measured from the underside of the head to the end of the item **(see illustration 2.11)**.

2.1 Internal hexagon/Allen (A), Torx (B) and splined (C) fasteners, with corresponding bits

2.2 External Torx (A), splined (B) and hexagon (C) fasteners, with corresponding sockets

- Certain fasteners on the motorcycle have a tensile marking on their heads, the higher the marking the stronger the fastener. High tensile fasteners generally carry a 10 or higher marking. Never replace a high tensile fastener with one of a lower tensile strength.

Washers (see illustration 2.3)

- Plain washers are used between a fastener head and a component to prevent damage to the component or to spread the load when torque is applied. Plain washers can also be used as spacers or shims in certain assemblies. Copper or aluminium plain washers are often used as sealing washers on drain plugs.

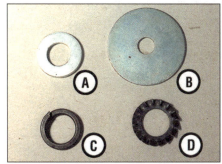

2.3 Plain washer (A), penny washer (B), spring washer (C) and serrated washer (D)

- The split-ring spring washer works by applying axial tension between the fastener head and component. If flattened, it is fatigued and must be renewed. If a plain (flat) washer is used on the fastener, position the spring washer between the fastener and the plain washer.
- Serrated star type washers dig into the fastener and component faces, preventing loosening. They are often used on electrical earth (ground) connections to the frame.
- Cone type washers (sometimes called Belleville) are conical and when tightened apply axial tension between the fastener head and component. They must be installed with the dished side against the component and often carry an OUTSIDE marking on their outer face. If flattened, they are fatigued and must be renewed.
- Tab washers are used to lock plain nuts or bolts on a shaft. A portion of the tab washer is bent up hard against one flat of the nut or bolt to prevent it loosening. Due to the tab washer being deformed in use, a new tab washer should be used every time it is disturbed.
- Wave washers are used to take up endfloat on a shaft. They provide light springing and prevent excessive side-to-side play of a component. Can be found on rocker arm shafts.

Nuts and split pins

- Conventional plain nuts are usually six-sided **(see illustration 2.4)**. They are sized by thread diameter and pitch. High tensile nuts carry a number on one end to denote their tensile strength.

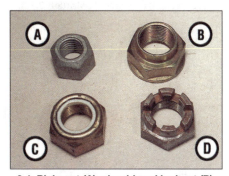

2.4 Plain nut (A), shouldered locknut (B), nylon insert nut (C) and castellated nut (D)

- Self-locking nuts either have a nylon insert, or two spring metal tabs, or a shoulder which is staked into a groove in the shaft - their advantage over conventional plain nuts is a resistance to loosening due to vibration. The nylon insert type can be used a number of times, but must be renewed when the friction of the nylon insert is reduced, ie when the nut spins freely on the shaft. The spring tab type can be reused unless the tabs are damaged. The shouldered type must be renewed every time it is disturbed.
- Split pins (cotter pins) are used to lock a castellated nut to a shaft or to prevent slackening of a plain nut. Common applications are wheel axles and brake torque arms. Because the split pin arms are deformed to lock around the nut a new split pin must always be used on installation - always fit the correct size split pin which will fit snugly in the shaft hole. Make sure the split pin arms are correctly located around the nut **(see illustrations 2.5 and 2.6)**.

2.5 Bend split pin (cotter pin) arms as shown (arrows) to secure a castellated nut

2.6 Bend split pin (cotter pin) arms as shown to secure a plain nut

Caution: If the castellated nut slots do not align with the shaft hole after tightening to the torque setting, tighten the nut until the next slot aligns with the hole - never slacken the nut to align its slot.

- R-pins (shaped like the letter R), or slip pins as they are sometimes called, are sprung and can be reused if they are otherwise in good condition. Always install R-pins with their closed end facing forwards **(see illustration 2.7)**.

REF•6 Tools and Workshop Tips

2.7 Correct fitting of R-pin. Arrow indicates forward direction

Circlips (see illustration 2.8)

● Circlips (sometimes called snap-rings) are used to retain components on a shaft or in a housing and have corresponding external or internal ears to permit removal. Parallel-sided (machined) circlips can be installed either way round in their groove, whereas stamped circlips (which have a chamfered edge on one face) must be installed with the chamfer facing away from the direction of thrust load **(see illustration 2.9)**.

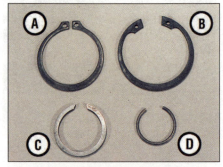

2.8 External stamped circlip (A), internal stamped circlip (B), machined circlip (C) and wire circlip (D)

● Always use circlip pliers to remove and install circlips; expand or compress them just enough to remove them. After installation, rotate the circlip in its groove to ensure it is securely seated. If installing a circlip on a splined shaft, always align its opening with a shaft channel to ensure the circlip ends are well supported and unlikely to catch **(see illustration 2.10)**.

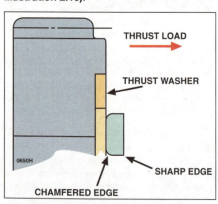

2.9 Correct fitting of a stamped circlip

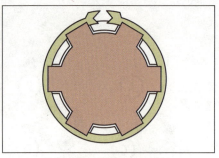

2.10 Align circlip opening with shaft channel

● Circlips can wear due to the thrust of components and become loose in their grooves, with the subsequent danger of becoming dislodged in operation. For this reason, renewal is advised every time a circlip is disturbed.
● Wire circlips are commonly used as piston pin retaining clips. If a removal tang is provided, long-nosed pliers can be used to dislodge them, otherwise careful use of a small flat-bladed screwdriver is necessary. Wire circlips should be renewed every time they are disturbed.

Thread diameter and pitch

● Diameter of a male thread (screw, bolt or stud) is the outside diameter of the threaded portion **(see illustration 2.11)**. Most motorcycle manufacturers use the ISO (International Standards Organisation) metric system expressed in millimetres, eg M6 refers to a 6 mm diameter thread. Sizing is the same for nuts, except that the thread diameter is measured across the valleys of the nut.
● Pitch is the distance between the peaks of the thread **(see illustration 2.11)**. It is expressed in millimetres, thus a common bolt size may be expressed as 6.0 x 1.0 mm (6 mm thread diameter and 1 mm pitch). Generally pitch increases in proportion to thread diameter, although there are always exceptions.
● Thread diameter and pitch are related for conventional fastener applications and the accompanying table can be used as a guide. Additionally, the AF (Across Flats), spanner or socket size dimension of the bolt or nut **(see illustration 2.11)** is linked to thread and pitch specification. Thread pitch can be measured with a thread gauge **(see illustration 2.12)**.

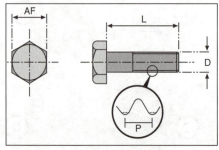

2.11 Fastener length (L), thread diameter (D), thread pitch (P) and head size (AF)

2.12 Using a thread gauge to measure pitch

AF size	Thread diameter x pitch (mm)
8 mm	M5 x 0.8
8 mm	M6 x 1.0
10 mm	M6 x 1.0
12 mm	M8 x 1.25
14 mm	M10 x 1.25
17 mm	M12 x 1.25

● The threads of most fasteners are of the right-hand type, ie they are turned clockwise to tighten and anti-clockwise to loosen. The reverse situation applies to left-hand thread fasteners, which are turned anti-clockwise to tighten and clockwise to loosen. Left-hand threads are used where rotation of a component might loosen a conventional right-hand thread fastener.

Seized fasteners

● Corrosion of external fasteners due to water or reaction between two dissimilar metals can occur over a period of time. It will build up sooner in wet conditions or in countries where salt is used on the roads during the winter. If a fastener is severely corroded it is likely that normal methods of removal will fail and result in its head being ruined. When you attempt removal, the fastener thread should be heard to crack free and unscrew easily - if it doesn't, stop there before damaging something.
● A smart tap on the head of the fastener will often succeed in breaking free corrosion which has occurred in the threads **(see illustration 2.13)**.
● An aerosol penetrating fluid (such as WD-40) applied the night beforehand may work its way down into the thread and ease removal. Depending on the location, you may be able to make up a Plasticine well around the fastener head and fill it with penetrating fluid.

2.13 A sharp tap on the head of a fastener will often break free a corroded thread

Tools and Workshop Tips

- If you are working on an engine internal component, corrosion will most likely not be a problem due to the well lubricated environment. However, components can be very tight and an impact driver is a useful tool in freeing them **(see illustration 2.14)**.

2.14 Using an impact driver to free a fastener

- Where corrosion has occurred between dissimilar metals (eg steel and aluminium alloy), the application of heat to the fastener head will create a disproportionate expansion rate between the two metals and break the seizure caused by the corrosion. Whether heat can be applied depends on the location of the fastener - any surrounding components likely to be damaged must first be removed **(see illustration 2.15)**. Heat can be applied using a paint stripper heat gun or clothes iron, or by immersing the component in boiling water - wear protective gloves to prevent scalding or burns to the hands.

2.15 Using heat to free a seized fastener

- As a last resort, it is possible to use a hammer and cold chisel to work the fastener head unscrewed **(see illustration 2.16)**. This will damage the fastener, but more importantly extreme care must be taken not to damage the surrounding component.

> **Caution:** Remember that the component being secured is generally of more value than the bolt, nut or screw - when the fastener is freed, do not unscrew it with force, instead work the fastener back and forth when resistance is felt to prevent thread damage.

2.16 Using a hammer and chisel to free a seized fastener

Broken fasteners and damaged heads

- If the shank of a broken bolt or screw is accessible you can grip it with self-locking grips. The knurled wheel type stud extractor tool or self-gripping stud puller tool is particularly useful for removing the long studs which screw into the cylinder mouth surface of the crankcase or bolts and screws from which the head has broken off **(see illustration 2.17)**. Studs can also be removed by locking two nuts together on the threaded end of the stud and using a spanner on the lower nut **(see illustration 2.18)**.

2.17 Using a stud extractor tool to remove a broken crankcase stud

2.18 Two nuts can be locked together to unscrew a stud from a component

- A bolt or screw which has broken off below or level with the casing must be extracted using a screw extractor set. Centre punch the fastener to centralise the drill bit, then drill a hole in the fastener **(see illustration 2.19)**. Select a drill bit which is approximately half to three-quarters the

2.19 When using a screw extractor, first drill a hole in the fastener . . .

diameter of the fastener and drill to a depth which will accommodate the extractor. Use the largest size extractor possible, but avoid leaving too small a wall thickness otherwise the extractor will merely force the fastener walls outwards wedging it in the casing thread.

- If a spiral type extractor is used, thread it anti-clockwise into the fastener. As it is screwed in, it will grip the fastener and unscrew it from the casing **(see illustration 2.20)**.

2.20 . . . then thread the extractor anti-clockwise into the fastener

- If a taper type extractor is used, tap it into the fastener so that it is firmly wedged in place. Unscrew the extractor (anti-clockwise) to draw the fastener out.

> **⚠ Warning:** Stud extractors are very hard and may break off in the fastener if care is not taken - ask an engineer about spark erosion if this happens.

- Alternatively, the broken bolt/screw can be drilled out and the hole retapped for an oversize bolt/screw or a diamond-section thread insert. It is essential that the drilling is carried out squarely and to the correct depth, otherwise the casing may be ruined - if in doubt, entrust the work to an engineer.
- Bolts and nuts with rounded corners cause the correct size spanner or socket to slip when force is applied. Of the types of spanner/socket available always use a six-point type rather than an eight or twelve-point type - better grip

REF•8 Tools and Workshop Tips

2.21 Comparison of surface drive ring spanner (left) with 12-point type (right)

is obtained. Surface drive spanners grip the middle of the hex flats, rather than the corners, and are thus good in cases of damaged heads **(see illustration 2.21)**.

● Slotted-head or Phillips-head screws are often damaged by the use of the wrong size screwdriver. Allen-head and Torx-head screws are much less likely to sustain damage. If enough of the screw head is exposed you can use a hacksaw to cut a slot in its head and then use a conventional flat-bladed screwdriver to remove it. Alternatively use a hammer and cold chisel to tap the head of the fastener around to slacken it. Always replace damaged fasteners with new ones, preferably Torx or Allen-head type.

HAYNES HINT

A dab of valve grinding compound between the screw head and screwdriver tip will often give a good grip.

Thread repair

● Threads (particularly those in aluminium alloy components) can be damaged by overtightening, being assembled with dirt in the threads, or from a component working loose and vibrating. Eventually the thread will fail completely, and it will be impossible to tighten the fastener.

● If a thread is damaged or clogged with old locking compound it can be renovated with a thread repair tool (thread chaser) **(see illustrations 2.22 and 2.23)**; special thread

2.22 A thread repair tool being used to correct an internal thread

2.23 A thread repair tool being used to correct an external thread

chasers are available for spark plug hole threads. The tool will not cut a new thread, but clean and true the original thread. Make sure that you use the correct diameter and pitch tool. Similarly, external threads can be cleaned up with a die or a thread restorer file **(see illustration 2.24)**.

2.24 Using a thread restorer file

● It is possible to drill out the old thread and retap the component to the next thread size. This will work where there is enough surrounding material and a new bolt or screw can be obtained. Sometimes, however, this is not possible - such as where the bolt/screw passes through another component which must also be suitably modified, also in cases where a spark plug or oil drain plug cannot be obtained in a larger diameter thread size.

● The diamond-section thread insert (often known by its popular trade name of Heli-Coil) is a simple and effective method of renewing the thread and retaining the original size. A kit can be purchased which contains the tap, insert and installing tool **(see illustration 2.25)**. Drill out the damaged thread with the size drill specified **(see illustration 2.26)**. Carefully retap the thread **(see illustration 2.27)**. Install the

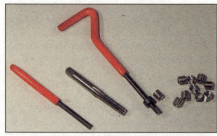

2.25 Obtain a thread insert kit to suit the thread diameter and pitch required

2.26 To install a thread insert, first drill out the original thread . . .

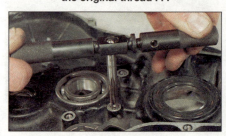

2.27 . . . tap a new thread . . .

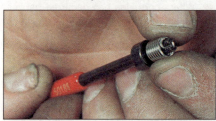

2.28 . . . fit insert on the installing tool . . .

2.29 . . . and thread into the component . . .

2.30 . . . break off the tang when complete

insert on the installing tool and thread it slowly into place using a light downward pressure **(see illustrations 2.28 and 2.29)**. When positioned between a 1/4 and 1/2 turn below the surface withdraw the installing tool and use the break-off tool to press down on the tang, breaking it off **(see illustration 2.30)**.

● There are epoxy thread repair kits on the market which can rebuild stripped internal threads, although this repair should not be used on high load-bearing components.

Tools and Workshop Tips

Thread locking and sealing compounds

● Locking compounds are used in locations where the fastener is prone to loosening due to vibration or on important safety-related items which might cause loss of control of the motorcycle if they fail. It is also used where important fasteners cannot be secured by other means such as lockwashers or split pins.

● Before applying locking compound, make sure that the threads (internal and external) are clean and dry with all old compound removed. Select a compound to suit the component being secured - a non-permanent general locking and sealing type is suitable for most applications, but a high strength type is needed for permanent fixing of studs in castings. Apply a drop or two of the compound to the first few threads of the fastener, then thread it into place and tighten to the specified torque. Do not apply excessive thread locking compound otherwise the thread may be damaged on subsequent removal.

● Certain fasteners are impregnated with a dry film type coating of locking compound on their threads. Always renew this type of fastener if disturbed.

● Anti-seize compounds, such as copper-based greases, can be applied to protect threads from seizure due to extreme heat and corrosion. A common instance is spark plug threads and exhaust system fasteners.

3 Measuring tools and gauges

Feeler gauges

● Feeler gauges (or blades) are used for measuring small gaps and clearances **(see illustration 3.1)**. They can also be used to measure endfloat (sideplay) of a component on a shaft where access is not possible with a dial gauge.

● Feeler gauge sets should be treated with care and not bent or damaged. They are etched with their size on one face. Keep them clean and very lightly oiled to prevent corrosion build-up.

3.1 Feeler gauges are used for measuring small gaps and clearances - thickness is marked on one face of gauge

● When measuring a clearance, select a gauge which is a light sliding fit between the two components. You may need to use two gauges together to measure the clearance accurately.

Micrometers

● A micrometer is a precision tool capable of measuring to 0.01 or 0.001 of a millimetre. It should always be stored in its case and not in the general toolbox. It must be kept clean and never dropped, otherwise its frame or measuring anvils could be distorted resulting in inaccurate readings.

● External micrometers are used for measuring outside diameters of components and have many more applications than internal micrometers. Micrometers are available in different size ranges, eg 0 to 25 mm, 25 to 50 mm, and upwards in 25 mm steps; some large micrometers have interchangeable anvils to allow a range of measurements to be taken. Generally the largest precision measurement you are likely to take on a motorcycle is the piston diameter.

● Internal micrometers (or bore micrometers) are used for measuring inside diameters, such as valve guides and cylinder bores. Telescoping gauges and small hole gauges are used in conjunction with an external micrometer, whereas the more expensive internal micrometers have their own measuring device.

External micrometer

Note: *The conventional analogue type instrument is described. Although much easier to read, digital micrometers are considerably more expensive.*

● Always check the calibration of the micrometer before use. With the anvils closed (0 to 25 mm type) or set over a test gauge (for

3.2 Check micrometer calibration before use

the larger types) the scale should read zero **(see illustration 3.2)**; make sure that the anvils (and test piece) are clean first. Any discrepancy can be adjusted by referring to the instructions supplied with the tool. Remember that the micrometer is a precision measuring tool - don't force the anvils closed, use the ratchet (4) on the end of the micrometer to close it. In this way, a measured force is always applied.

● To use, first make sure that the item being measured is clean. Place the anvil of the micrometer (1) against the item and use the thimble (2) to bring the spindle (3) lightly into contact with the other side of the item **(see illustration 3.3)**. Don't tighten the thimble down because this will damage the micrometer - instead use the ratchet (4) on the end of the micrometer. The ratchet mechanism applies a measured force preventing damage to the instrument.

● The micrometer is read by referring to the linear scale on the sleeve and the annular scale on the thimble. Read off the sleeve first to obtain the base measurement, then add the fine measurement from the thimble to obtain the overall reading. The linear scale on the sleeve represents the measuring range of the micrometer (eg 0 to 25 mm). The annular scale

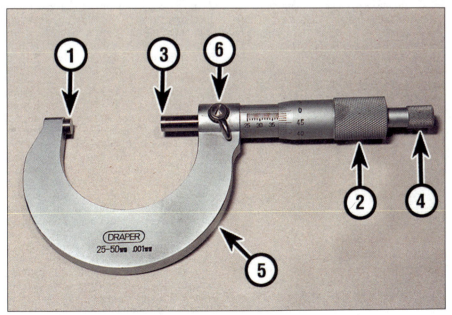

3.3 Micrometer component parts

1 Anvil	3 Spindle	5 Frame
2 Thimble	4 Ratchet	6 Locking lever

REF•10 Tools and Workshop Tips

on the thimble will be in graduations of 0.01 mm (or as marked on the frame) - one full revolution of the thimble will move 0.5 mm on the linear scale. Take the reading where the datum line on the sleeve intersects the thimble's scale. Always position the eye directly above the scale otherwise an inaccurate reading will result.

In the example shown the item measures 2.95 mm (see illustration 3.4):

Linear scale	2.00 mm
Linear scale	0.50 mm
Annular scale	0.45 mm
Total figure	**2.95 mm**

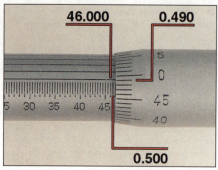

3.5 Micrometer reading of 46.99 mm on linear and annular scales . . .

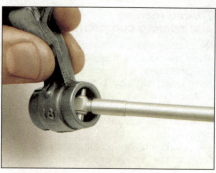

3.7 Expand the telescoping gauge in the bore, lock its position . . .

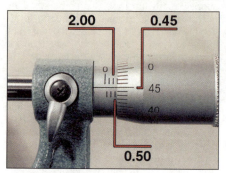

3.4 Micrometer reading of 2.95 mm

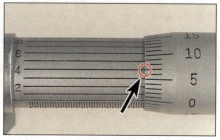

3.6 . . . and 0.004 mm on vernier scale

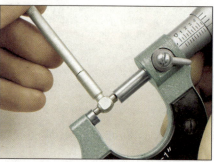

3.8 . . . then measure the gauge with a micrometer

Most micrometers have a locking lever (6) on the frame to hold the setting in place, allowing the item to be removed from the micrometer.
● Some micrometers have a vernier scale on their sleeve, providing an even finer measurement to be taken, in 0.001 increments of a millimetre. Take the sleeve and thimble measurement as described above, then check which graduation on the vernier scale aligns with that of the annular scale on the thimble **Note:** *The eye must be perpendicular to the scale when taking the vernier reading - if necessary rotate the body of the micrometer to ensure this.* Multiply the vernier scale figure by 0.001 and add it to the base and fine measurement figures.

In the example shown the item measures 46.994 mm (see illustrations 3.5 and 3.6):

Linear scale (base)	46.000 mm
Linear scale (base)	00.500 mm
Annular scale (fine)	00.490 mm
Vernier scale	00.004 mm
Total figure	**46.994 mm**

Internal micrometer

● Internal micrometers are available for measuring bore diameters, but are expensive and unlikely to be available for home use. It is suggested that a set of telescoping gauges and small hole gauges, both of which must be used with an external micrometer, will suffice for taking internal measurements on a motorcycle.
● Telescoping gauges can be used to measure internal diameters of components. Select a gauge with the correct size range, make sure its ends are clean and insert it into the bore. Expand the gauge, then lock its position and withdraw it from the bore (see illustration 3.7). Measure across the gauge ends with a micrometer (see illustration 3.8).
● Very small diameter bores (such as valve guides) are measured with a small hole gauge. Once adjusted to a slip-fit inside the component, its position is locked and the gauge withdrawn for measurement with a micrometer (see illustrations 3.9 and 3.10).

Vernier caliper

Note: *The conventional linear and dial gauge type instruments are described. Digital types are easier to read, but are far more expensive.*
● The vernier caliper does not provide the precision of a micrometer, but is versatile in being able to measure internal and external diameters. Some types also incorporate a depth gauge. It is ideal for measuring clutch plate friction material and spring free lengths.
● To use the conventional linear scale vernier, slacken off the vernier clamp screws (1) and set its jaws over (2), or inside (3), the item to be measured (see illustration 3.11). Slide the jaw into contact, using the thumbwheel (4) for fine movement of the sliding scale (5) then tighten the clamp screws (1). Read off the main scale (6) where the zero on the sliding scale (5) intersects it, taking the whole number to the left of the zero; this provides the base measurement. View along the sliding scale and select the division which

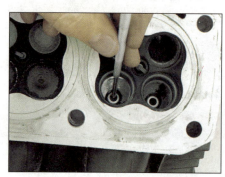

3.9 Expand the small hole gauge in the bore, lock its position . . .

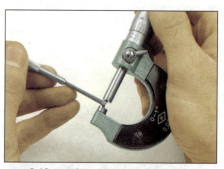

3.10 . . . then measure the gauge with a micrometer

lines up exactly with any of the divisions on the main scale, noting that the divisions usually represents 0.02 of a millimetre. Add this fine measurement to the base measurement to obtain the total reading.

Tools and Workshop Tips REF•11

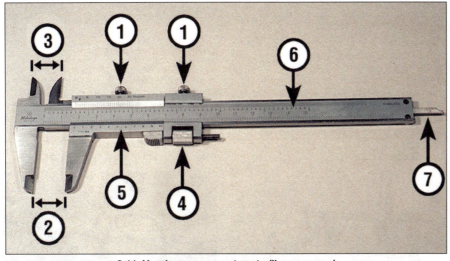

3.11 Vernier component parts (linear gauge)

1 Clamp screws
2 External jaws
3 Internal jaws
4 Thumbwheel
5 Sliding scale
6 Main scale
7 Depth gauge

In the example shown the item measures 55.92 mm **(see illustration 3.12)**:

Base measurement	55.00 mm
Fine measurement	00.92 mm
Total figure	**55.92 mm**

● Some vernier calipers are equipped with a dial gauge for fine measurement. Before use, check that the jaws are clean, then close them fully and check that the dial gauge reads zero. If necessary adjust the gauge ring accordingly. Slacken the vernier clamp screw (1) and set its jaws over (2), or inside (3), the item to be measured **(see illustration 3.13)**. Slide the jaws into contact, using the thumbwheel (4) for fine movement. Read off the main scale (5) where the edge of the sliding scale (6) intersects it, taking the whole number to the left of the zero; this provides the base measurement. Read off the needle position on the dial gauge (7) scale to provide the fine measurement; each division represents 0.05 of a millimetre. Add this fine measurement to the base measurement to obtain the total reading.

In the example shown the item measures 55.95 mm **(see illustration 3.14)**:

Base measurement	55.00 mm
Fine measurement	00.95 mm
Total figure	**55.95 mm**

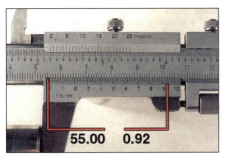

3.12 Vernier gauge reading of 55.92 mm

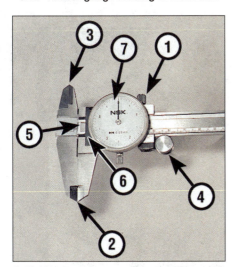

3.13 Vernier component parts (dial gauge)

1 Clamp screw
2 External jaws
3 Internal jaws
4 Thumbwheel
5 Main scale
6 Sliding scale
7 Dial gauge

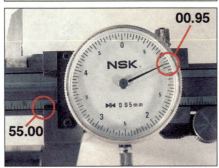

3.14 Vernier gauge reading of 55.95 mm

Plastigauge

● Plastigauge is a plastic material which can be compressed between two surfaces to measure the oil clearance between them. The width of the compressed Plastigauge is measured against a calibrated scale to determine the clearance.

● Common uses of Plastigauge are for measuring the clearance between crankshaft journal and main bearing inserts, between crankshaft journal and big-end bearing inserts, and between camshaft and bearing surfaces. The following example describes big-end oil clearance measurement.

● Handle the Plastigauge material carefully to prevent distortion. Using a sharp knife, cut a length which corresponds with the width of the bearing being measured and place it carefully across the journal so that it is parallel with the shaft **(see illustration 3.15)**. Carefully install both bearing shells and the connecting rod. Without rotating the rod on the journal tighten its bolts or nuts (as applicable) to the specified torque. The connecting rod and bearings are then disassembled and the crushed Plastigauge examined.

3.15 Plastigauge placed across shaft journal

● Using the scale provided in the Plastigauge kit, measure the width of the material to determine the oil clearance **(see illustration 3.16)**. Always remove all traces of Plastigauge after use using your fingernails.

Caution: Arriving at the correct clearance demands that the assembly is torqued correctly, according to the settings and sequence (where applicable) provided by the motorcycle manufacturer.

3.16 Measuring the width of the crushed Plastigauge

Tools and Workshop Tips

Dial gauge or DTI (Dial Test Indicator)

● A dial gauge can be used to accurately measure small amounts of movement. Typical uses are measuring shaft runout or shaft endfloat (sideplay) and setting piston position for ignition timing on two-strokes. A dial gauge set usually comes with a range of different probes and adapters and mounting equipment.

● The gauge needle must point to zero when at rest. Rotate the ring around its periphery to zero the gauge.

● Check that the gauge is capable of reading the extent of movement in the work. Most gauges have a small dial set in the face which records whole millimetres of movement as well as the fine scale around the face periphery which is calibrated in 0.01 mm divisions. Read off the small dial first to obtain the base measurement, then add the measurement from the fine scale to obtain the total reading.

In the example shown the gauge reads 1.48 mm (see illustration 3.17):

Base measurement	1.00 mm
Fine measurement	0.48 mm
Total figure	**1.48 mm**

3.17 Dial gauge reading of 1.48 mm

● If measuring shaft runout, the shaft must be supported in vee-blocks and the gauge mounted on a stand perpendicular to the shaft. Rest the tip of the gauge against the centre of the shaft and rotate the shaft slowly whilst watching the gauge reading (see illustration 3.18). Take several measurements along the length of the shaft and record the maximum gauge reading as the amount of runout in the shaft. Note: *The reading obtained will be total runout at that point - some manufacturers specify that the runout figure is halved to compare with their specified runout limit.*

● Endfloat (sideplay) measurement requires that the gauge is mounted securely to the surrounding component with its probe touching the end of the shaft. Using hand pressure, push and pull on the shaft noting the maximum endfloat recorded on the gauge (see illustration 3.19).

3.19 Using a dial gauge to measure shaft endfloat

● A dial gauge with suitable adapters can be used to determine piston position BTDC on two-stroke engines for the purposes of ignition timing. The gauge, adapter and suitable length probe are installed in the place of the spark plug and the gauge zeroed at TDC. If the piston position is specified as 1.14 mm BTDC, rotate the engine back to 2.00 mm BTDC, then slowly forwards to 1.14 mm BTDC.

Cylinder compression gauges

● A compression gauge is used for measuring cylinder compression. Either the rubber-cone type or the threaded adapter type can be used. The latter is preferred to ensure a perfect seal against the cylinder head. A 0 to 300 psi (0 to 20 Bar) type gauge (for petrol/gasoline engines) will be suitable for motorcycles.

● The spark plug is removed and the gauge either held hard against the cylinder head (cone type) or the gauge adapter screwed into the cylinder head (threaded type) (see illustration 3.20). Cylinder compression is measured with the engine turning over, but not running - carry out the compression test as described in *Fault Finding Equipment*. The gauge will hold the reading until manually released.

Oil pressure gauge

● An oil pressure gauge is used for measuring engine oil pressure. Most gauges come with a set of adapters to fit the thread of the take-off point (see illustration 3.21). If the take-off point specified by the motorcycle manufacturer is an external oil pipe union, make sure that the specified replacement union is used to prevent oil starvation.

3.21 Oil pressure gauge and take-off point adapter (arrow)

● Oil pressure is measured with the engine running (at a specific rpm) and often the manufacturer will specify pressure limits for a cold and hot engine.

Straight-edge and surface plate

● If checking the gasket face of a component for warpage, place a steel rule or precision straight-edge across the gasket face and measure any gap between the straight-edge and component with feeler gauges (see illustration 3.22). Check diagonally across the component and between mounting holes (see illustration 3.23).

3.22 Use a straight-edge and feeler gauges to check for warpage

3.18 Using a dial gauge to measure shaft runout

3.20 Using a rubber-cone type cylinder compression gauge

3.23 Check for warpage in these directions

Tools and Workshop Tips

- Checking individual components for warpage, such as clutch plain (metal) plates, requires a perfectly flat plate or piece or plate glass and feeler gauges.

4 Torque and leverage

What is torque?

- Torque describes the twisting force about a shaft. The amount of torque applied is determined by the distance from the centre of the shaft to the end of the lever and the amount of force being applied to the end of the lever; distance multiplied by force equals torque.
- The manufacturer applies a measured torque to a bolt or nut to ensure that it will not slacken in use and to hold two components securely together without movement in the joint. The actual torque setting depends on the thread size, bolt or nut material and the composition of the components being held.
- Too little torque may cause the fastener to loosen due to vibration, whereas too much torque will distort the joint faces of the component or cause the fastener to shear off. Always stick to the specified torque setting.

Using a torque wrench

- Check the calibration of the torque wrench and make sure it has a suitable range for the job. Torque wrenches are available in Nm (Newton-metres), kgf m (kilograms-force metre), lbf ft (pounds-feet), lbf in (inch-pounds). Do not confuse lbf ft with lbf in.
- Adjust the tool to the desired torque on the scale (see illustration 4.1). If your torque wrench is not calibrated in the units specified, carefully convert the figure (see *Conversion Factors*). A manufacturer sometimes gives a torque setting as a range (8 to 10 Nm) rather than a single figure - in this case set the tool midway between the two settings. The same torque may be expressed as 9 Nm ± 1 Nm. Some torque wrenches have a method of locking the setting so that it isn't inadvertently altered during use.

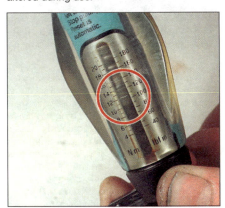

4.1 Set the torque wrench index mark to the setting required, in this case 12 Nm

- Install the bolts/nuts in their correct location and secure them lightly. Their threads must be clean and free of any old locking compound. Unless specified the threads and flange should be dry - oiled threads are necessary in certain circumstances and the manufacturer will take this into account in the specified torque figure. Similarly, the manufacturer may also specify the application of thread-locking compound.
- Tighten the fasteners in the specified sequence until the torque wrench clicks, indicating that the torque setting has been reached. Apply the torque again to double-check the setting. Where different thread diameter fasteners secure the component, as a rule tighten the larger diameter ones first.
- When the torque wrench has been finished with, release the lock (where applicable) and fully back off its setting to zero - do not leave the torque wrench tensioned. Also, do not use a torque wrench for slackening a fastener.

Angle-tightening

- Manufacturers often specify a figure in degrees for final tightening of a fastener. This usually follows tightening to a specific torque setting.
- A degree disc can be set and attached to the socket (see illustration 4.2) or a protractor can be used to mark the angle of movement on the bolt/nut head and the surrounding casting (see illustration 4.3).

4.2 Angle tightening can be accomplished with a torque-angle gauge . . .

4.3 . . . or by marking the angle on the surrounding component

Loosening sequences

- Where more than one bolt/nut secures a component, loosen each fastener evenly a little at a time. In this way, not all the stress of the joint is held by one fastener and the components are not likely to distort.
- If a tightening sequence is provided, work in the REVERSE of this, but if not, work from the outside in, in a criss-cross sequence (see illustration 4.4).

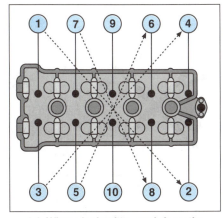

4.4 When slackening, work from the outside inwards

Tightening sequences

- If a component is held by more than one fastener it is important that the retaining bolts/nuts are tightened evenly to prevent uneven stress build-up and distortion of sealing faces. This is especially important on high-compression joints such as the cylinder head.
- A sequence is usually provided by the manufacturer, either in a diagram or actually marked in the casting. If not, always start in the centre and work outwards in a criss-cross pattern (see illustration 4.5). Start off by securing all bolts/nuts finger-tight, then set the torque wrench and tighten each fastener by a small amount in sequence until the final torque is reached. By following this practice,

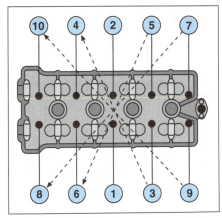

4.5 When tightening, work from the inside outwards

REF•14 Tools and Workshop Tips

the joint will be held evenly and will not be distorted. Important joints, such as the cylinder head and big-end fasteners often have two- or three-stage torque settings.

Applying leverage

● Use tools at the correct angle. Position a socket wrench or spanner on the bolt/nut so that you pull it towards you when loosening. If this can't be done, push the spanner without curling your fingers around it **(see illustration 4.6)** - the spanner may slip or the fastener loosen suddenly, resulting in your fingers being crushed against a component.

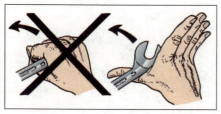

4.6 If you can't pull on the spanner to loosen a fastener, push with your hand open

● Additional leverage is gained by extending the length of the lever. The best way to do this is to use a breaker bar instead of the regular length tool, or to slip a length of tubing over the end of the spanner or socket wrench.
● If additional leverage will not work, the fastener head is either damaged or firmly corroded in place (see *Fasteners*).

5 Bearings

Bearing removal and installation
Drivers and sockets

● Before removing a bearing, always inspect the casing to see which way it must be driven out - some casings will have retaining plates or a cast step. Also check for any identifying markings on the bearing and if installed to a certain depth, measure this at this stage. Some roller bearings are sealed on one side - take note of the original fitted position.
● Bearings can be driven out of a casing using a bearing driver tool (with the correct size head) or a socket of the correct diameter. Select the driver head or socket so that it contacts the outer race of the bearing, not the balls/rollers or inner race. Always support the casing around the bearing housing with wood blocks, otherwise there is a risk of fracture. The bearing is driven out with a few blows on the driver or socket from a heavy mallet. Unless access is severely restricted (as with wheel bearings), a pin-punch is not recommended unless it is moved around the bearing to keep it square in its housing.

● The same equipment can be used to install bearings. Make sure the bearing housing is supported on wood blocks and line up the bearing in its housing. Fit the bearing as noted on removal - generally they are installed with their marked side facing outwards. Tap the bearing squarely into its housing using a driver or socket which bears only on the bearing's outer race - contact with the bearing balls/rollers or inner race will destroy it **(see illustrations 5.1 and 5.2)**.
● Check that the bearing inner race and balls/rollers rotate freely.

5.1 Using a bearing driver against the bearing's outer race

5.2 Using a large socket against the bearing's outer race

Pullers and slide-hammers

● Where a bearing is pressed on a shaft a puller will be required to extract it **(see illustration 5.3)**. Make sure that the puller clamp or legs fit securely behind the bearing and are unlikely to slip out. If pulling a bearing

5.3 This bearing puller clamps behind the bearing and pressure is applied to the shaft end to draw the bearing off

off a gear shaft for example, you may have to locate the puller behind a gear pinion if there is no access to the race and draw the gear pinion off the shaft as well **(see illustration 5.4)**.

Caution: Ensure that the puller's centre bolt locates securely against the end of the shaft and will not slip when pressure is applied. Also ensure that puller does not damage the shaft end.

5.4 Where no access is available to the rear of the bearing, it is sometimes possible to draw off the adjacent component

● Operate the puller so that its centre bolt exerts pressure on the shaft end and draws the bearing off the shaft.
● When installing the bearing on the shaft, tap only on the bearing's inner race - contact with the balls/rollers or outer race with destroy the bearing. Use a socket or length of tubing as a drift which fits over the shaft end **(see illustration 5.5)**.

5.5 When installing a bearing on a shaft use a piece of tubing which bears only on the bearing's inner race

● Where a bearing locates in a blind hole in a casing, it cannot be driven or pulled out as described above. A slide-hammer with knife-edged bearing puller attachment will be required. The puller attachment passes through the bearing and when tightened expands to fit firmly behind the bearing **(see illustration 5.6)**. By operating the slide-hammer part of the tool the bearing is jarred out of its housing **(see illustration 5.7)**.
● It is possible, if the bearing is of reasonable weight, for it to drop out of its housing if the casing is heated as described opposite. If this

Tools and Workshop Tips

5.6 Expand the bearing puller so that it locks behind the bearing . . .

5.7 . . . attach the slide hammer to the bearing puller

method is attempted, first prepare a work surface which will enable the casing to be tapped face down to help dislodge the bearing - a wood surface is ideal since it will not damage the casing's gasket surface. Wearing protective gloves, tap the heated casing several times against the work surface to dislodge the bearing under its own weight **(see illustration 5.8)**.

5.8 Tapping a casing face down on wood blocks can often dislodge a bearing

● Bearings can be installed in blind holes using the driver or socket method described above.

Drawbolts

● Where a bearing or bush is set in the eye of a component, such as a suspension linkage arm or connecting rod small-end, removal by drift may damage the component. Furthermore, a rubber bushing in a shock absorber eye cannot successfully be driven out of position. If access is available to a engineering press, the task is straightforward. If not, a drawbolt can be fabricated to extract the bearing or bush.

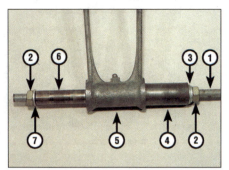

5.9 Drawbolt component parts assembled on a suspension arm

1. Bolt or length of threaded bar
2. Nuts
3. Washer (external diameter greater than tubing internal diameter)
4. Tubing (internal diameter sufficient to accommodate bearing)
5. Suspension arm with bearing
6. Tubing (external diameter slightly smaller than bearing)
7. Washer (external diameter slightly smaller than bearing)

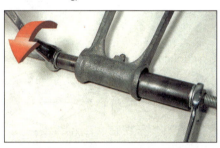

5.10 Drawing the bearing out of the suspension arm

● To extract the bearing/bush you will need a long bolt with nut (or piece of threaded bar with two nuts), a piece of tubing which has an internal diameter larger than the bearing/bush, another piece of tubing which has an external diameter slightly smaller than the bearing/bush, and a selection of washers **(see illustrations 5.9 and 5.10)**. Note that the pieces of tubing must be of the same length, or longer, than the bearing/bush.

● The same kit (without the pieces of tubing) can be used to draw the new bearing/bush back into place **(see illustration 5.11)**.

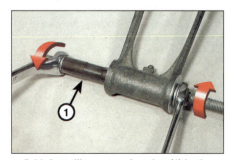

5.11 Installing a new bearing (1) in the suspension arm

Temperature change

● If the bearing's outer race is a tight fit in the casing, the aluminium casing can be heated to release its grip on the bearing. Aluminium will expand at a greater rate than the steel bearing outer race. There are several ways to do this, but avoid any localised extreme heat (such as a blow torch) - aluminium alloy has a low melting point.

● Approved methods of heating a casing are using a domestic oven (heated to 100°C) or immersing the casing in boiling water **(see illustration 5.12)**. Low temperature range localised heat sources such as a paint stripper heat gun or clothes iron can also be used **(see illustration 5.13)**. Alternatively, soak a rag in boiling water, wring it out and wrap it around the bearing housing.

> ⚠ **Warning:** *All of these methods require care in use to prevent scalding and burns to the hands. Wear protective gloves when handling hot components.*

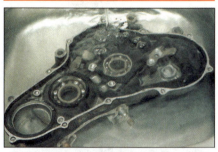

5.12 A casing can be immersed in a sink of boiling water to aid bearing removal

5.13 Using a localised heat source to aid bearing removal

● If heating the whole casing note that plastic components, such as the neutral switch, may suffer - remove them beforehand.

● After heating, remove the bearing as described above. You may find that the expansion is sufficient for the bearing to fall out of the casing under its own weight or with a light tap on the driver or socket.

● If necessary, the casing can be heated to aid bearing installation, and this is sometimes the recommended procedure if the motorcycle manufacturer has designed the housing and bearing fit with this intention.

REF•16 Tools and Workshop Tips

- Installation of bearings can be eased by placing them in a freezer the night before installation. The steel bearing will contract slightly, allowing easy insertion in its housing. This is often useful when installing steering head outer races in the frame.

Bearing types and markings

- Plain shell bearings, ball bearings, needle roller bearings and tapered roller bearings will all be found on motorcycles (see illustrations 5.14 and 5.15). The ball and roller types are usually caged between an inner and outer race, but uncaged variations may be found.

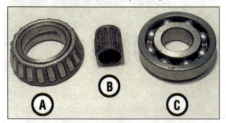

5.14 Shell bearings are either plain or grooved. They are usually identified by colour code (arrow)

5.15 Tapered roller bearing (A), needle roller bearing (B) and ball journal bearing (C)

- Shell bearings (often called inserts) are usually found at the crankshaft main and connecting rod big-end where they are good at coping with high loads. They are made of a phosphor-bronze material and are impregnated with self-lubricating properties.
- Ball bearings and needle roller bearings consist of a steel inner and outer race with the balls or rollers between the races. They require constant lubrication by oil or grease and are good at coping with axial loads. Taper roller bearings consist of rollers set in a tapered cage set on the inner race; the outer race is separate. They are good at coping with axial loads and prevent movement along the shaft - a typical application is in the steering head.
- Bearing manufacturers produce bearings to ISO size standards and stamp one face of the bearing to indicate its internal and external diameter, load capacity and type (see illustration 5.16).
- Metal bushes are usually of phosphor-bronze material. Rubber bushes are used in suspension mounting eyes. Fibre bushes have also been used in suspension pivots.

5.16 Typical bearing marking

Bearing fault finding

- If a bearing outer race has spun in its housing, the housing material will be damaged. You can use a bearing locking compound to bond the outer race in place if damage is not too severe.
- Shell bearings will fail due to damage of their working surface, as a result of lack of lubrication, corrosion or abrasive particles in the oil (see illustration 5.17). Small particles of dirt in the oil may embed in the bearing material whereas larger particles will score the bearing and shaft journal. If a number of short journeys are made, insufficient heat will be generated to drive off condensation which has built up on the bearings.

5.17 Typical bearing failures

- Ball and roller bearings will fail due to lack of lubrication or damage to the balls or rollers. Tapered-roller bearings can be damaged by overloading them. Unless the bearing is sealed on both sides, wash it in paraffin (kerosene) to remove all old grease then allow it to dry. Make a visual inspection looking to dented balls or rollers, damaged cages and worn or pitted races (see illustration 5.18).
- A ball bearing can be checked for wear by listening to it when spun. Apply a film of light oil to the bearing and hold it close to the ear - hold the outer race with one hand and spin the inner race with the other hand (see illustration 5.19). The bearing should be almost silent when spun; if it grates or rattles it is worn.

5.18 Example of ball journal bearing with damaged balls and cages

5.19 Hold outer race and listen to inner race when spun

6 Oil seals

Oil seal removal and installation

- Oil seals should be renewed every time a component is dismantled. This is because the seal lips will become set to the sealing surface and will not necessarily reseal.
- Oil seals can be prised out of position using a large flat-bladed screwdriver (see illustration 6.1). In the case of crankcase seals, check first that the seal is not lipped on the inside, preventing its removal with the crankcases joined.

6.1 Prise out oil seals with a large flat-bladed screwdriver

- New seals are usually installed with their marked face (containing the seal reference code) outwards and the spring side towards the fluid being retained. In certain cases, such as a two-stroke engine crankshaft seal, a double lipped seal may be used due to there being fluid or gas on each side of the joint.

Tools and Workshop Tips

● Use a bearing driver or socket which bears only on the outer hard edge of the seal to install it in the casing - tapping on the inner edge will damage the sealing lip.

Oil seal types and markings

● Oil seals are usually of the single-lipped type. Double-lipped seals are found where a liquid or gas is on both sides of the joint.
● Oil seals can harden and lose their sealing ability if the motorcycle has been in storage for a long period - renewal is the only solution.
● Oil seal manufacturers also conform to the ISO markings for seal size - these are moulded into the outer face of the seal **(see illustration 6.2)**.

6.2 These oil seal markings indicate inside diameter, outside diameter and seal thickness

7 Gaskets and sealants

Types of gasket and sealant

● Gaskets are used to seal the mating surfaces between components and keep lubricants, fluids, vacuum or pressure contained within the assembly. Aluminium gaskets are sometimes found at the cylinder joints, but most gaskets are paper-based. If the mating surfaces of the components being joined are undamaged the gasket can be installed dry, although a dab of sealant or grease will be useful to hold it in place during assembly.
● RTV (Room Temperature Vulcanising) silicone rubber sealants cure when exposed to moisture in the atmosphere. These sealants are good at filling pits or irregular gasket faces, but will tend to be forced out of the joint under very high torque. They can be used to replace a paper gasket, but first make sure that the width of the paper gasket is not essential to the shimming of internal components. RTV sealants should not be used on components containing petrol (gasoline).
● Non-hardening, semi-hardening and hard setting liquid gasket compounds can be used with a gasket or between a metal-to-metal joint. Select the sealant to suit the application: universal non-hardening sealant can be used on virtually all joints; semi-hardening on joint faces which are rough or damaged; hard setting sealant on joints which require a permanent bond and are subjected to high temperature and pressure. **Note:** *Check first if the paper gasket has a bead of sealant impregnated in its surface before applying additional sealant.*
● When choosing a sealant, make sure it is suitable for the application, particularly if being applied in a high-temperature area or in the vicinity of fuel. Certain manufacturers produce sealants in either clear, silver or black colours to match the finish of the engine. This has a particular application on motorcycles where much of the engine is exposed.
● Do not over-apply sealant. That which is squeezed out on the outside of the joint can be wiped off, whereas an excess of sealant on the inside can break off and clog oilways.

Breaking a sealed joint

● Age, heat, pressure and the use of hard setting sealant can cause two components to stick together so tightly that they are difficult to separate using finger pressure alone. Do not resort to using levers unless there is a pry point provided for this purpose **(see illustration 7.1)** or else the gasket surfaces will be damaged.
● Use a soft-faced hammer **(see illustration 7.2)** or a wood block and conventional hammer to strike the component near the mating surface. Avoid hammering against cast extremities since they may break off. If this method fails, try using a wood wedge between the two components.

Caution: If the joint will not separate, double-check that you have removed all the fasteners.

7.1 If a pry point is provided, apply gently pressure with a flat-bladed screwdriver

7.2 Tap around the joint with a soft-faced mallet if necessary - don't strike cooling fins

Removal of old gasket and sealant

● Paper gaskets will most likely come away complete, leaving only a few traces stuck on

HAYNES HiNT

Most components have one or two hollow locating dowels between the two gasket faces. If a dowel cannot be removed, do not resort to gripping it with pliers - it will almost certainly be distorted. Install a close-fitting socket or Phillips screwdriver into the dowel and then grip the outer edge of the dowel to free it.

the sealing faces of the components. It is imperative that all traces are removed to ensure correct sealing of the new gasket.
● Very carefully scrape all traces of gasket away making sure that the sealing surfaces are not gouged or scored by the scraper **(see illustrations 7.3, 7.4 and 7.5)**. Stubborn deposits can be removed by spraying with an aerosol gasket remover. Final preparation of

7.3 Paper gaskets can be scraped off with a gasket scraper tool . . .

7.4 . . . a knife blade . . .

7.5 . . . or a household scraper

REF•18 Tools and Workshop Tips

7.6 Fine abrasive paper is wrapped around a flat file to clean up the gasket face

7.7 A kitchen scourer can be used on stubborn deposits

the gasket surface can be made with very fine abrasive paper or a plastic kitchen scourer **(see illustrations 7.6 and 7.7)**.
● Old sealant can be scraped or peeled off components, depending on the type originally used. Note that gasket removal compounds are available to avoid scraping the components clean; make sure the gasket remover suits the type of sealant used.

8 Chains

Breaking and joining final drive chains

● Drive chains for all but small bikes are continuous and do not have a clip-type connecting link. The chain must be broken using a chain breaker tool and the new chain securely riveted together using a new soft rivet-type link. Never use a clip-type connecting link instead of a rivet-type link, except in an emergency. Various chain breaking and riveting tools are available, either as separate tools or combined as illustrated in the accompanying photographs - read the instructions supplied with the tool carefully.

> **Warning: The need to rivet the new link pins correctly cannot be overstressed - loss of control of the motorcycle is very likely to result if the chain breaks in use.**

● Rotate the chain and look for the soft link. The soft link pins look like they have been

8.1 Tighten the chain breaker to push the pin out of the link . . .

8.2 . . . withdraw the pin, remove the tool . . .

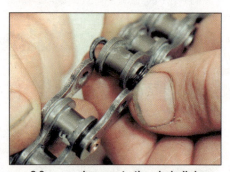

8.3 . . . and separate the chain link

deeply centre-punched instead of peened over like all the other pins **(see illustration 8.9)** and its sideplate may be a different colour. Position the soft link midway between the sprockets and assemble the chain breaker tool over one of the soft link pins **(see illustration 8.1)**. Operate the tool to push the pin out through the chain **(see illustration 8.2)**. On an O-ring chain, remove the O-rings **(see illustration 8.3)**. Carry out the same procedure on the other soft link pin.

> **Caution: Certain soft link pins (particularly on the larger chains) may require their ends to be filed or ground off before they can be pressed out using the tool.**

● Check that you have the correct size and strength (standard or heavy duty) new soft link - do not reuse the old link. Look for the size marking on the chain sideplates **(see illustration 8.10)**.
● Position the chain ends so that they are engaged over the rear sprocket. On an O-ring

8.4 Insert the new soft link, with O-rings, through the chain ends . . .

8.5 . . . install the O-rings over the pin ends . . .

8.6 . . . followed by the sideplate

chain, install a new O-ring over each pin of the link and insert the link through the two chain ends **(see illustration 8.4)**. Install a new O-ring over the end of each pin, followed by the sideplate (with the chain manufacturer's marking facing outwards) **(see illustrations 8.5 and 8.6)**. On an unsealed chain, insert the link through the two chain ends, then install the sideplate with the chain manufacturer's marking facing outwards.
● Note that it may not be possible to install the sideplate using finger pressure alone. If using a joining tool, assemble it so that the plates of the tool clamp the link and press the sideplate over the pins **(see illustration 8.7)**. Otherwise, use two small sockets placed over

8.7 Push the sideplate into position using a clamp

Tools and Workshop Tips

8.8 Assemble the chain riveting tool over one pin at a time and tighten it fully

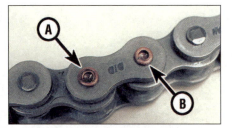

8.9 Pin end correctly riveted (A), pin end unriveted (B)

the rivet ends and two pieces of the wood between a G-clamp. Operate the clamp to press the sideplate over the pins.
● Assemble the joining tool over one pin (following the maker's instructions) and tighten the tool down to spread the pin end securely **(see illustrations 8.8 and 8.9)**. Do the same on the other pin.

> ⚠ **Warning: Check that the pin ends are secure and that there is no danger of the sideplate coming loose. If the pin ends are cracked the soft link must be renewed.**

Final drive chain sizing

● Chains are sized using a three digit number, followed by a suffix to denote the chain type **(see illustration 8.10)**. Chain type is either standard or heavy duty (thicker sideplates), and also unsealed or O-ring/X-ring type.
● The first digit of the number relates to the pitch of the chain, ie the distance from the centre of one pin to the centre of the next pin **(see illustration 8.11)**. Pitch is expressed in eighths of an inch, as follows:

8.10 Typical chain size and type marking

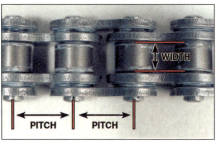

8.11 Chain dimensions

| Sizes commencing with a 4 (eg 428) have a pitch of 1/2 inch (12.7 mm) |
| Sizes commencing with a 5 (eg 520) have a pitch of 5/8 inch (15.9 mm) |
| Sizes commencing with a 6 (eg 630) have a pitch of 3/4 inch (19.1 mm) |

● The second and third digits of the chain size relate to the width of the rollers, again in imperial units, eg the 525 shown has 5/16 inch (7.94 mm) rollers **(see illustration 8.11)**.

9 Hoses

Clamping to prevent flow

● Small-bore flexible hoses can be clamped to prevent fluid flow whilst a component is worked on. Whichever method is used, ensure that the hose material is not permanently distorted or damaged by the clamp.
a) A brake hose clamp available from auto accessory shops **(see illustration 9.1)**.
b) A wingnut type hose clamp **(see illustration 9.2)**.

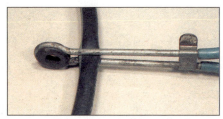

9.1 Hoses can be clamped with an automotive brake hose clamp . . .

9.2 . . . a wingnut type hose clamp . . .

c) Two sockets placed each side of the hose and held with straight-jawed self-locking grips **(see illustration 9.3)**.
d) Thick card each side of the hose held between straight-jawed self-locking grips **(see illustration 9.4)**.

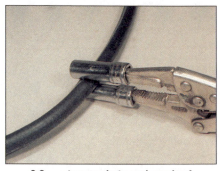

9.3 . . . two sockets and a pair of self-locking grips . . .

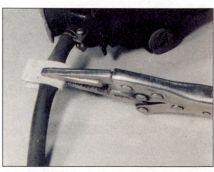

9.4 . . . or thick card and self-locking grips

Freeing and fitting hoses

● Always make sure the hose clamp is moved well clear of the hose end. Grip the hose with your hand and rotate it whilst pulling it off the union. If the hose has hardened due to age and will not move, slit it with a sharp knife and peel its ends off the union **(see illustration 9.5)**.
● Resist the temptation to use grease or soap on the unions to aid installation; although it helps the hose slip over the union it will equally aid the escape of fluid from the joint. It is preferable to soften the hose ends in hot water and wet the inside surface of the hose with water or a fluid which will evaporate.

9.5 Cutting a coolant hose free with a sharp knife

Conversion Factors

Length (distance)
Inches (in)	x 25.4	= Millimetres (mm)	x 0.0394	=	Inches (in)
Feet (ft)	x 0.305	= Metres (m)	x 3.281	=	Feet (ft)
Miles	x 1.609	= Kilometres (km)	x 0.621	=	Miles

Volume (capacity)
Cubic inches (cu in; in³)	x 16.387	= Cubic centimetres (cc; cm³)	x 0.061	=	Cubic inches (cu in; in³)
Imperial pints (Imp pt)	x 0.568	= Litres (l)	x 1.76	=	Imperial pints (Imp pt)
Imperial quarts (Imp qt)	x 1.137	= Litres (l)	x 0.88	=	Imperial quarts (Imp qt)
Imperial quarts (Imp qt)	x 1.201	= US quarts (US qt)	x 0.833	=	Imperial quarts (Imp qt)
US quarts (US qt)	x 0.946	= Litres (l)	x 1.057	=	US quarts (US qt)
Imperial gallons (Imp gal)	x 4.546	= Litres (l)	x 0.22	=	Imperial gallons (Imp gal)
Imperial gallons (Imp gal)	x 1.201	= US gallons (US gal)	x 0.833	=	Imperial gallons (Imp gal)
US gallons (US gal)	x 3.785	= Litres (l)	x 0.264	=	US gallons (US gal)

Mass (weight)
Ounces (oz)	x 28.35	= Grams (g)	x 0.035	=	Ounces (oz)
Pounds (lb)	x 0.454	= Kilograms (kg)	x 2.205	=	Pounds (lb)

Force
Ounces-force (ozf; oz)	x 0.278	= Newtons (N)	x 3.6	=	Ounces-force (ozf; oz)
Pounds-force (lbf; lb)	x 4.448	= Newtons (N)	x 0.225	=	Pounds-force (lbf; lb)
Newtons (N)	x 0.1	= Kilograms-force (kgf; kg)	x 9.81	=	Newtons (N)

Pressure
Pounds-force per square inch (psi; lbf/in²; lb/in²)	x 0.070	= Kilograms-force per square centimetre (kgf/cm²; kg/cm²)	x 14.223	=	Pounds-force per square inch (psi; lbf/in²; lb/in²)
Pounds-force per square inch (psi; lbf/in²; lb/in²)	x 0.068	= Atmospheres (atm)	x 14.696	=	Pounds-force per square inch (psi; lbf/in²; lb/in²)
Pounds-force per square inch (psi; lbf/in²; lb/in²)	x 0.069	= Bars	x 14.5	=	Pounds-force per square inch (psi; lbf/in²; lb/in²)
Pounds-force per square inch (psi; lbf/in²; lb/in²)	x 6.895	= Kilopascals (kPa)	x 0.145	=	Pounds-force per square inch (psi; lbf/in²; lb/in²)
Kilopascals (kPa)	x 0.01	= Kilograms-force per square centimetre (kgf/cm²; kg/cm²)	x 98.1	=	Kilopascals (kPa)
Millibar (mbar)	x 100	= Pascals (Pa)	x 0.01	=	Millibar (mbar)
Millibar (mbar)	x 0.0145	= Pounds-force per square inch (psi; lbf/in²; lb/in²)	x 68.947	=	Millibar (mbar)
Millibar (mbar)	x 0.75	= Millimetres of mercury (mmHg)	x 1.333	=	Millibar (mbar)
Millibar (mbar)	x 0.401	= Inches of water (inH$_2$O)	x 2.491	=	Millibar (mbar)
Millimetres of mercury (mmHg)	x 0.535	= Inches of water (inH$_2$O)	x 1.868	=	Millimetres of mercury (mmHg)
Inches of water (inH$_2$O)	x 0.036	= Pounds-force per square inch (psi; lbf/in²; lb/in²)	x 27.68	=	Inches of water (inH$_2$O)

Torque (moment of force)
Pounds-force inches (lbf in; lb in)	x 1.152	= Kilograms-force centimetre (kgf cm; kg cm)	x 0.868	=	Pounds-force inches (lbf in; lb in)
Pounds-force inches (lbf in; lb in)	x 0.113	= Newton metres (Nm)	x 8.85	=	Pounds-force inches (lbf in; lb in)
Pounds-force inches (lbf in; lb in)	x 0.083	= Pounds-force feet (lbf ft; lb ft)	x 12	=	Pounds-force inches (lbf in; lb in)
Pounds-force feet (lbf ft; lb ft)	x 0.138	= Kilograms-force metres (kgf m; kg m)	x 7.233	=	Pounds-force feet (lbf ft; lb ft)
Pounds-force feet (lbf ft; lb ft)	x 1.356	= Newton metres (Nm)	x 0.738	=	Pounds-force feet (lbf ft; lb ft)
Newton metres (Nm)	x 0.102	= Kilograms-force metres (kgf m; kg m)	x 9.804	=	Newton metres (Nm)

Power
Horsepower (hp)	x 745.7	= Watts (W)	x 0.0013	=	Horsepower (hp)

Velocity (speed)
Miles per hour (miles/hr; mph)	x 1.609	= Kilometres per hour (km/hr; kph)	x 0.621	=	Miles per hour (miles/hr; mph)

Fuel consumption*
Miles per gallon (mpg)	x 0.354	= Kilometres per litre (km/l)	x 2.825	=	Miles per gallon (mpg)

Temperature

Degrees Fahrenheit = (°C x 1.8) + 32   Degrees Celsius (Degrees Centigrade; °C) = (°F - 32) x 0.56

It is common practice to convert from miles per gallon (mpg) to litres/100 kilometres (l/100km), where mpg x l/100 km = 282

Motorcycle Chemicals and Lubricants

A number of chemicals and lubricants are available for use in motorcycle maintenance and repair. They include a wide variety of products ranging from cleaning solvents and degreasers to lubricants and protective sprays for rubber, plastic and vinyl.

● **Contact point/spark plug cleaner** is a solvent used to clean oily film and dirt from points, grime from electrical connectors and oil deposits from spark plugs. It is oil free and leaves no residue. It can also be used to remove gum and varnish from carburettor jets and other orifices.

● **Carburettor cleaner** is similar to contact point/spark plug cleaner but it usually has a stronger solvent and may leave a slight oily reside. It is not recommended for cleaning electrical components or connections.

● **Brake system cleaner** is used to remove grease or brake fluid from brake system components (where clean surfaces are absolutely necessary and petroleum-based solvents cannot be used); it also leaves no residue.

● **Silicone-based lubricants** are used to protect rubber parts such as hoses and grommets, and are used as lubricants for hinges and locks.

● **Multi-purpose grease** is an all purpose lubricant used wherever grease is more practical than a liquid lubricant such as oil. Some multi-purpose grease is coloured white and specially formulated to be more resistant to water than ordinary grease.

● **Gear oil** (sometimes called gear lube) is a specially designed oil used in transmissions and final drive units, as well as other areas where high friction, high temperature lubrication is required. It is available in a number of viscosities (weights) for various applications.

● **Motor oil**, of course, is the lubricant specially formulated for use in the engine. It normally contains a wide variety of additives to prevent corrosion and reduce foaming and wear. Motor oil comes in various weights (viscosity ratings) of from 5 to 80. The recommended weight of the oil depends on the seasonal temperature and the demands on the engine. Light oil is used in cold climates and under light load conditions; heavy oil is used in hot climates and where high loads are encountered. Multi-viscosity oils are designed to have characteristics of both light and heavy oils and are available in a number of weights from 5W-20 to 20W-50.

● **Petrol additives** perform several functions, depending on their chemical makeup. They usually contain solvents that help dissolve gum and varnish that build up on carburettor and inlet parts. They also serve to break down carbon deposits that form on the inside surfaces of the combustion chambers. Some additives contain upper cylinder lubricants for valves and piston rings.

● **Brake and clutch fluid** is a specially formulated hydraulic fluid that can withstand the heat and pressure encountered in brake/clutch systems. Care must be taken that this fluid does not come in contact with painted surfaces or plastics. An opened container should always be resealed to prevent contamination by water or dirt.

● **Chain lubricants** are formulated especially for use on motorcycle final drive chains. A good chain lube should adhere well and have good penetrating qualities to be effective as a lubricant inside the chain and on the side plates, pins and rollers. Most chain lubes are either the foaming type or quick drying type and are usually marketed as sprays. Take care to use a lubricant marked as being suitable for O-ring chains.

● **Degreasers** are heavy duty solvents used to remove grease and grime that may accumulate on engine and frame components. They can be sprayed or brushed on and, depending on the type, are rinsed with either water or solvent.

● **Solvents** are used alone or in combination with degreasers to clean parts and assemblies during repair and overhaul. The home mechanic should use only solvents that are non-flammable and that do not produce irritating fumes.

● **Gasket sealing compounds** may be used in conjunction with gaskets, to improve their sealing capabilities, or alone, to seal metal-to-metal joints. Many gasket sealers can withstand extreme heat, some are impervious to petrol and lubricants, while others are capable of filling and sealing large cavities. Depending on the intended use, gasket sealers either dry hard or stay relatively soft and pliable. They are usually applied by hand, with a brush, or are sprayed on the gasket sealing surfaces.

● **Thread locking compound** is an adhesive locking compound that prevents threaded fasteners from loosening because of vibration. It is available in a variety of types for different applications.

● **Moisture dispersants** are usually sprays that can be used to dry out electrical components such as the fuse block and wiring connectors. Some types can also be used as treatment for rubber and as a lubricant for hinges, cables and locks.

● **Waxes and polishes** are used to help protect painted and plated surfaces from the weather. Different types of paint may require the use of different types of wax polish. Some polishes utilise a chemical or abrasive cleaner to help remove the top layer of oxidised (dull) paint on older vehicles. In recent years, many non-wax polishes (that contain a wide variety of chemicals such as polymers and silicones) have been introduced. These non-wax polishes are usually easier to apply and last longer than conventional waxes and polishes.

REF•22 MOT Test Checks

About the MOT Test

In the UK, all vehicles more than three years old are subject to an annual test to ensure that they meet minimum safety requirements. A current test certificate must be issued before a machine can be used on public roads, and is required before a road fund licence can be issued. Riding without a current test certificate will also invalidate your insurance.

For most owners, the MOT test is an annual cause for anxiety, and this is largely due to owners not being sure what needs to be checked prior to submitting the motorcycle for testing. The simple answer is that a fully roadworthy motorcycle will have no difficulty in passing the test.

This is a guide to getting your motorcycle through the MOT test. Obviously it will not be possible to examine the motorcycle to the same standard as the professional MOT tester, particularly in view of the equipment required for some of the checks. However, working through the following procedures will enable you to identify any problem areas before submitting the motorcycle for the test.

It has only been possible to summarise the test requirements here, based on the regulations in force at the time of printing. Test standards are becoming increasingly stringent, although there are some exemptions for older vehicles. More information about the MOT test can be obtained from the TSO publications, *How Safe is your Motorcycle* and *The MOT Inspection Manual for Motorcycle Testing*.

Many of the checks require that one of the wheels is raised off the ground. If the motorcycle doesn't have a centre stand, note that an auxiliary stand will be required. Additionally, the help of an assistant may prove useful.

Certain exceptions apply to machines under 50 cc, machines without a lighting system, and Classic bikes - if in doubt about any of the requirements listed below seek confirmation from an MOT tester prior to submitting the motorcycle for the test.

Check that the frame number is clearly visible.

> **HAYNES HiNT** *If a component is in borderline condition, the tester has discretion in deciding whether to pass or fail it. If the motorcycle presented is clean and evidently well cared for, the tester may be more inclined to pass a borderline component than if the motorcycle is scruffy and apparently neglected.*

Electrical System

Lights, turn signals, horn and reflector

✔ With the ignition on, check the operation of the following electrical components. **Note:** *The electrical components on certain small-capacity machines are powered by the generator, requiring that the engine is run for this check.*

a) Headlight and tail light. Check that both illuminate in the low and high beam switch positions.
b) Position lights. Check that the front position (or sidelight) and tail light illuminate in this switch position.
c) Turn signals. Check that all flash at the correct rate, and that the warning light(s) function correctly. Check that the turn signal switch works correctly.
d) Hazard warning system (where fitted). Check that all four turn signals flash in this switch position.
e) Brake stop light. Check that the light comes on when the front and rear brakes are independently applied. Models first used on or after 1st April 1986 must have a brake light switch on each brake.
f) Horn. Check that the sound is continuous and of reasonable volume.

✔ Check that there is a red reflector on the rear of the machine, either mounted separately or as part of the tail light lens.
✔ Check the condition of the headlight, tail light and turn signal lenses.

Headlight beam height

✔ The MOT tester will perform a headlight beam height check using specialised beam setting equipment **(see illustration 1)**. This equipment will not be available to the home mechanic, but if you suspect that the headlight is incorrectly set or may have been maladjusted in the past, you can perform a rough test as follows.

✔ Position the bike in a straight line facing a brick wall. The bike must be off its stand, upright and with a rider seated. Measure the height from the ground to the centre of the headlight and mark a horizontal line on the wall at this height. Position the motorcycle 3.8 metres from the wall and draw a vertical line up the wall central to the centreline of the motorcycle. Switch to dipped beam and check that the beam pattern falls slightly lower than the horizontal line and to the left of the vertical line **(see illustration 2)**.

Headlight beam height checking equipment

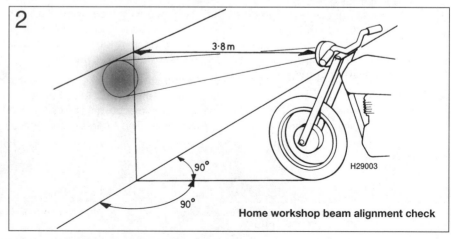

Home workshop beam alignment check

MOT Test Checks REF•23

Exhaust System and Final Drive

Exhaust

✔ Check that the exhaust mountings are secure and that the system does not foul any of the rear suspension components.

✔ Start the motorcycle. When the revs are increased, check that the exhaust is neither holed nor leaking from any of its joints. On a linked system, check that the collector box is not leaking due to corrosion.

✔ Note that the exhaust decibel level ("loudness" of the exhaust) is assessed at the discretion of the tester. If the motorcycle was first used on or after 1st January 1985 the silencer must carry the BSAU 193 stamp, or a marking relating to its make and model, or be of OE (original equipment) manufacture. If the silencer is marked NOT FOR ROAD USE, RACING USE ONLY or similar, it will fail the MOT.

Final drive

✔ On chain or belt drive machines, check that the chain/belt is in good condition and does not have excessive slack. Also check that the sprocket is securely mounted on the rear wheel hub. Check that the chain/belt guard is in place.

✔ On shaft drive bikes, check for oil leaking from the drive unit and fouling the rear tyre.

Steering and Suspension

Steering

✔ With the front wheel raised off the ground, rotate the steering from lock to lock. The handlebar or switches must not contact the fuel tank or be close enough to trap the rider's hand. Problems can be caused by damaged lock stops on the lower yoke and frame, or by the fitting of non-standard handlebars.

✔ When performing the lock to lock check, also ensure that the steering moves freely without drag or notchiness. Steering movement can be impaired by poorly routed cables, or by overtight head bearings or worn bearings. The tester will perform a check of the steering head bearing lower race by mounting the front wheel on a surface plate, then performing a lock to lock check with the weight of the machine on the lower bearing (see illustration 3).

✔ Grasp the fork sliders (lower legs) and attempt to push and pull on the forks (see illustration 4). Any play in the steering head bearings will be felt. Note that in extreme cases, wear of the front fork bushes can be misinterpreted for head bearing play.

✔ Check that the handlebars are securely mounted.

✔ Check that the handlebar grip rubbers are secure. They should by bonded to the bar left end and to the throttle cable pulley on the right end.

Front wheel mounted on a surface plate for steering head bearing lower race check

Front suspension

✔ With the motorcycle off the stand, hold the front brake on and pump the front forks up and down (see illustration 5). Check that they are adequately damped.

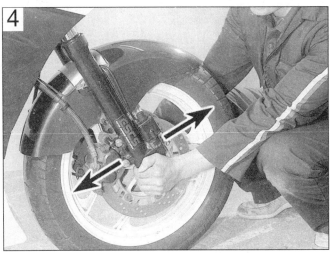

Checking the steering head bearings for freeplay

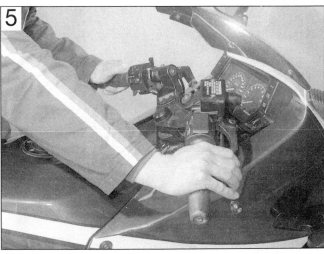

Hold the front brake on and pump the front forks up and down to check operation

REF•24 MOT Test Checks

Inspect the area around the fork dust seal for oil leakage (arrow)

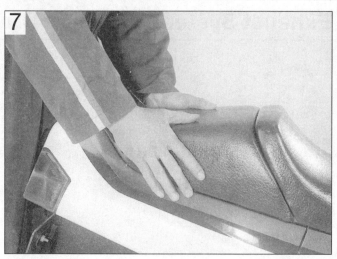

Bounce the rear of the motorcycle to check rear suspension operation

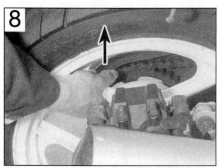

Checking for rear suspension linkage play

✔ Inspect the area above and around the front fork oil seals **(see illustration 6)**. There should be no sign of oil on the fork tube (stanchion) nor leaking down the slider (lower leg). On models so equipped, check that there is no oil leaking from the anti-dive units.
✔ On models with swingarm front suspension, check that there is no freeplay in the linkage when moved from side to side.

Rear suspension

✔ With the motorcycle off the stand and an assistant supporting the motorcycle by its handlebars, bounce the rear suspension **(see illustration 7)**. Check that the suspension components do not foul on any of the cycle parts and check that the shock absorber(s) provide adequate damping.
✔ Visually inspect the shock absorber(s) and check that there is no sign of oil leakage from its damper. This is somewhat restricted on certain single shock models due to the location of the shock absorber.
✔ With the rear wheel raised off the ground, grasp the wheel at the highest point and attempt to pull it up **(see illustration 8)**. Any play in the swingarm pivot or suspension linkage bearings will be felt as movement.
Note: *Do not confuse play with actual suspension movement.* Failure to lubricate suspension linkage bearings can lead to bearing failure **(see illustration 9)**.
✔ With the rear wheel raised off the ground, grasp the swingarm ends and attempt to move the swingarm from side to side and forwards and backwards - any play indicates wear of the swingarm pivot bearings **(see illustration 10)**.

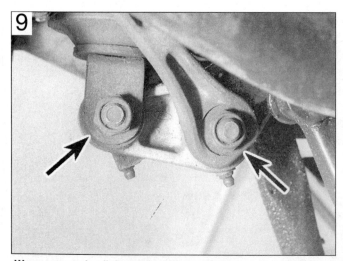

Worn suspension linkage pivots (arrows) are usually the cause of play in the rear suspension

Grasp the swingarm at the ends to check for play in its pivot bearings

MOT Test Checks

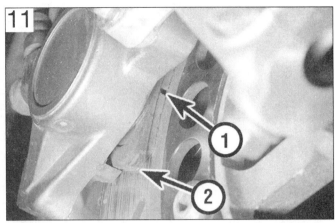

Brake pad wear can usually be viewed without removing the caliper. Most pads have wear indicator grooves (1) and some also have indicator tangs (2)

On drum brakes, check the angle of the operating lever with the brake fully applied. Most drum brakes have a wear indicator pointer and scale.

Brakes, Wheels and Tyres

Brakes

✔ With the wheel raised off the ground, apply the brake then free it off, and check that the wheel is about to revolve freely without brake drag.

✔ On disc brakes, examine the disc itself. Check that it is securely mounted and not cracked.

✔ On disc brakes, view the pad material through the caliper mouth and check that the pads are not worn down beyond the limit **(see illustration 11)**.

✔ On drum brakes, check that when the brake is applied the angle between the operating lever and cable or rod is not too great **(see illustration 12)**. Check also that the operating lever doesn't foul any other components.

✔ On disc brakes, examine the flexible hoses from top to bottom. Have an assistant hold the brake on so that the fluid in the hose is under pressure, and check that there is no sign of fluid leakage, bulges or cracking. If there are any metal brake pipes or unions, check that these are free from corrosion and damage. Where a brake-linked anti-dive system is fitted, check the hoses to the anti-dive in a similar manner.

✔ Check that the rear brake torque arm is secure and that its fasteners are secured by self-locking nuts or castellated nuts with split-pins or R-pins **(see illustration 13)**.

✔ On models with ABS, check that the self-check warning light in the instrument panel works.

✔ The MOT tester will perform a test of the motorcycle's braking efficiency based on a calculation of rider and motorcycle weight. Although this cannot be carried out at home, you can at least ensure that the braking systems are properly maintained. For hydraulic disc brakes, check the fluid level, lever/pedal feel (bleed of air if its spongy) and pad material. For drum brakes, check adjustment, cable or rod operation and shoe lining thickness.

Wheels and tyres

✔ Check the wheel condition. Cast wheels should be free from cracks and if of the built-up design, all fasteners should be secure. Spoked wheels should be checked for broken, corroded, loose or bent spokes.

✔ With the wheel raised off the ground, spin the wheel and visually check that the tyre and wheel run true. Check that the tyre does not foul the suspension or mudguards.

✔ With the wheel raised off the ground, grasp the wheel and attempt to move it about the axle (spindle) **(see illustration 14)**. Any play felt here indicates wheel bearing failure.

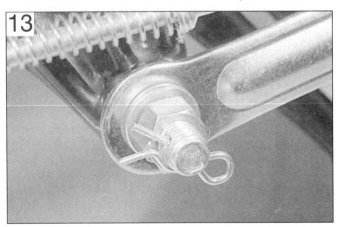

Brake torque arm must be properly secured at both ends

Check for wheel bearing play by trying to move the wheel about the axle (spindle)

REF•26 MOT Test Checks

Checking the tyre tread depth

Tyre direction of rotation arrow can be found on tyre sidewall

Castellated type wheel axle (spindle) nut must be secured by a split pin or R-pin

Two straightedges are used to check wheel alignment

✔ Check the tyre tread depth, tread condition and sidewall condition **(see illustration 15)**.
✔ Check the tyre type. Front and rear tyre types must be compatible and be suitable for road use. Tyres marked NOT FOR ROAD USE, COMPETITION USE ONLY or similar, will fail the MOT.

✔ If the tyre sidewall carries a direction of rotation arrow, this must be pointing in the direction of normal wheel rotation **(see illustration 16)**.
✔ Check that the wheel axle (spindle) nuts (where applicable) are properly secured. A self-locking nut or castellated nut with a split-pin or R-pin can be used **(see illustration 17)**.
✔ Wheel alignment is checked with the motorcycle off the stand and a rider seated. With the front wheel pointing straight ahead, two perfectly straight lengths of metal or wood and placed against the sidewalls of both tyres **(see illustration 18)**. The gap each side of the front tyre must be equidistant on both sides. Incorrect wheel alignment may be due to a cocked rear wheel (often as the result of poor chain adjustment) or in extreme cases, a bent frame.

General checks and condition

✔ Check the security of all major fasteners, bodypanels, seat, fairings (where fitted) and mudguards.

✔ Check that the rider and pillion footrests, handlebar levers and brake pedal are securely mounted.

✔ Check for corrosion on the frame or any load-bearing components. If severe, this may affect the structure, particularly under stress.

Sidecars

A motorcycle fitted with a sidecar requires additional checks relating to the stability of the machine and security of attachment and swivel joints, plus specific wheel alignment (toe-in) requirements. Additionally, tyre and lighting requirements differ from conventional motorcycle use. Owners are advised to check MOT test requirements with an official test centre.

Storage

Preparing for storage

Before you start

If repairs or an overhaul is needed, see that this is carried out now rather than left until you want to ride the bike again.

Give the bike a good wash and scrub all dirt from its underside. Make sure the bike dries completely before preparing for storage.

Engine

● Remove the spark plug(s) and lubricate the cylinder bores with approximately a teaspoon of motor oil using a spout-type oil can **(see illustration 1)**. Reinstall the spark plug(s). Crank the engine over a couple of times to coat the piston rings and bores with oil. If the bike has a kickstart, use this to turn the engine over. If not, flick the kill switch to the OFF position and crank the engine over on the starter **(see illustration 2)**. If the nature on the ignition system prevents the starter operating with the kill switch in the OFF position, remove the spark plugs and fit them back in their caps; ensure that the plugs are earthed (grounded) against the cylinder head when the starter is operated **(see illustration 3)**.

 Warning: It is important that the plugs are earthed (grounded) away from the spark plug holes otherwise there is a risk of atomised fuel from the cylinders igniting.

 On a single cylinder four-stroke engine, you can seal the combustion chamber completely by positioning the piston at TDC on the compression stroke.

● Drain the carburettor(s) otherwise there is a risk of jets becoming blocked by gum deposits from the fuel **(see illustration 4)**.

● If the bike is going into long-term storage, consider adding a fuel stabiliser to the fuel in the tank. If the tank is drained completely, corrosion of its internal surfaces may occur if left unprotected for a long period. The tank can be treated with a rust preventative especially for this purpose. Alternatively, remove the tank and pour half a litre of motor oil into it, install the filler cap and shake the tank to coat its internals with oil before draining off the excess. The same effect can also be achieved by spraying WD40 or a similar water-dispersant around the inside of the tank via its flexible nozzle.

● Make sure the cooling system contains the correct mix of antifreeze. Antifreeze also contains important corrosion inhibitors.

● The air intakes and exhaust can be sealed off by covering or plugging the openings. Ensure that you do not seal in any condensation; run the engine until it is hot,

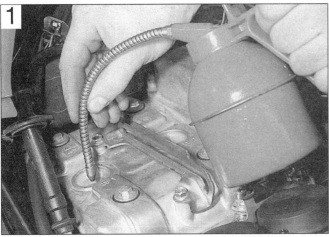

Squirt a drop of motor oil into each cylinder

Flick the kill switch to OFF . . .

. . . and ensure that the metal bodies of the plugs (arrows) are earthed against the cylinder head

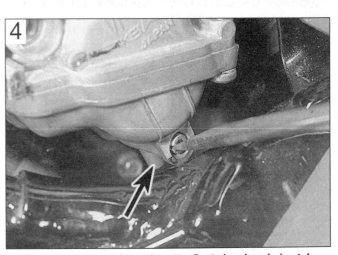

Connect a hose to the carburettor float chamber drain stub (arrow) and unscrew the drain screw

REF•28 Storage

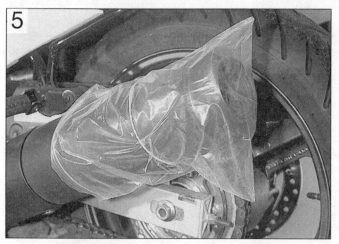

Exhausts can be sealed off with a plastic bag

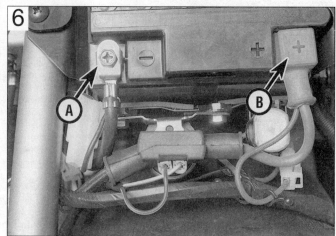

Disconnect the negative lead (A) first, followed by the positive lead (B)

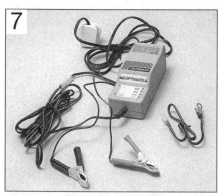

Use a suitable battery charger - this kit also assess battery condition

then switch off and allow to cool. Tape a piece of thick plastic over the silencer end(s) **(see illustration 5)**. Note that some advocate pouring a tablespoon of motor oil into the silencer(s) before sealing them off.

Battery
● Remove it from the bike - in extreme cases of cold the battery may freeze and crack its case **(see illustration 6)**.

● Check the electrolyte level and top up if necessary (conventional refillable batteries). Clean the terminals.
● Store the battery off the motorcycle and away from any sources of fire. Position a wooden block under the battery if it is to sit on the ground.
● Give the battery a trickle charge for a few hours every month **(see illustration 7)**.

Tyres
● Place the bike on its centrestand or an auxiliary stand which will support the motorcycle in an upright position. Position wood blocks under the tyres to keep them off the ground and to provide insulation from damp. If the bike is being put into long-term storage, ideally both tyres should be off the ground; not only will this protect the tyres, but will also ensure that no load is placed on the steering head or wheel bearings.
● Deflate each tyre by 5 to 10 psi, no more or the beads may unseat from the rim, making subsequent inflation difficult on tubeless tyres.

Pivots and controls
● Lubricate all lever, pedal, stand and footrest pivot points. If grease nipples are fitted to the rear suspension components, apply lubricant to the pivots.
● Lubricate all control cables.

Cycle components
● Apply a wax protectant to all painted and plastic components. Wipe off any excess, but don't polish to a shine. Where fitted, clean the screen with soap and water.
● Coat metal parts with Vaseline (petroleum jelly). When applying this to the fork tubes, do not compress the forks otherwise the seals will rot from contact with the Vaseline.
● Apply a vinyl cleaner to the seat.

Storage conditions
● Aim to store the bike in a shed or garage which does not leak and is free from damp.
● Drape an old blanket or bedspread over the bike to protect it from dust and direct contact with sunlight (which will fade paint). This also hides the bike from prying eyes. Beware of tight-fitting plastic covers which may allow condensation to form and settle on the bike.

Getting back on the road

Engine and transmission
● Change the oil and replace the oil filter. If this was done prior to storage, check that the oil hasn't emulsified - a thick whitish substance which occurs through condensation.
● Remove the spark plugs. Using a spout-type oil can, squirt a few drops of oil into the cylinder(s). This will provide initial lubrication as the piston rings and bores comes back into contact. Service the spark plugs, or fit new ones, and install them in the engine.

● Check that the clutch isn't stuck on. The plates can stick together if left standing for some time, preventing clutch operation. Engage a gear and try rocking the bike back and forth with the clutch lever held against the handlebar. If this doesn't work on cable-operated clutches, hold the clutch lever back against the handlebar with a strong elastic band or cable tie for a couple of hours **(see illustration 8)**.
● If the air intakes or silencer end(s) were blocked off, remove the bung or cover used.
● If the fuel tank was coated with a rust

Hold clutch lever back against the handlebar with elastic bands or a cable tie

Storage

preventative, oil or a stabiliser added to the fuel, drain and flush the tank and dispose of the fuel sensibly. If no action was taken with the fuel tank prior to storage, it is advised that the old fuel is disposed of since it will go off over a period of time. Refill the fuel tank with fresh fuel.

Frame and running gear

- Oil all pivot points and cables.
- Check the tyre pressures. They will definitely need inflating if pressures were reduced for storage.
- Lubricate the final drive chain (where applicable).
- Remove any protective coating applied to the fork tubes (stanchions) since this may well destroy the fork seals. If the fork tubes weren't protected and have picked up rust spots, remove them with very fine abrasive paper and refinish with metal polish.
- Check that both brakes operate correctly. Apply each brake hard and check that it's not possible to move the motorcycle forwards, then check that the brake frees off again once released. Brake caliper pistons can stick due to corrosion around the piston head, or on the sliding caliper types, due to corrosion of the slider pins. If the brake doesn't free after repeated operation, take the caliper off for examination. Similarly drum brakes can stick due to a seized operating cam, cable or rod linkage.
- If the motorcycle has been in long-term storage, renew the brake fluid and clutch fluid (where applicable).
- Depending on where the bike has been stored, the wiring, cables and hoses may have been nibbled by rodents. Make a visual check and investigate disturbed wiring loom tape.

Battery

- If the battery has been previously removal and given top up charges it can simply be reconnected. Remember to connect the positive cable first and the negative cable last.
- On conventional refillable batteries, if the battery has not received any attention, remove it from the motorcycle and check its electrolyte level. Top up if necessary then charge the battery. If the battery fails to hold a charge and a visual checks show heavy white sulphation of the plates, the battery is probably defective and must be renewed. This is particularly likely if the battery is old. Confirm battery condition with a specific gravity check.
- On sealed (MF) batteries, if the battery has not received any attention, remove it from the motorcycle and charge it according to the information on the battery case - if the battery fails to hold a charge it must be renewed.

Starting procedure

- If a kickstart is fitted, turn the engine over a couple of times with the ignition OFF to distribute oil around the engine. If no kickstart is fitted, flick the engine kill switch OFF and the ignition ON and crank the engine over a couple of times to work oil around the upper cylinder components. If the nature of the ignition system is such that the starter won't work with the kill switch OFF, remove the spark plugs, fit them back into their caps and earth (ground) their bodies on the cylinder head. Reinstall the spark plugs afterwards.
- Switch the kill switch to RUN, operate the choke and start the engine. If the engine won't start don't continue cranking the engine - not only will this flatten the battery, but the starter motor will overheat. Switch the ignition off and try again later. If the engine refuses to start, go through the fault finding procedures in this manual. **Note:** *If the bike has been in storage for a long time, old fuel or a carburettor blockage may be the problem. Gum deposits in carburettors can block jets - if a carburettor cleaner doesn't prove successful the carburettors must be dismantled for cleaning.*
- Once the engine has started, check that the lights, turn signals and horn work properly.
- Treat the bike gently for the first ride and check all fluid levels on completion. Settle the bike back into the maintenance schedule.

REF•30 Fault Finding

This Section provides an easy reference-guide to the more common faults that are likely to afflict your machine. Obviously, the opportunities are almost limitless for faults to occur as a result of obscure failures, and to try and cover all eventualities would require a book. Indeed, a number have been written on the subject.

Successful troubleshooting is not a mysterious 'black art' but the application of a bit of knowledge combined with a systematic and logical approach to the problem. Approach any troubleshooting by first accurately identifying the symptom and then checking through the list of possible causes, starting with the simplest or most obvious and progressing in stages to the most complex.

Take nothing for granted, but above all apply liberal quantities of common sense.

The main symptom of a fault is given in the text as a major heading below which are listed the various systems or areas which may contain the fault. Details of each possible cause for a fault and the remedial action to be taken are given, in brief, in the paragraphs below each heading. Further information should be sought in the relevant Chapter.

1 Engine doesn't start or is difficult to start
- [ ] Starter motor doesn't rotate
- [ ] Starter motor rotates but engine does not turn over
- [ ] Starter works but engine won't turn over (seized)
- [ ] No fuel flow
- [ ] Engine flooded
- [ ] No spark or weak spark
- [ ] Compression low
- [ ] Stalls after starting
- [ ] Rough idle

2 Poor running at low speed
- [ ] Spark weak
- [ ] Fuel/air mixture incorrect
- [ ] Compression low
- [ ] Poor acceleration

3 Poor running or no power at high speed
- [ ] Firing incorrect
- [ ] Fuel/air mixture incorrect
- [ ] Compression low
- [ ] Knocking or pinking
- [ ] Miscellaneous causes

4 Overheating
- [ ] Engine overheats
- [ ] Firing incorrect
- [ ] Fuel/air mixture incorrect
- [ ] Compression too high
- [ ] Engine load excessive
- [ ] Lubrication inadequate
- [ ] Miscellaneous causes

5 Clutch problems
- [ ] Clutch slipping
- [ ] Clutch not disengaging completely

6 Gearchanging problems
- [ ] Doesn't go into gear, or lever doesn't return
- [ ] Jumps out of gear
- [ ] Overselects

7 Abnormal engine noise
- [ ] Knocking or pinking
- [ ] Piston slap or rattling
- [ ] Valve noise
- [ ] Other noise

8 Abnormal driveline noise
- [ ] Clutch noise
- [ ] Transmission noise
- [ ] Final drive noise

9 Abnormal frame and suspension noise
- [ ] Front end noise
- [ ] Shock absorber noise
- [ ] Brake noise

10 Oil pressure warning light comes on
- [ ] Engine lubrication system
- [ ] Electrical system

11 Excessive exhaust smoke
- [ ] White smoke
- [ ] Black smoke
- [ ] Brown smoke

12 Poor handling or stability
- [ ] Handlebar hard to turn
- [ ] Handlebar shakes or vibrates excessively
- [ ] Handlebar pulls to one side
- [ ] Poor shock absorbing qualities

13 Braking problems
- [ ] Brakes are spongy, don't hold
- [ ] Brake lever or pedal pulsates
- [ ] Brakes drag

14 Electrical problems
- [ ] Battery dead or weak
- [ ] Battery overcharged

Fault Finding REF•31

1 Engine doesn't start or is difficult to start

Starter motor doesn't rotate
- [ ] Engine kill switch OFF.
- [ ] Fuse blown. Check main fuse (Chapter 9).
- [ ] Battery voltage low. Check and recharge battery (Chapter 9).
- [ ] Starter motor defective. Make sure the wiring to the starter is secure. Make sure the starter relay clicks when the start button is pushed. If the relay clicks, then the fault is in the wiring or motor.
- [ ] Starter relay faulty. Check it according to the procedure in Chapter 9.
- [ ] Starter switch not contacting. The contacts could be wet, corroded or dirty. Disassemble and clean the switch (Chapter 9).
- [ ] Wiring open or shorted. Check all wiring connections and harnesses to make sure that they are dry, tight and not corroded. Also check for broken or frayed wires that can cause a short to earth (see wiring diagram, Chapter 9).
- [ ] Ignition (main) switch defective. Check the switch according to the procedure in Chapter 9. Renew the switch if it is defective.
- [ ] Engine kill switch defective. Check for wet, dirty or corroded contacts. Clean or renew the switch as necessary (Chapter 9).
- [ ] Faulty neutral, sidestand (L, N, R models) or clutch switch. Check the wiring to each switch and the switch itself according to the procedures in Chapter 9.

Starter motor rotates but engine does not turn over
- [ ] Starter motor clutch defective. Inspect and repair or renew (Chapter 2).
- [ ] Damaged idler or starter gears. Inspect and renew the damaged parts (Chapter 2).

Starter works but engine won't turn over (seized)
- [ ] Seized engine caused by one or more internally damaged components. Failure due to wear, abuse or lack of lubrication. Damage can include seized valves, followers, camshafts, pistons, crankshaft, connecting rod bearings, or transmission gears or bearings. Refer to Chapter 2 for engine disassembly.

No fuel flow
- [ ] No fuel in tank.
- [ ] Fuel tank breather hose obstructed.
- [ ] Fuel tap filter or in-line filter clogged. Remove the tap and clean it and the filters (Chapter 4).
- [ ] Fuel line clogged. Pull the fuel line loose and carefully blow through it.
- [ ] Float needle valve clogged. For all of the valves to be clogged, either a very bad batch of fuel with an unusual additive has been used, or some other foreign material has entered the tank. Many times after a machine has been stored for many months without running, the fuel turns to a varnish-like liquid and forms deposits on the inlet needle valves and jets. The carburettors should be removed and overhauled if draining the float chambers doesn't solve the problem.

Engine flooded
- [ ] Float height too high. Check as described in Chapter 4.
- [ ] Float needle valve worn or stuck open. A piece of dirt, rust or other debris can cause the valve to seat improperly, causing excess fuel to be admitted to the float chamber. In this case, the float chamber should be cleaned and the needle valve and seat inspected. If the needle and seat are worn, then the leaking will persist and the parts should be renewed (Chapter 4).
- [ ] Starting technique incorrect. Under normal circumstances (ie, if all the carburettor functions are sound) the machine should start with little or no throttle. When the engine is cold, the choke should be operated and the engine started without opening the throttle. When the engine is at operating temperature, only a very slight amount of throttle should be necessary. If the engine is flooded, turn the fuel tap OFF and hold the throttle open while cranking the engine. This will allow additional air to reach the cylinders. Remember to turn the fuel tap back ON after the engine starts.

No spark or weak spark
- [ ] Ignition switch OFF.
- [ ] Engine kill switch turned to the OFF position.
- [ ] Battery voltage low. Check and recharge the battery as necessary (Chapter 9).
- [ ] Spark plugs dirty, defective or worn out. Locate reason for fouled plugs using spark plug condition chart at the end of this manual and follow the plug maintenance procedures (Chapter 1).
- [ ] Spark plug caps or secondary (HT) wiring faulty. Check condition. Renew if cracks or deterioration are evident (Chapter 5).
- [ ] Spark plug caps not making good contact. Make sure that the plug caps fit snugly over the plug ends.
- [ ] Ignition control unit defective (Chapter 5).
- [ ] Pulse generator coil(s) defective. Check, referring to Chapter 5 for details.
- [ ] Ignition HT coils defective. Check the coils, referring to Chapter 5.
- [ ] Ignition or kill switch shorted. This is usually caused by water, corrosion, damage or excessive wear. The switches can be disassembled and cleaned with electrical contact cleaner. If cleaning does not help, renew the switches (Chapter 9).
- [ ] Wiring shorted or broken between:
    a) Ignition (main) switch and engine kill switch (or blown fuse)
    b) Ignition control unit and engine kill switch
    c) Ignition control unit and ignition HT coils
    d) Ignition HT coils and spark plugs
    e) Ignition control unit and pulse generator
- [ ] Make sure that all wiring connections are clean, dry and tight. Look for chafed and broken wires (Chapters 5 and 9).

Compression low
- [ ] Spark plugs loose. Remove the plugs and inspect their threads. Reinstall and tighten securely.
- [ ] Cylinder head not sufficiently tightened down. If the cylinder head is suspected of being loose, then there's a chance that the gasket or head is damaged if the problem has persisted for any length of time. The head nuts should be tightened to the proper torque in the correct sequence (Chapter 2).
- [ ] Improper valve clearance. This means that the valve is not closing completely and compression pressure is leaking past the valve. Check and adjust the valve clearances (Chapter 1).
- [ ] Cylinder and/or piston worn. Excessive wear will cause compression pressure to leak past the rings. This is usually accompanied by worn rings as well. A top-end overhaul is necessary (Chapter 2).
- [ ] Piston rings worn, weak, broken, or sticking. Broken or sticking piston rings usually indicate a lubrication or carburation problem that causes excess carbon deposits or seizures to form on the pistons and rings. Top-end overhaul is necessary (Chapter 2).
- [ ] Piston ring-to-groove clearance excessive. This is caused by excessive wear of the piston ring lands. Piston renewal is necessary (Chapter 2).
- [ ] Cylinder head gasket damaged. If a head is allowed to become loose, or if excessive carbon build-up on the piston crown and combustion chamber causes extremely high compression, the head gasket may leak. Retorquing the head is not always sufficient to restore the seal, so gasket renewal is necessary (Chapter 2).
- [ ] Cylinder head warped. This is caused by overheating or improperly tightened head nuts. Machine shop resurfacing or head renewal is necessary (Chapter 2).

Fault Finding

1 Engine doesn't start or is difficult to start (continued)

- [ ] Valve spring broken or weak. Caused by component failure or wear; the springs must be renewed (Chapter 2).
- [ ] Valve not seating properly. This is caused by a bent valve (from over-revving or improper valve adjustment), burned valve or seat (improper carburation) or an accumulation of carbon deposits on the seat (from carburation or lubrication problems). The valves must be cleaned and/or renewed and the seats serviced if possible (Chapter 2).

Stalls after starting

- [ ] Improper choke action. Make sure the choke linkage shaft is getting a full stroke and staying in the out position (Chapter 4).
- [ ] Ignition malfunction. See Chapter 5.
- [ ] Carburettor malfunction. See Chapter 4.
- [ ] Fuel contaminated. The fuel can be contaminated with either dirt or water, or can change chemically if the machine is allowed to sit for several months or more. Drain the tank and float chambers (Chapter 4).

- [ ] Inlet air leak. Check for loose carburettor-to-inlet manifold connections, loose or missing vacuum gauge adapter screws or caps, or loose carburettor tops (Chapter 4).
- [ ] Engine idle speed incorrect. Turn idle adjusting screw until the engine idles at the specified rpm (Chapter 1).

Rough idle

- [ ] Ignition malfunction. See Chapter 5.
- [ ] Idle speed incorrect. See Chapter 1.
- [ ] Carburettors not synchronised. Adjust carburettors with vacuum gauge or manometer set as described in Chapter 1.
- [ ] Carburettor malfunction. See Chapter 4.
- [ ] Fuel contaminated. The fuel can be contaminated with either dirt or water, or can change chemically if the machine is allowed to sit for several months or more. Drain the tank and float chambers (Chapter 4).
- [ ] Inlet air leak. Check for loose carburettor-to-inlet manifold connections, loose or missing vacuum gauge adapter screws or caps, or loose carburettor tops (Chapter 4).
- [ ] Air filter clogged. Renew the air filter element (Chapter 1).

2 Poor running at low speeds

Spark weak

- [ ] Battery voltage low. Check and recharge battery (Chapter 9).
- [ ] Spark plugs fouled, defective or worn out. Refer to Chapter 1 for spark plug maintenance.
- [ ] Spark plug cap or HT wiring defective. Refer to Chapters 1 and 5 for details on the ignition system.
- [ ] Spark plug caps not making contact.
- [ ] Incorrect spark plugs. Wrong type, heat range or cap configuration. Check and install correct plugs listed in Chapter 1.
- [ ] Ignition control unit defective. See Chapter 5.
- [ ] Pulse generator defective. See Chapter 5.
- [ ] Ignition HT coils defective. See Chapter 5.

Fuel/air mixture incorrect

- [ ] Pilot screws out of adjustment (Chapter 4).
- [ ] Pilot jet or air passage clogged. Remove and overhaul the carburettors (Chapter 4).
- [ ] Air bleed holes clogged. Remove carburettor and blow out all passages (Chapter 4).
- [ ] Air filter clogged, poorly sealed or missing (Chapter 1).
- [ ] Air filter housing poorly sealed. Look for cracks, holes or loose clamps and renew or repair defective parts.
- [ ] Fuel level too high or too low. Check the float height (Chapter 4).
- [ ] Fuel tank breather hose obstructed.
- [ ] Carburettor inlet manifolds loose. Check for cracks, breaks, tears or loose clamps. Renew the rubber inlet manifold joints if split or perished.

Compression low

- [ ] Spark plugs loose. Remove the plugs and inspect their threads. Reinstall and tighten securely.
- [ ] Cylinder head not sufficiently tightened down. If the cylinder head is suspected of being loose, then there's a chance that the gasket and head are damaged if the problem has persisted for any length of time. The head nuts should be tightened to the proper torque in the correct sequence (Chapter 2).
- [ ] Improper valve clearance. This means that the valve is not closing completely and compression pressure is leaking past the valve. Check and adjust the valve clearances (Chapter 1).
- [ ] Cylinder and/or piston worn. Excessive wear will cause compression pressure to leak past the rings. This is usually accompanied by worn rings as well. A top-end overhaul is necessary (Chapter 2).
- [ ] Piston rings worn, weak, broken, or sticking. Broken or sticking piston rings usually indicate a lubrication or carburation problem that causes excess carbon deposits or seizures to form on the pistons and rings. Top-end overhaul is necessary (Chapter 2).
- [ ] Piston ring-to-groove clearance excessive. This is caused by excessive wear of the piston ring lands. Piston renewal is necessary (Chapter 2).
- [ ] Cylinder head gasket damaged. If a head is allowed to become loose, or if excessive carbon build-up on the piston crown and combustion chamber causes extremely high compression, the head gasket may leak. Retorquing the head is not always sufficient to restore the seal, so gasket renewal is necessary (Chapter 2).
- [ ] Cylinder head warped. This is caused by overheating or improperly tightened head nuts. Machine shop resurfacing or head renewal is necessary (Chapter 2).
- [ ] Valve spring broken or weak. Caused by component failure or wear; the springs must be renewed (Chapter 2).
- [ ] Valve not seating properly. This is caused by a bent valve (from over-revving or improper valve adjustment), burned valve or seat (improper carburation) or an accumulation of carbon deposits on the seat (from carburation, lubrication problems). The valves must be cleaned and/or renewed and the seats serviced if possible (Chapter 2).

Poor acceleration

- [ ] Carburettors leaking or dirty. Overhaul the carburettors (Chapter 4).
- [ ] Timing not advancing. The pulse generator or the ignition control module may be defective. If so, they must be renewed, as they can't be repaired.
- [ ] Carburettors not synchronised. Adjust them with a vacuum gauge set or manometer (Chapter 1).
- [ ] Engine oil viscosity too high. Using a heavier oil than that recommended in Chapter 1 can damage the oil pump or lubrication system and cause drag on the engine.
- [ ] Brakes dragging. Usually caused by debris which has entered the brake piston seals, or from a warped disc or bent axle. Repair as necessary (Chapter 7).

Fault Finding REF•33

3 Poor running or no power at high speed

Firing incorrect
- [ ] Air filter restricted. Clean or renew filter (Chapter 1).
- [ ] Spark plugs fouled, defective or worn out. See Chapter 1 for spark plug maintenance.
- [ ] Spark plug caps or HT wiring defective. See Chapters 1 and 5 for details of the ignition system.
- [ ] Spark plug caps not in good contact. See Chapter 5.
- [ ] Incorrect spark plugs. Wrong type, heat range or cap configuration. Check and install correct plugs listed in Chapter 1.
- [ ] Ignition control unit defective. See Chapter 5.
- [ ] Ignition HT coils defective. See Chapter 5.

Fuel/air mixture incorrect
- [ ] Main jet clogged. Dirt, water or other contaminants can clog the main jets. Clean the fuel tap filter, the in-line filter, the float chamber area, and the jets and carburettor orifices (Chapter 4).
- [ ] Main jet wrong size. The standard jetting is for sea level atmospheric pressure and oxygen content.
- [ ] Air bleed holes clogged. Remove and overhaul carburettors (Chapter 4).
- [ ] Air filter clogged, poorly sealed, or missing (Chapter 1).
- [ ] Air filter housing poorly sealed. Look for cracks, holes or loose clamps, and renew defective parts.
- [ ] Fuel level too high or too low. Check the float height (Chapter 4).
- [ ] Fuel tank breather hose obstructed.
- [ ] Carburettor inlet manifolds loose. Check for cracks, breaks, tears or loose clamps. Renew the rubber inlet manifolds if they are split or perished (Chapter 4).

Compression low
- [ ] Spark plugs loose. Remove the plugs and inspect their threads. Reinstall and tighten securely (Chapter 1).
- [ ] Cylinder head not sufficiently tightened down. If the cylinder head is suspected of being loose, then there's a chance that the gasket and head are damaged if the problem has persisted for any length of time. The head nuts should be tightened to the proper torque in the correct sequence (Chapter 2).
- [ ] Improper valve clearance. This means that the valve is not closing completely and compression pressure is leaking past the valve. Check and adjust the valve clearances (Chapter 1).
- [ ] Cylinder and/or piston worn. Excessive wear will cause compression pressure to leak past the rings. This is usually accompanied by worn rings as well. A top-end overhaul is necessary (Chapter 2).
- [ ] Piston rings worn, weak, broken, or sticking. Broken or sticking piston rings usually indicate a lubrication or carburation problem that causes excess carbon deposits or seizures to form on the pistons and rings. Top-end overhaul is necessary (Chapter 2).
- [ ] Piston ring-to-groove clearance excessive. This is caused by excessive wear of the piston ring lands. Piston renewal is necessary (Chapter 2).
- [ ] Cylinder head gasket damaged. If a head is allowed to become loose, or if excessive carbon build-up on the piston crown and combustion chamber causes extremely high compression, the head gasket may leak. Retorquing the head is not always sufficient to restore the seal, so gasket renewal is necessary (Chapter 2).
- [ ] Cylinder head warped. This is caused by overheating or improperly tightened head nuts. Machine shop resurfacing or head renewal is necessary (Chapter 2).
- [ ] Valve spring broken or weak. Caused by component failure or wear; the springs must be renewed (Chapter 2).
- [ ] Valve not seating properly. This is caused by a bent valve (from over-revving or improper valve adjustment), burned valve or seat (improper carburation) or an accumulation of carbon deposits on the seat (from carburation or lubrication problems). The valves must be cleaned and/or renewed and the seats serviced if possible (Chapter 2).

Knocking or pinking
- [ ] Carbon build-up in combustion chamber. Use of a fuel additive that will dissolve the adhesive bonding the carbon particles to the crown and chamber is the easiest way to remove the build-up. Otherwise, the cylinder head will have to be removed and decarbonised (Chapter 2).
- [ ] Incorrect or poor quality fuel. Old or improper grades of fuel can cause detonation. This causes the piston to rattle, thus the knocking or pinking sound. Drain old fuel and always use the recommended fuel grade.
- [ ] Spark plug heat range incorrect. Uncontrolled detonation indicates the plug heat range is too hot. The plug in effect becomes a glow plug, raising cylinder temperatures. Install the proper heat range plug (Chapter 1).
- [ ] Improper air/fuel mixture. This will cause the cylinders to run hot, which leads to detonation. Clogged jets or an air leak can cause this imbalance. See Chapter 4.

Miscellaneous causes
- [ ] Throttle valve doesn't open fully. Adjust the throttle grip freeplay (Chapter 1).
- [ ] Clutch slipping. May be caused by loose or worn clutch components. Refer to Chapter 2 for clutch overhaul procedures.
- [ ] Timing not advancing. Check ignition timing (Chapter 5).
- [ ] Engine oil viscosity too high. Using a heavier oil than the one recommended in Chapter 1 can damage the oil pump or lubrication system and cause drag on the engine.
- [ ] Brakes dragging. Usually caused by debris which has entered the brake piston seals, or from a warped disc or bent axle. Repair as necessary.

4 Overheating

Engine overheats
- ☐ Coolant level low. Check and add coolant (Chapter 1).
- ☐ Leak in cooling system. Check cooling system hoses and radiator for leaks and other damage. Repair or renew parts as necessary (Chapter 3).
- ☐ Thermostat sticking open or closed. Check as described in Chapter 3.
- ☐ Faulty radiator cap. Remove the cap and have it pressure tested.
- ☐ Coolant passages clogged. Drain and flush the entire system, then refill with fresh coolant (Chapter 1).
- ☐ Water pump defective. Remove the pump and check the components (Chapter 3).
- ☐ Clogged radiator fins. Clean them by blowing compressed air through the fins from the backside.
- ☐ Cooling fan or fan switch fault (Chapter 3).

Firing incorrect
- ☐ Spark plugs fouled, defective or worn out. See Chapter 1 for spark plug maintenance.
- ☐ Incorrect spark plugs.
- ☐ Ignition control unit defective. See Chapter 5.
- ☐ Faulty ignition HT coils (Chapter 5).

Fuel/air mixture incorrect
- ☐ Main jet clogged. Dirt, water and other contaminants can clog the main jets. Clean the fuel tap filter, the fuel pump in-line filter, the float chamber area and the jets and carburettor orifices (Chapter 4).
- ☐ Main jet wrong size. The standard jetting is for sea level atmospheric pressure and oxygen content.
- ☐ Air filter clogged, poorly sealed or missing (Chapter 1).
- ☐ Air filter housing poorly sealed. Look for cracks, holes or loose clamps and renew or repair.
- ☐ Fuel level too low. Check float height (Chapter 4).
- ☐ Fuel tank breather hose obstructed.
- ☐ Carburettor inlet manifolds loose. Check for cracks, breaks, tears or loose clamps. Renew the rubber inlet manifold joints if split or perished.

Compression too high
- ☐ Carbon build-up in combustion chamber. Use of a fuel additive that will dissolve the adhesive bonding the carbon particles to the piston crown and chamber is the easiest way to remove the build-up. Otherwise, the cylinder head will have to be removed and decarbonised (Chapter 2).
- ☐ Improperly machined head surface or installation of incorrect gasket during engine assembly.

Engine load excessive
- ☐ Clutch slipping. Can be caused by damaged, loose or worn clutch components. Refer to Chapter 2 for overhaul procedures.
- ☐ Engine oil level too high. The addition of too much oil will cause pressurisation of the crankcase and inefficient engine operation. Check Specifications and drain to proper level (Chapter 1).
- ☐ Engine oil viscosity too high. Using a heavier oil than the one recommended in Chapter 1 can damage the oil pump or lubrication system as well as cause drag on the engine.
- ☐ Brakes dragging. Usually caused by debris which has entered the brake piston seals, or from a warped disc or bent axle. Repair as necessary.

Lubrication inadequate
- ☐ Engine oil level too low. Friction caused by intermittent lack of lubrication or from oil that is overworked can cause overheating. The oil provides a definite cooling function in the engine. Check the oil level (Daily (pre-ride) checks).
- ☐ Poor quality engine oil or incorrect viscosity or type. Oil is rated not only according to viscosity but also according to type. Some oils are not rated high enough for use in this engine. Refer to Chapter 1 Specifications and change to the correct oil.

Miscellaneous causes
- ☐ Modification to exhaust system. Most aftermarket exhaust systems cause the engine to run leaner, which make them run hotter. When installing an accessory exhaust system, always rejet the carburettors.

5 Clutch problems

Clutch slipping
- ☐ Insufficient clutch cable freeplay. Check and adjust (Chapter 1).
- ☐ Friction plates worn or warped. Overhaul the clutch assembly (Chapter 2).
- ☐ Plain plates warped (Chapter 2).
- ☐ Clutch springs broken or weak. Old or heat-damaged (from slipping clutch) springs should be renewed (Chapter 2).
- ☐ Clutch release mechanism defective. Renew any defective parts (Chapter 2).
- ☐ Clutch centre or housing unevenly worn. This causes improper engagement of the plates. Renew the damaged or worn parts (Chapter 2).

Clutch not disengaging completely
- ☐ Excessive clutch cable freeplay. Check and adjust (Chapter 1).
- ☐ Clutch plates warped or damaged. This will cause clutch drag, which in turn will cause the machine to creep. Overhaul the clutch assembly (Chapter 2).
- ☐ Clutch spring tension uneven. Usually caused by a sagged or broken spring. Check and renew the springs as a set (Chapter 2).
- ☐ Engine oil deteriorated. Old, thin, worn out oil will not provide proper lubrication for the plates, causing the clutch to drag. Renew the oil and filter (Chapter 1).
- ☐ Engine oil viscosity too high. Using a heavier oil than recommended in Chapter 1 can cause the plates to stick together, putting a drag on the engine. Change to the correct weight oil (Chapter 1).
- ☐ Clutch housing bush seized on mainshaft. Lack of lubrication, severe wear or damage can cause the guide to seize on the shaft. Overhaul of the clutch, and perhaps transmission, may be necessary to repair the damage (Chapter 2).
- ☐ Clutch release mechanism defective. Overhaul the clutch cover components (Chapter 2).
- ☐ Loose clutch centre nut. Causes housing and centre misalignment putting a drag on the engine. Engagement adjustment continually varies. Overhaul the clutch assembly (Chapter 2).

Fault Finding REF•35

6 Gearchanging problems

Doesn't go into gear or lever doesn't return
- [ ] Clutch not disengaging. See above.
- [ ] Selector fork(s) bent or seized. Often caused by dropping the machine or from lack of lubrication. Overhaul the transmission (Chapter 2).
- [ ] Gear(s) stuck on shaft. Most often caused by a lack of lubrication or excessive wear in transmission bearings and bushes. Overhaul the transmission (Chapter 2).
- [ ] Selector drum binding. Caused by lubrication failure or excessive wear. Renew the drum and bearing (Chapter 2).
- [ ] Gearchange lever return spring weak or broken (Chapter 2).
- [ ] Gearchange lever broken. Splines stripped out of lever or shaft, caused by allowing the lever to get loose or from dropping the machine. Renew necessary parts (Chapter 2).
- [ ] Gearchange mechanism stopper arm broken or worn. Full engagement and rotary movement of selector drum results. Renew the arm (Chapter 2).
- [ ] Stopper arm spring broken. Allows arm to float, causing sporadic selector operation. Renew the spring (Chapter 2).

Jumps out of gear
- [ ] Selector fork(s) worn. Overhaul the transmission (Chapter 2).
- [ ] Gear groove(s) worn. Overhaul the transmission (Chapter 2).
- [ ] Gear dogs or dog slots worn or damaged. The gears should be inspected and renewed. No attempt should be made to service the worn parts.

Overselects
- [ ] Stopper arm spring weak or broken (Chapter 2).
- [ ] Gearchange shaft return spring post broken or distorted (Chapter 2).

7 Abnormal engine noise

Knocking or pinking
- [ ] Carbon build-up in combustion chamber. Use of a fuel additive that will dissolve the adhesive bonding the carbon particles to the piston crown and chamber is the easiest way to remove the build-up. Otherwise, the cylinder head will have to be removed and decarbonised (Chapter 2).
- [ ] Incorrect or poor quality fuel. Old or improper fuel can cause detonation. This causes the pistons to rattle, thus the knocking or pinking sound. Drain the old fuel and always use the recommended grade fuel (Chapter 4).
- [ ] Spark plug heat range incorrect. Uncontrolled detonation indicates that the plug heat range is too hot. The plug in effect becomes a glow plug, raising cylinder temperatures. Install the proper heat range plug (Chapter 1).
- [ ] Improper air/fuel mixture. This will cause the cylinders to run hot and lead to detonation. Clogged jets or an air leak can cause this imbalance. See Chapter 4.

Piston slap or rattling
- [ ] Cylinder-to-piston clearance excessive. Caused by improper assembly. Inspect and overhaul top-end parts (Chapter 2).
- [ ] Connecting rod bent. Caused by over-revving, trying to start a badly flooded engine or from ingesting a foreign object into the combustion chamber. Renew the damaged parts (Chapter 2).
- [ ] Piston pin or piston pin bore worn or seized from wear or lack of lubrication. Renew damaged parts (Chapter 2).
- [ ] Piston ring(s) worn, broken or sticking. Overhaul the top-end (Chapter 2).
- [ ] Piston seizure damage. Usually from lack of lubrication or overheating. Renew the pistons and bore the cylinders, as necessary (Chapter 2).
- [ ] Connecting rod upper or lower end clearance excessive. Caused by excessive wear or lack of lubrication. Renew worn parts.

Valve noise
- [ ] Incorrect valve clearances. Adjust the clearances by referring to Chapter 1.
- [ ] Valve spring broken or weak. Check and renew weak valve springs (Chapter 2).
- [ ] Camshaft or cylinder head worn or damaged. Lack of lubrication at high rpm is usually the cause of damage. Insufficient oil or failure to change the oil at the recommended intervals are the chief causes. Since there are no replaceable bearings in the head, the head itself will have to be renewed if there is excessive wear or damage (Chapter 2).

Other noise
- [ ] Cylinder head gasket leaking.
- [ ] Exhaust pipe leaking at cylinder head connection. Caused by improper fit of pipe(s) or loose exhaust flange. All exhaust fasteners should be tightened evenly and carefully. Failure to do this will lead to a leak.
- [ ] Crankshaft runout excessive. Caused by a bent crankshaft (from over-revving) or damage from an upper cylinder component failure.
- [ ] Engine mounting bolts loose. Tighten all engine mount bolts (Chapter 2).
- [ ] Crankshaft bearings worn (Chapter 2).
- [ ] Camshaft drive gear assembly defective. Renew according to the procedure in Chapter 2.

8 Abnormal driveline noise

Clutch noise
- [ ] Clutch housing/friction plate clearance excessive (Chapter 2).
- [ ] Loose or damaged clutch pressure plate and/or bolts (Chapter 2).

Transmission noise
- [ ] Bearings worn. Also includes the possibility that the shafts are worn. Overhaul the transmission (Chapter 2).
- [ ] Gears worn or chipped (Chapter 2).
- [ ] Metal chips jammed in gear teeth. Probably pieces from a broken clutch, gear or selector mechanism that were picked up by the gears. This will cause early bearing failure (Chapter 2).
- [ ] Engine oil level too low. Causes a howl from transmission. Also affects engine power and clutch operation (Chapter 1).

Final drive noise
- [ ] Chain not adjusted properly (Chapter 1).
- [ ] Front or rear sprocket loose. Tighten fasteners (Chapter 6).
- [ ] Sprockets worn. Renew both sprockets and chain as a set (Chapter 6).
- [ ] Rear sprocket warped. Renew both sprockets and chain as a set (Chapter 6).
- [ ] Cush drive dampers worn (Chapter 6).

REF•36 Fault Finding

9 Abnormal frame and suspension noise

Front end noise

☐ Low fluid level or improper viscosity oil in forks. This can sound like spurting and is usually accompanied by irregular fork action (Chapter 6).
☐ Spring weak or broken. Makes a clicking or scraping sound. Fork oil, when drained, will have a lot of metal particles in it (Chapter 6).
☐ Steering head bearings loose or damaged. Clicks when braking. Check and adjust or renew as necessary (Chapters 1 and 6).
☐ Fork yokes loose. Make sure all clamp pinch bolts are tightened to the specified torque (Chapter 6).
☐ Fork tube bent. Good possibility if machine has been dropped. Renew the tube (Chapter 6).
☐ Front axle bolt or axle clamp bolts loose. Tighten them to the specified torque (Chapter 7).
☐ Worn wheel bearings. Check (Chapter 1) and renew (Chapter 7).

Shock absorber noise

☐ Fluid level incorrect. Indicates a leak caused by defective seal. Shock will be covered with oil. Renew shock or seek advice on repair from a suspension specialist (Chapter 6).
☐ Defective shock absorber with internal damage. This is in the body of the shock and can't be remedied. The shock must be renewed (Chapter 6).
☐ Bent or damaged shock body. Renew the shock (Chapter 6).
☐ Loose or worn suspension linkage components. Check and renew bearings where possible (Chapter 6).

Brake noise

☐ Squeal caused by pad shim not installed or positioned correctly (where fitted) (Chapter 7).
☐ Squeal caused by dust on brake pads. Usually found in combination with glazed pads. Clean using brake cleaning solvent (Chapter 7).
☐ Contamination of brake pads. Oil, brake fluid or dirt causing brake to chatter or squeal. Clean or renew pads (Chapter 7).
☐ Pads glazed. Caused by excessive heat from prolonged use or from contamination. Do not use sandpaper, emery cloth, carborundum cloth or any other abrasive to roughen the pad surfaces as abrasives will stay in the pad material and damage the disc. A very fine flat file can be used, but pad renewal is suggested as a cure (Chapter 7).
☐ Disc warped. Can cause a chattering, clicking or intermittent squeal. Usually accompanied by a pulsating lever and uneven braking. Renew the disc (Chapter 7).
☐ Worn wheel bearings. Check (Chapter 1) and renew (Chapter 7).

10 Oil pressure warning light comes on

Engine lubrication system

☐ Engine oil pump defective, blocked oil strainer gauze or failed relief valve. Carry out oil pressure check (Chapter 2).
☐ Engine oil level low. Inspect for leak or other problem causing low oil level and add recommended oil (Daily (pre-ride) checks).
☐ Engine oil viscosity too low. Very old, thin oil or an improper weight of oil used in the engine. Change to correct oil (Chapter 1).
☐ Camshaft or journals worn. Excessive wear causing drop in oil pressure. Renew camshaft and/or cylinder head. Abnormal wear could be caused by oil starvation at high rpm from low oil level or improper weight or type of oil (Chapter 1).
☐ Crankshaft and/or bearings worn. Same problems as above. Check and renew crankshaft and/or bearings (Chapter 2).

Electrical system

☐ Oil pressure switch defective. Check the switch according to the procedure in Chapter 9. Renew it if it is defective.
☐ Oil pressure warning light circuit defective. Check for pinched, shorted, disconnected or damaged wiring (Chapter 9).

11 Excessive exhaust smoke

White smoke

☐ Piston oil ring worn. The ring may be broken or damaged, causing oil from the crankcase to be pulled past the piston into the combustion chamber. Renew the rings (Chapter 2).
☐ Cylinders worn, cracked, or scored. Caused by overheating or oil starvation. The cylinders will have to be rebored and new pistons installed.
☐ Valve oil seal damaged or worn. Renew oil seals (Chapter 2).
☐ Valve guide worn. Have the guides renewed by an engineer (Chapter 2).
☐ Engine oil level too high, which causes the oil to be forced past the rings. Drain oil to the proper level (Daily (pre-ride) checks).
☐ Head gasket broken between oil return and cylinder. Causes oil to be pulled into the combustion chamber. Renew the head gasket and check the head for warpage (Chapter 2).
☐ Abnormal crankcase pressurisation, which forces oil past the rings. Clogged breather is usually the cause.

Black smoke

☐ Air filter clogged. Clean or renew the element (Chapter 1).
☐ Main jet too large or loose. Compare the jet size to the Specifications (Chapter 4).
☐ Choke cable or linkage shaft stuck, causing fuel to be pulled through choke circuit (Chapter 4).
☐ Fuel level too high. Check and adjust the float height(s) as necessary (Chapter 4).
☐ Float needle valve held off needle seat. Clean the float chambers and fuel line and renew the needles and seats if necessary (Chapter 4).

Brown smoke

☐ Main jet too small or clogged. Lean condition caused by wrong size main jet or by a restricted orifice. Clean float chambers and jets and compare jet size to Specifications (Chapter 4).
☐ Fuel flow insufficient. Float needle valve stuck closed due to chemical reaction with old fuel. Float height incorrect. Restricted fuel line. Clean line and float chamber and adjust floats if necessary.
☐ Carburettor inlet manifold clamps loose (Chapter 4).
☐ Air filter poorly sealed or not installed (Chapter 1).

Fault Finding

12 Poor handling or stability

Handlebar hard to turn

- [ ] Steering head bearing adjuster nut too tight. Check adjustment as described in Chapter 1.
- [ ] Bearings damaged. Roughness can be felt as the bars are turned from side-to-side. Renew bearings (Chapter 6).
- [ ] Races dented or worn. Denting results from wear in only one position (eg, straight ahead), from a collision or hitting a pothole or from dropping the machine. Renew bearings (Chapter 6).
- [ ] Steering stem lubrication inadequate. Causes are grease getting hard from age or being washed out by high pressure car washes. Disassemble steering head and repack bearings (Chapter 6).
- [ ] Steering stem bent. Caused by a collision, hitting a pothole or by dropping the machine. Renew damaged part. Don't try to straighten the steering stem (Chapter 6).
- [ ] Front tyre air pressure too low (Chapter 1).

Handlebar shakes or vibrates excessively

- [ ] Tyres worn or out of balance (Chapter 7).
- [ ] Swingarm bearings worn. Renew worn bearings (Chapter 6).
- [ ] Wheel rim(s) warped or damaged. Inspect wheels for runout (Chapter 7).
- [ ] Wheel bearings worn. Worn front or rear wheel bearings can cause poor tracking. Worn front bearings will cause wobble (Chapters 1 and 7).
- [ ] Handlebar clamp bolts loose. Tighten them to the specified torque (Chapter 6).
- [ ] Fork yoke bolts loose. Tighten them to the specified torque (Chapter 6).
- [ ] Engine mounting bolts loose. Will cause excessive vibration with increased engine rpm (Chapter 2).

Handlebar pulls to one side

- [ ] Frame bent. Definitely suspect this if the machine has been dropped. May or may not be accompanied by cracking near the bend. Renew the frame (Chapter 6).
- [ ] Wheels out of alignment. Caused by improper location of axle spacers or from bent steering stem or frame (Chapter 6).
- [ ] Swingarm bent or twisted as the result of impact damage. Renew the swingarm (Chapter 6).
- [ ] Steering stem bent. Caused by impact damage or by dropping the motorcycle. Renew the steering stem (Chapter 6).
- [ ] Fork tube bent. Disassemble the forks and renew the damaged parts (Chapter 6).
- [ ] Fork oil level uneven. Check and add or drain as necessary (Chapters 1 and 6).

Poor shock absorbing qualities

- [ ] Too hard:
    - a) *Front fork oil level excessive (Chapter 6).*
    - b) *Front fork oil viscosity too high. Use a lighter oil (see the Specifications in Chapter 6).*
    - c) *Front fork tube bent. Causes a harsh, sticking feeling (Chapter 6).*
    - d) *Front fork internal damage (Chapter 6).*
    - e) *Incorrect adjustment setting (Chapter 6).*
    - f) *Rear shock shaft or body bent or damaged (Chapter 6).*
    - g) *Rear shock internal damage.*
    - h) *Tyre pressure too high (Chapter 1).*
- [ ] Too soft:
    - a) *Front fork or rear shock oil insufficient and/or leaking (Chapter 6).*
    - b) *Front fork oil level too low (Chapter 6).*
    - c) *Front fork oil viscosity too light (Chapter 6).*
    - d) *Front fork springs weak or broken (Chapter 6).*
    - e) *Shock internal damage or leakage (Chapter 6).*

13 Braking problems

Brakes are spongy, don't hold

- [ ] Air in brake line. Caused by inattention to master cylinder fluid level or by leakage. Locate problem and bleed brakes (Chapter 7).
- [ ] Pads or disc worn (Chapters 1 and 7).
- [ ] Brake fluid leak. Locate problem and renew faulty component (Chapter 7).
- [ ] Contaminated pads. Caused by contamination with oil, grease, brake fluid, etc. Renew pads. Clean disc thoroughly with brake cleaner (Chapter 7).
- [ ] Brake fluid deteriorated. Fluid is old or contaminated. Drain system, replenish with new fluid and bleed the system (Chapter 7).
- [ ] Master cylinder internal parts worn or damaged causing fluid to bypass. Overhaul master cylinder (Chapter 7).
- [ ] Master cylinder bore scratched by foreign material or broken spring. Repair or renew master cylinder (Chapter 7).
- [ ] Disc warped. Renew disc (Chapter 7).

Brake lever or pedal pulsates

- [ ] Disc warped. Renew disc (Chapter 7).
- [ ] Axle bent. Renew axle (Chapter 7).
- [ ] Brake caliper bolts loose (Chapter 7).
- [ ] Brake caliper sliders damaged or sticking, causing caliper to bind. Lubricate the sliders or renew them if they are corroded or bent (Chapter 7).
- [ ] Wheel warped or otherwise damaged (Chapter 7).
- [ ] Wheel bearings damaged or worn (Chapter 7).

Brakes drag

- [ ] Master cylinder piston seized. Caused by wear or damage to piston or cylinder bore (Chapter 7).
- [ ] Lever balky or stuck. Check pivot and lubricate (Chapter 7).
- [ ] Brake caliper binds. Caused by inadequate lubrication of caliper slider pins (Chapter 7).
- [ ] Brake caliper piston seized in bore. Caused by wear or ingestion of dirt past deteriorated seals (Chapter 7).
- [ ] Brake pad damaged. Pad material separated from backing plate. Usually caused by faulty manufacturing process or from contact with chemicals. Renew pads (Chapter 7).
- [ ] Pads improperly installed (Chapter 7).

14 Electrical problems

Battery dead or weak

- [ ] Battery faulty. Caused by sulphated plates which are shorted through sedimentation. Also, broken battery terminal making only occasional contact (Chapter 9).
- [ ] Battery cables making poor contact (Chapter 9).
- [ ] Load excessive. Caused by addition of high wattage lights or other electrical accessories.
- [ ] Ignition (main) switch defective. Switch either earths internally or fails to shut off system. Renew the switch (Chapter 9).
- [ ] Regulator/rectifier defective (Chapter 9).
- [ ] Alternator stator coil open or shorted (Chapter 9).
- [ ] Wiring faulty. Wiring earthed or connections loose in ignition, charging or lighting circuits (Chapter 9).

Battery overcharged

- [ ] Regulator/rectifier defective. Overcharging is noticed when battery gets excessively warm (Chapter 9).
- [ ] Battery defective. Renew battery (Chapter 9).
- [ ] Battery amperage too low, wrong type or size. Install manufacturer's specified amp-hour battery to handle charging load (Chapter 9).

Fault Finding Equipment REF•39

Checking engine compression

● Low compression will result in exhaust smoke, heavy oil consumption, poor starting and poor performance. A compression test will provide useful information about an engine's condition and if performed regularly, can give warning of trouble before any other symptoms become apparent.
● A compression gauge will be required, along with an adapter to suit the spark plug hole thread size. Note that the screw-in type gauge/adapter set up is preferable to the rubber cone type.
● Before carrying out the test, first check the valve clearances as described in Chapter 1.

1 Run the engine until it reaches normal operating temperature, then stop it and remove the spark plug(s), taking care not to scald your hands on the hot components.
2 Install the gauge adapter and compression gauge in No. 1 cylinder spark plug hole (see illustration 1).

Screw the compression gauge adapter into the spark plug hole, then screw the gauge into the adapter

3 On kickstart-equipped motorcycles, make sure the ignition switch is OFF, then open the throttle fully and kick the engine over a couple of times until the gauge reading stabilises.
4 On motorcycles with electric start only, the procedure will differ depending on the nature of the ignition system. Flick the engine kill switch (engine stop switch) to OFF and turn the ignition switch ON; open the throttle fully and crank the engine over on the starter motor for a couple of revolutions until the gauge reading stabilises. If the starter will not operate with the kill switch OFF, turn the ignition switch OFF and refer to the next paragraph.
5 Install the spark plugs back into their suppressor caps and arrange the plug electrodes so that their metal bodies are earthed (grounded) against the cylinder head; this is essential to prevent damage to the ignition system as the engine is spun over (see illustration 2). Position the plugs well

All spark plugs must be earthed (grounded) against the cylinder head

away from the plug holes otherwise there is a risk of atomised fuel escaping from the combustion chambers and igniting. As a safety precaution, cover the top of the valve cover with rag. Now turn the ignition switch ON and kill switch ON, open the throttle fully and crank the engine over on the starter motor for a couple of revolutions until the gauge reading stabilises.
6 After one or two revolutions the pressure should build up to a maximum figure and then stabilise. Take a note of this reading and on multi-cylinder engines repeat the test on the remaining cylinders.
7 The correct pressures are given in Chapter 2 Specifications. If the results fall within the specified range and on multi-cylinder engines all are relatively equal, the engine is in good condition. If there is a marked difference between the readings, or if the readings are lower than specified, inspection of the top-end components will be required.
8 Low compression pressure may be due to worn cylinder bores, pistons or rings, failure of the cylinder head gasket, worn valve seals, or poor valve seating.
9 To distinguish between cylinder/piston wear and valve leakage, pour a small quantity of oil into the bore to temporarily seal the piston rings, then repeat the compression tests (see illustration 3). If the readings show

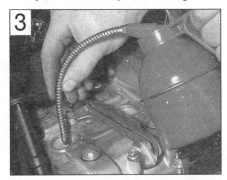

Bores can be temporarily sealed with a squirt of motor oil

a noticeable increase in pressure this confirms that the cylinder bore, piston, or rings are worn. If, however, no change is indicated, the cylinder head gasket or valves should be examined.
10 High compression pressure indicates excessive carbon build-up in the combustion chamber and on the piston crown. If this is the case the cylinder head should be removed and the deposits removed. Note that excessive carbon build-up is less likely with the used on modern fuels.

Checking battery open-circuit voltage

 Warning: The gases produced by the battery are explosive - never smoke or create any sparks in the vicinity of the battery. Never allow the electrolyte to contact your skin or clothing - if it does, wash it off and seek immediate medical attention.

Fault Finding Equipment

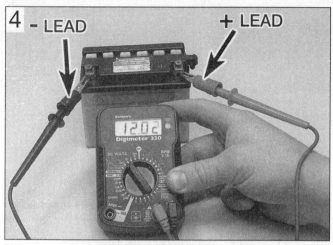

Measuring open-circuit battery voltage

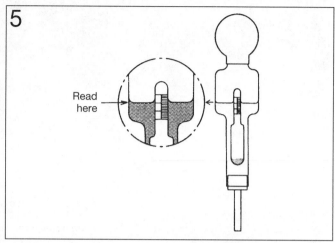

Float-type hydrometer for measuring battery specific gravity

- Before any electrical fault is investigated the battery should be checked.
- You'll need a dc voltmeter or multimeter to check battery voltage. Check that the leads are inserted in the correct terminals on the meter, red lead to positive (+ve), black lead to negative (-ve). Incorrect connections can damage the meter.
- A sound fully-charged 12 volt battery should produce between 12.3 and 12.6 volts across its terminals (12.8 volts for a maintenance-free battery). On machines with a 6 volt battery, voltage should be between 6.1 and 6.3 volts.

1 Set a multimeter to the 0 to 20 volts dc range and connect its probes across the battery terminals. Connect the meter's positive (+ve) probe, usually red, to the battery positive (+ve) terminal, followed by the meter's negative (-ve) probe, usually black, to the battery negative terminal (-ve) **(see illustration 4)**.

2 If battery voltage is low (below 10 volts on a 12 volt battery or below 4 volts on a six volt battery), charge the battery and test the voltage again. If the battery repeatedly goes flat, investigate the motorcycle's charging system.

Checking battery specific gravity (SG)

⚠ *Warning: The gases produced by the battery are explosive - never smoke or create any sparks in the vicinity of the battery. Never allow the electrolyte to contact your skin or clothing - if it does, wash it off and seek immediate medical attention.*

- The specific gravity check gives an indication of a battery's state of charge.
- A hydrometer is used for measuring specific gravity. Make sure you purchase one which has a small enough hose to insert in the aperture of a motorcycle battery.
- Specific gravity is simply a measure of the electrolyte's density compared with that of water. Water has an SG of 1.000 and fully-charged battery electrolyte is about 26% heavier, at 1.260.
- Specific gravity checks are not possible on maintenance-free batteries. Testing the open-circuit voltage is the only means of determining their state of charge.

1 To measure SG, remove the battery from the motorcycle and remove the first cell cap. Draw

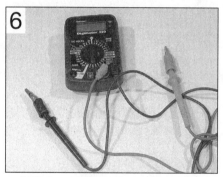

Digital multimeter can be used for all electrical tests

some electrolyte into the hydrometer and note the reading **(see illustration 5)**. Return the electrolyte to the cell and install the cap.

2 The reading should be in the region of 1.260 to 1.280. If SG is below 1.200 the battery needs charging. Note that SG will vary with temperature; it should be measured at 20°C (68°F). Add 0.007 to the reading for every 10°C above 20°C, and subtract 0.007 from the reading for every 10°C below 20°C. Add 0.004 to the reading for every 10°F above 68°F, and subtract 0.004 from the reading for every 10°F below 68°F.

3 When the check is complete, rinse the hydrometer thoroughly with clean water.

Checking for continuity

- The term continuity describes the uninterrupted flow of electricity through an electrical circuit. A continuity check will determine whether an **open-circuit** situation exists.
- Continuity can be checked with an ohmmeter, multimeter, continuity tester or battery and bulb test circuit **(see illustrations 6, 7 and 8)**.

Battery-powered continuity tester

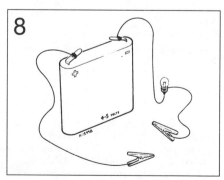

Battery and bulb test circuit

Fault Finding Equipment

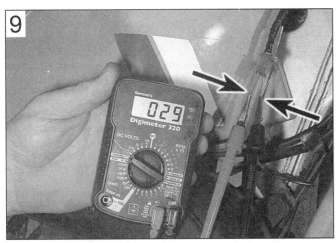

Continuity check of front brake light switch using a meter - note split pins used to access connector terminals

Continuity check of rear brake light switch using a continuity tester

- All of these instruments are self-powered by a battery, therefore the checks are made with the ignition OFF.
- As a safety precaution, always disconnect the battery negative (-ve) lead before making checks, particularly if ignition switch checks are being made.
- If using a meter, select the appropriate ohms scale and check that the meter reads infinity (∞). Touch the meter probes together and check that meter reads zero; where necessary adjust the meter so that it reads zero.
- After using a meter, always switch it OFF to conserve its battery.

Switch checks

1 If a switch is at fault, trace its wiring up to the wiring connectors. Separate the wire connectors and inspect them for security and condition. A build-up of dirt or corrosion here will most likely be the cause of the problem - clean up and apply a water dispersant such as WD40.

2 If using a test meter, set the meter to the ohms x 10 scale and connect its probes across the wires from the switch **(see illustration 9)**. Simple ON/OFF type switches, such as brake light switches, only have two wires whereas combination switches, like the ignition switch, have many internal links. Study the wiring diagram to ensure that you are connecting across the correct pair of wires. Continuity (low or no measurable resistance - 0 ohms) should be indicated with the switch ON and no continuity (high resistance) with it OFF.

3 Note that the polarity of the test probes doesn't matter for continuity checks, although care should be taken to follow specific test procedures if a diode or solid-state component is being checked.

4 A continuity tester or battery and bulb circuit can be used in the same way. Connect its probes as described above **(see illustration 10)**. The light should come on to indicate continuity in the ON switch position, but should extinguish in the OFF position.

Wiring checks

- Many electrical faults are caused by damaged wiring, often due to incorrect routing or chaffing on frame components.
- Loose, wet or corroded wire connectors can also be the cause of electrical problems, especially in exposed locations.

1 A continuity check can be made on a single length of wire by disconnecting it at each end and connecting a meter or continuity tester across both ends of the wire **(see illustration 11)**.

2 Continuity (low or no resistance - 0 ohms) should be indicated if the wire is good. If no continuity (high resistance) is shown, suspect a broken wire.

Checking for voltage

- A voltage check can determine whether current is reaching a component.
- Voltage can be checked with a dc voltmeter, multimeter set on the dc volts scale, test light or buzzer **(see illustrations 12 and 13)**. A meter has the advantage of being able to measure actual voltage.
- When using a meter, check that its leads are inserted in the correct terminals on the meter, red to positive (+ve), black to negative (-ve). Incorrect connections can damage the meter.
- A voltmeter (or multimeter set to the dc volts scale) should always be connected in parallel (across the load). Connecting it in series will destroy the meter.
- Voltage checks are made with the ignition ON.

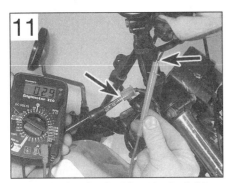

Continuity check of front brake light switch sub-harness

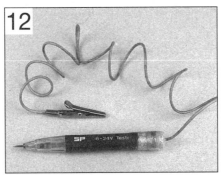

A simple test light can be used for voltage checks

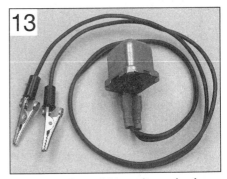

A buzzer is useful for voltage checks

Fault Finding Equipment

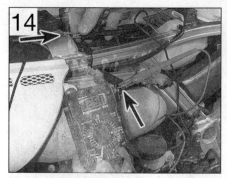

Checking for voltage at the rear brake light power supply wire using a meter . . .

1 First identify the relevant wiring circuit by referring to the wiring diagram at the end of this manual. If other electrical components share the same power supply (ie are fed from the same fuse), take note whether they are working correctly - this is useful information in deciding where to start checking the circuit.

2 If using a meter, check first that the meter leads are plugged into the correct terminals on the meter (see above). Set the meter to the dc volts function, at a range suitable for the battery voltage. Connect the meter red probe (+ve) to the power supply wire and the black probe to a good metal earth (ground) on the motorcycle's frame or directly to the battery negative (-ve) terminal **(see illustration 14)**. Battery voltage should be shown on the meter

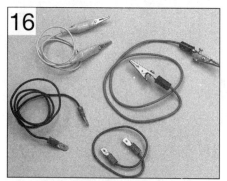

A selection of jumper wires for making earth (ground) checks

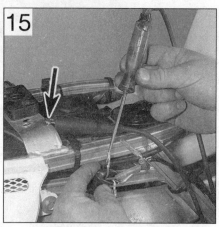

. . . or a test light - note the earth connection to the frame (arrow)

with the ignition switched ON.

3 If using a test light or buzzer, connect its positive (+ve) probe to the power supply terminal and its negative (-ve) probe to a good earth (ground) on the motorcycle's frame or directly to the battery negative (-ve) terminal **(see illustration 15)**. With the ignition ON, the test light should illuminate or the buzzer sound.

4 If no voltage is indicated, work back towards the fuse continuing to check for voltage. When you reach a point where there is voltage, you know the problem lies between that point and your last check point.

Checking the earth (ground)

● Earth connections are made either directly to the engine or frame (such as sensors, neutral switch etc. which only have a positive feed) or by a separate wire into the earth circuit of the wiring harness. Alternatively a short earth wire is sometimes run directly from the component to the motorcycle's frame.
● Corrosion is often the cause of a poor earth connection.
● If total failure is experienced, check the security of the main earth lead from the negative (-ve) terminal of the battery and also the main earth (ground) point on the wiring harness. If corroded, dismantle the connection and clean all surfaces back to bare metal.

1 To check the earth on a component, use an insulated jumper wire to temporarily bypass its earth connection **(see illustration 16)**. Connect one end of the jumper wire between the earth terminal or metal body of the component and the other end to the motorcycle's frame.

2 If the circuit works with the jumper wire installed, the original earth circuit is faulty. Check the wiring for open-circuits or poor connections. Clean up direct earth connections, removing all traces of corrosion and remake the joint. Apply petroleum jelly to the joint to prevent future corrosion.

Tracing a short-circuit

● A short-circuit occurs where current shorts to earth (ground) bypassing the circuit components. This usually results in a blown fuse.

● A short-circuit is most likely to occur where the insulation has worn through due to wiring chafing on a component, allowing a direct path to earth (ground) on the frame.

1 Remove any bodypanels necessary to access the circuit wiring.
2 Check that all electrical switches in the circuit are OFF, then remove the circuit fuse and connect a test light, buzzer or voltmeter (set to the dc scale) across the fuse terminals. No voltage should be shown.
3 Move the wiring from side to side whilst observing the test light or meter. When the test light comes on, buzzer sounds or meter shows voltage, you have found the cause of the short. It will usually shown up as damaged or burned insulation.
4 Note that the same test can be performed on each component in the circuit, even the switch.

Technical Terms Explained

A

ABS (Anti-lock braking system) A system, usually electronically controlled, that senses incipient wheel lockup during braking and relieves hydraulic pressure at wheel which is about to skid.

Aftermarket Components suitable for the motorcycle, but not produced by the motorcycle manufacturer.

Allen key A hexagonal wrench which fits into a recessed hexagonal hole.

Alternating current (ac) Current produced by an alternator. Requires converting to direct current by a rectifier for charging purposes.

Alternator Converts mechanical energy from the engine into electrical energy to charge the battery and power the electrical system.

Ampere (amp) A unit of measurement for the flow of electrical current. Current = Volts ÷ Ohms.

Ampere-hour (Ah) Measure of battery capacity.

Angle-tightening A torque expressed in degrees. Often follows a conventional tightening torque for cylinder head or main bearing fasteners **(see illustration)**.

Angle-tightening cylinder head bolts

Antifreeze A substance (usually ethylene glycol) mixed with water, and added to the cooling system, to prevent freezing of the coolant in winter. Antifreeze also contains chemicals to inhibit corrosion and the formation of rust and other deposits that would tend to clog the radiator and coolant passages and reduce cooling efficiency.

Anti-dive System attached to the fork lower leg (slider) to prevent fork dive when braking hard.

Anti-seize compound A coating that reduces the risk of seizing on fasteners that are subjected to high temperatures, such as exhaust clamp bolts and nuts.

API American Petroleum Institute. A quality standard for 4-stroke motor oils.

Asbestos A natural fibrous mineral with great heat resistance, commonly used in the composition of brake friction materials. Asbestos is a health hazard and the dust created by brake systems should never be inhaled or ingested.

ATF Automatic Transmission Fluid. Often used in front forks.

ATU Automatic Timing Unit. Mechanical device for advancing the ignition timing on early engines.

ATV All Terrain Vehicle. Often called a Quad.

Axial play Side-to-side movement.

Axle A shaft on which a wheel revolves. Also known as a spindle.

B

Backlash The amount of movement between meshed components when one component is held still. Usually applies to gear teeth.

Ball bearing A bearing consisting of a hardened inner and outer race with hardened steel balls between the two races.

Bearings Used between two working surfaces to prevent wear of the components and a build-up of heat. Four types of bearing are commonly used on motorcycles: plain shell bearings, ball bearings, tapered roller bearings and needle roller bearings.

Bevel gears Used to turn the drive through 90°. Typical applications are shaft final drive and camshaft drive **(see illustration)**.

Bevel gears are used to turn the drive through 90°

BHP Brake Horsepower. The British measurement for engine power output. Power output is now usually expressed in kilowatts (kW).

Bias-belted tyre Similar construction to radial tyre, but with outer belt running at an angle to the wheel rim.

Big-end bearing The bearing in the end of the connecting rod that's attached to the crankshaft.

Bleeding The process of removing air from an hydraulic system via a bleed nipple or bleed screw.

Bottom-end A description of an engine's crankcase components and all components contained there-in.

BTDC Before Top Dead Centre in terms of piston position. Ignition timing is often expressed in terms of degrees or millimetres BTDC.

Bush A cylindrical metal or rubber component used between two moving parts.

Burr Rough edge left on a component after machining or as a result of excessive wear.

C

Cam chain The chain which takes drive from the crankshaft to the camshaft(s).

Canister The main component in an evaporative emission control system (California market only); contains activated charcoal granules to trap vapours from the fuel system rather than allowing them to vent to the atmosphere.

Castellated Resembling the parapets along the top of a castle wall. For example, a castellated wheel axle or spindle nut.

Catalytic converter A device in the exhaust system of some machines which converts certain pollutants in the exhaust gases into less harmful substances.

Charging system Description of the components which charge the battery, ie the alternator, rectifier and regulator.

Circlip A ring-shaped clip used to prevent endwise movement of cylindrical parts and shafts. An internal circlip is installed in a groove in a housing; an external circlip fits into a groove on the outside of a cylindrical piece such as a shaft. Also known as a snap-ring.

Clearance The amount of space between two parts. For example, between a piston and a cylinder, between a bearing and a journal, etc.

Coil spring A spiral of elastic steel found in various sizes throughout a vehicle, for example as a springing medium in the suspension and in the valve train.

Compression Reduction in volume, and increase in pressure and temperature, of a gas, caused by squeezing it into a smaller space.

Compression damping Controls the speed the suspension compresses when hitting a bump.

Compression ratio The relationship between cylinder volume when the piston is at top dead centre and cylinder volume when the piston is at bottom dead centre.

Continuity The uninterrupted path in the flow of electricity. Little or no measurable resistance.

Continuity tester Self-powered bleeper or test light which indicates continuity.

Cp Candlepower. Bulb rating commonly found on US motorcycles.

Crossply tyre Tyre plies arranged in a criss-cross pattern. Usually four or six plies used, hence 4PR or 6PR in tyre size codes.

Cush drive Rubber damper segments fitted between the rear wheel and final drive sprocket to absorb transmission shocks **(see illustration)**.

Cush drive rubbers dampen out transmission shocks

D

Degree disc Calibrated disc for measuring piston position. Expressed in degrees.

Dial gauge Clock-type gauge with adapters for measuring runout and piston position. Expressed in mm or inches.

Diaphragm The rubber membrane in a master cylinder or carburettor which seals the upper chamber.

Diaphragm spring A single sprung plate often used in clutches.

Direct current (dc) Current produced by a dc generator.

Technical Terms Explained

Decarbonisation The process of removing carbon deposits - typically from the combustion chamber, valves and exhaust port/system.
Detonation Destructive and damaging explosion of fuel/air mixture in combustion chamber instead of controlled burning.
Diode An electrical valve which only allows current to flow in one direction. Commonly used in rectifiers and starter interlock systems.
Disc valve (or rotary valve) A induction system used on some two-stroke engines.
Double-overhead camshaft (DOHC) An engine that uses two overhead camshafts, one for the intake valves and one for the exhaust valves.
Drivebelt A toothed belt used to transmit drive to the rear wheel on some motorcycles. A drivebelt has also been used to drive the camshafts. Drivebelts are usually made of Kevlar.
Driveshaft Any shaft used to transmit motion. Commonly used when referring to the final driveshaft on shaft drive motorcycles.

E

Earth return The return path of an electrical circuit, utilising the motorcycle's frame.
ECU (Electronic Control Unit) A computer which controls (for instance) an ignition system, or an anti-lock braking system.
EGO Exhaust Gas Oxygen sensor. Sometimes called a Lambda sensor.
Electrolyte The fluid in a lead-acid battery.
EMS (Engine Management System) A computer controlled system which manages the fuel injection and the ignition systems in an integrated fashion.
Endfloat The amount of lengthways movement between two parts. As applied to a crankshaft, the distance that the crankshaft can move side-to-side in the crankcase.
Endless chain A chain having no joining link. Common use for cam chains and final drive chains.
EP (Extreme Pressure) Oil type used in locations where high loads are applied, such as between gear teeth.
Evaporative emission control system Describes a charcoal filled canister which stores fuel vapours from the tank rather than allowing them to vent to the atmosphere. Usually only fitted to California models and referred to as an EVAP system.
Expansion chamber Section of two-stroke engine exhaust system so designed to improve engine efficiency and boost power.

F

Feeler blade or gauge A thin strip or blade of hardened steel, ground to an exact thickness, used to check or measure clearances between parts.
Final drive Description of the drive from the transmission to the rear wheel. Usually by chain or shaft, but sometimes by belt.
Firing order The order in which the engine cylinders fire, or deliver their power strokes, beginning with the number one cylinder.
Flooding Term used to describe a high fuel level in the carburettor float chambers, leading to fuel overflow. Also refers to excess fuel in the combustion chamber due to incorrect starting technique.
Free length The no-load state of a component when measured. Clutch, valve and fork spring lengths are measured at rest, without any preload.
Freeplay The amount of travel before any action takes place. The looseness in a linkage, or an assembly of parts, between the initial application of force and actual movement. For example, the distance the rear brake pedal moves before the rear brake is actuated.
Fuel injection The fuel/air mixture is metered electronically and directed into the engine intake ports (indirect injection) or into the cylinders (direct injection). Sensors supply information on engine speed and conditions.
Fuel/air mixture The charge of fuel and air going into the engine. See **Stoichiometric ratio**.
Fuse An electrical device which protects a circuit against accidental overload. The typical fuse contains a soft piece of metal which is calibrated to melt at a predetermined current flow (expressed as amps) and break the circuit.

G

Gap The distance the spark must travel in jumping from the centre electrode to the side electrode in a spark plug. Also refers to the distance between the ignition rotor and the pickup coil in an electronic ignition system.
Gasket Any thin, soft material - usually cork, cardboard, asbestos or soft metal - installed between two metal surfaces to ensure a good seal. For instance, the cylinder head gasket seals the joint between the block and the cylinder head.
Gauge An instrument panel display used to monitor engine conditions. A gauge with a movable pointer on a dial or a fixed scale is an analogue gauge. A gauge with a numerical readout is called a digital gauge.
Gear ratios The drive ratio of a pair of gears in a gearbox, calculated on their number of teeth.
Glaze-busting see **Honing**
Grinding Process for renovating the valve face and valve seat contact area in the cylinder head.
Gudgeon pin The shaft which connects the connecting rod small-end with the piston. Often called a piston pin or wrist pin.

H

Helical gears Gear teeth are slightly curved and produce less gear noise that straight-cut gears. Often used for primary drives.

Installing a Helicoil thread insert in a cylinder head

Helicoil A thread insert repair system. Commonly used as a repair for stripped spark plug threads **(see illustration)**.
Honing A process used to break down the glaze on a cylinder bore (also called glaze-busting). Can also be carried out to roughen a rebored cylinder to aid ring bedding-in.
HT (High Tension) Description of the electrical circuit from the secondary winding of the ignition coil to the spark plug.
Hydraulic A liquid filled system used to transmit pressure from one component to another. Common uses on motorcycles are brakes and clutches.
Hydrometer An instrument for measuring the specific gravity of a lead-acid battery.
Hygroscopic Water absorbing. In motorcycle applications, braking efficiency will be reduced if DOT 3 or 4 hydraulic fluid absorbs water from the air - care must be taken to keep new brake fluid in tightly sealed containers.

I

lbf ft Pounds-force feet. An imperial unit of torque. Sometimes written as ft-lbs.
lbf in Pound-force inch. An imperial unit of torque, applied to components where a very low torque is required. Sometimes written as in-lbs.
IC Abbreviation for Integrated Circuit.
Ignition advance Means of increasing the timing of the spark at higher engine speeds. Done by mechanical means (ATU) on early engines or electronically by the ignition control unit on later engines.
Ignition timing The moment at which the spark plug fires, expressed in the number of crankshaft degrees before the piston reaches the top of its stroke, or in the number of millimetres before the piston reaches the top of its stroke.
Infinity (∞) Description of an open-circuit electrical state, where no continuity exists.
Inverted forks (upside down forks) The sliders or lower legs are held in the yokes and the fork tubes or stanchions are connected to the wheel axle (spindle). Less unsprung weight and stiffer construction than conventional forks.

J

JASO Quality standard for 2-stroke oils.
Joule The unit of electrical energy.
Journal The bearing surface of a shaft.

K

Kickstart Mechanical means of turning the engine over for starting purposes. Only usually fitted to mopeds, small capacity motorcycles and off-road motorcycles.
Kill switch Handebar-mounted switch for emergency ignition cut-out. Cuts the ignition circuit on all models, and additionally prevent starter motor operation on others.
km Symbol for kilometre.
kmh Abbreviation for kilometres per hour.

L

Lambda (λ) sensor A sensor fitted in the exhaust system to measure the exhaust gas oxygen content (excess air factor).

Technical Terms Explained

Lapping see **Grinding**.
LCD Abbreviation for Liquid Crystal Display.
LED Abbreviation for Light Emitting Diode.
Liner A steel cylinder liner inserted in a aluminium alloy cylinder block.
Locknut A nut used to lock an adjustment nut, or other threaded component, in place.
Lockstops The lugs on the lower triple clamp (yoke) which abut those on the frame, preventing handlebar-to-fuel tank contact.
Lockwasher A form of washer designed to prevent an attaching nut from working loose.
LT Low Tension Description of the electrical circuit from the power supply to the primary winding of the ignition coil.

M

Main bearings The bearings between the crankshaft and crankcase.
Maintenance-free (MF) battery A sealed battery which cannot be topped up.
Manometer Mercury-filled calibrated tubes used to measure intake tract vacuum. Used to synchronise carburettors on multi-cylinder engines.
Micrometer A precision measuring instrument that measures component outside diameters **(see illustration)**.

Tappet shims are measured with a micrometer

MON (Motor Octane Number) A measure of a fuel's resistance to knock.
Monograde oil An oil with a single viscosity, eg SAE80W.
Monoshock A single suspension unit linking the swingarm or suspension linkage to the frame.
mph Abbreviation for miles per hour.
Multigrade oil Having a wide viscosity range (eg 10W40). The W stands for Winter, thus the viscosity ranges from SAE10 when cold to SAE40 when hot.
Multimeter An electrical test instrument with the capability to measure voltage, current and resistance. Some meters also incorporate a continuity tester and buzzer.

N

Needle roller bearing Inner race of caged needle rollers and hardened outer race. Examples of uncaged needle rollers can be found on some engines. Commonly used in rear suspension applications and in two-stroke engines.
Nm Newton metres.
NOx Oxides of Nitrogen. A common toxic pollutant emitted by petrol engines at higher temperatures.

O

Octane The measure of a fuel's resistance to knock.
OE (Original Equipment) Relates to components fitted to a motorcycle as standard or replacement parts supplied by the motorcycle manufacturer.
Ohm The unit of electrical resistance. Ohms = Volts ÷ Current.
Ohmmeter An instrument for measuring electrical resistance.
Oil cooler System for diverting engine oil outside of the engine to a radiator for cooling purposes.
Oil injection A system of two-stroke engine lubrication where oil is pump-fed to the engine in accordance with throttle position.
Open-circuit An electrical condition where there is a break in the flow of electricity - no continuity (high resistance).
O-ring A type of sealing ring made of a special rubber-like material; in use, the O-ring is compressed into a groove to provide the sealing action.
Oversize (OS) Term used for piston and ring size options fitted to a rebored cylinder.
Overhead cam (sohc) engine An engine with single camshaft located on top of the cylinder head.
Overhead valve (ohv) engine An engine with the valves located in the cylinder head, but with the camshaft located in the engine block or crankcase.
Oxygen sensor A device installed in the exhaust system which senses the oxygen content in the exhaust and converts this information into an electric current. Also called a Lambda sensor.

P

Plastigauge A thin strip of plastic thread, available in different sizes, used for measuring clearances. For example, a strip of Plastigauge is laid across a bearing journal. The parts are assembled and dismantled; the width of the crushed strip indicates the clearance between journal and bearing.
Polarity Either negative or positive earth (ground), determined by which battery lead is connected to the frame (earth return). Modern motorcycles are usually negative earth.
Pre-ignition A situation where the fuel/air mixture ignites before the spark plug fires. Often due to a hot spot in the combustion chamber caused by carbon build-up. Engine has a tendency to 'run-on'.
Pre-load (suspension) The amount a spring is compressed when in the unloaded state. Preload can be applied by gas, spacer or mechanical adjuster.
Premix The method of engine lubrication on older two-stroke engines. Engine oil is mixed with the petrol in the fuel tank in a specific ratio. The fuel/oil mix is sometimes referred to as "petroil".
Primary drive Description of the drive from the crankshaft to the clutch. Usually by gear or chain.
PS Pfedestärke - a German interpretation of BHP.
PSI Pounds-force per square inch. Imperial measurement of tyre pressure and cylinder pressure measurement.
PTFE Polytetrafluoroethylene. A low friction substance.

Pulse secondary air injection system A process of promoting the burning of excess fuel present in the exhaust gases by routing fresh air into the exhaust ports.

Q

Quartz halogen bulb Tungsten filament surrounded by a halogen gas. Typically used for the headlight **(see illustration)**.

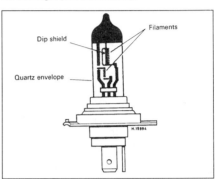

Quartz halogen headlight bulb construction

R

Rack-and-pinion A pinion gear on the end of a shaft that mates with a rack (think of a geared wheel opened up and laid flat). Sometimes used in clutch operating systems.
Radial play Up and down movement about a shaft.
Radial ply tyres Tyre plies run across the tyre (from bead to bead) and around the circumference of the tyre. Less resistant to tread distortion than other tyre types.
Radiator A liquid-to-air heat transfer device designed to reduce the temperature of the coolant in a liquid cooled engine.
Rake A feature of steering geometry - the angle of the steering head in relation to the vertical **(see illustration)**.

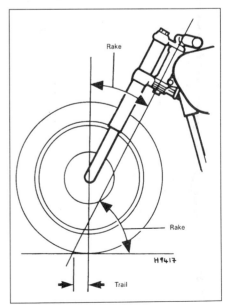

Steering geometry

Technical Terms Explained

Rebore Providing a new working surface to the cylinder bore by boring out the old surface. Necessitates the use of oversize piston and rings.
Rebound damping A means of controlling the oscillation of a suspension unit spring after it has been compressed. Resists the spring's natural tendency to bounce back after being compressed.
Rectifier Device for converting the ac output of an alternator into dc for battery charging.
Reed valve An induction system commonly used on two-stroke engines.
Regulator Device for maintaining the charging voltage from the generator or alternator within a specified range.
Relay A electrical device used to switch heavy current on and off by using a low current auxiliary circuit.
Resistance Measured in ohms. An electrical component's ability to pass electrical current.
RON (Research Octane Number) A measure of a fuel's resistance to knock.
rpm revolutions per minute.
Runout The amount of wobble (in-and-out movement) of a wheel or shaft as it's rotated. The amount a shaft rotates `out-of-true'. The out-of-round condition of a rotating part.

S

SAE (Society of Automotive Engineers) A standard for the viscosity of a fluid.
Sealant A liquid or paste used to prevent leakage at a joint. Sometimes used in conjunction with a gasket.
Service limit Term for the point where a component is no longer useable and must be renewed.
Shaft drive A method of transmitting drive from the transmission to the rear wheel.
Shell bearings Plain bearings consisting of two shell halves. Most often used as big-end and main bearings in a four-stroke engine. Often called bearing inserts.
Shim Thin spacer, commonly used to adjust the clearance or relative positions between two parts. For example, shims inserted into or under tappets or followers to control valve clearances. Clearance is adjusted by changing the thickness of the shim.
Short-circuit An electrical condition where current shorts to earth (ground) bypassing the circuit components.
Skimming Process to correct warpage or repair a damaged surface, eg on brake discs or drums.
Slide-hammer A special puller that screws into or hooks onto a component such as a shaft or bearing; a heavy sliding handle on the shaft bottoms against the end of the shaft to knock the component free.
Small-end bearing The bearing in the upper end of the connecting rod at its joint with the gudgeon pin.
Spalling Damage to camshaft lobes or bearing journals shown as pitting of the working surface.
Specific gravity (SG) The state of charge of the electrolyte in a lead-acid battery. A measure of the electrolyte's density compared with water.
Straight-cut gears Common type gear used on gearbox shafts and for oil pump and water pump drives.
Stanchion The inner sliding part of the front forks, held by the yokes. Often called a fork tube.
Stoichiometric ratio The optimum chemical air/fuel ratio for a petrol engine, said to be 14.7 parts of air to 1 part of fuel.
Sulphuric acid The liquid (electrolyte) used in a lead-acid battery. Poisonous and extremely corrosive.
Surface grinding (lapping) Process to correct a warped gasket face, commonly used on cylinder heads.

T

Tapered-roller bearing Tapered inner race of caged needle rollers and separate tapered outer race. Examples of taper roller bearings can be found on steering heads.
Tappet A cylindrical component which transmits motion from the cam to the valve stem, either directly or via a pushrod and rocker arm. Also called a cam follower.
TCS Traction Control System. An electronically-controlled system which senses wheel spin and reduces engine speed accordingly.
TDC Top Dead Centre denotes that the piston is at its highest point in the cylinder.
Thread-locking compound Solution applied to fastener threads to prevent slackening. Select type to suit application.
Thrust washer A washer positioned between two moving components on a shaft. For example, between gear pinions on gearshaft.
Timing chain See **Cam Chain**.
Timing light Stroboscopic lamp for carrying out ignition timing checks with the engine running.
Top-end A description of an engine's cylinder block, head and valve gear components.
Torque Turning or twisting force about a shaft.
Torque setting A prescribed tightness specified by the motorcycle manufacturer to ensure that the bolt or nut is secured correctly. Undertightening can result in the bolt or nut coming loose or a surface not being sealed. Overtightening can result in stripped threads, distortion or damage to the component being retained.
Torx key A six-point wrench.
Tracer A stripe of a second colour applied to a wire insulator to distinguish that wire from another one with the same colour insulator. For example, Br/W is often used to denote a brown insulator with a white tracer.
Trail A feature of steering geometry. Distance from the steering head axis to the tyre's central contact point.
Triple clamps The cast components which extend from the steering head and support the fork stanchions or tubes. Often called fork yokes.
Turbocharger A centrifugal device, driven by exhaust gases, that pressurises the intake air. Normally used to increase the power output from a given engine displacement.
TWI Abbreviation for Tyre Wear Indicator. Indicates the location of the tread depth indicator bars on tyres.

U

Universal joint or U-joint (UJ) A double-pivoted connection for transmitting power from a driving to a driven shaft through an angle. Typically found in shaft drive assemblies.
Unsprung weight Anything not supported by the bike's suspension (ie the wheel, tyres, brakes, final drive and bottom (moving) part of the suspension).

V

Vacuum gauges Clock-type gauges for measuring intake tract vacuum. Used for carburettor synchronisation on multi-cylinder engines.
Valve A device through which the flow of liquid, gas or vacuum may be stopped, started or regulated by a moveable part that opens, shuts or partially obstructs one or more ports or passageways. The intake and exhaust valves in the cylinder head are of the poppet type.
Valve clearance The clearance between the valve tip (the end of the valve stem) and the rocker arm or tappet/follower. The valve clearance is measured when the valve is closed. The correct clearance is important - if too small the valve won't close fully and will burn out, whereas if too large noisy operation will result.
Valve lift The amount a valve is lifted off its seat by the camshaft lobe.
Valve timing The exact setting for the opening and closing of the valves in relation to piston position.
Vernier caliper A precision measuring instrument that measures inside and outside dimensions. Not quite as accurate as a micrometer, but more convenient.
VIN Vehicle Identification Number. Term for the bike's engine and frame numbers.
Viscosity The thickness of a liquid or its resistance to flow.
Volt A unit for expressing electrical "pressure" in a circuit. Volts = current x ohms.

W

Water pump A mechanically-driven device for moving coolant around the engine.
Watt A unit for expressing electrical power. Watts = volts x current.
Wear limit see **Service limit**
Wet liner A liquid-cooled engine design where the pistons run in liners which are directly surrounded by coolant **(see illustration)**.

Wet liner arrangement

Wheelbase Distance from the centre of the front wheel to the centre of the rear wheel.
Wiring harness or loom Describes the electrical wires running the length of the motorcycle and enclosed in tape or plastic sheathing. Wiring coming off the main harness is usually referred to as a sub harness.
Woodruff key A key of semi-circular or square section used to locate a gear to a shaft. Often used to locate the alternator rotor on the crankshaft.
Wrist pin Another name for gudgeon or piston pin.

Index

Note: References throughout this index are in the form – "Chapter number" • "page number"

A

About this manual - 0•7
Acknowledgements - 0•7
Air ducts removal and installation - 8•5
Air filter
 cleaning - 1•11
 housing removal and installation - 4•3
 renewal - 1•23
Air mixture adjustment - 4•4
Alternator check, removal and installation - 9•22
Asbestos - 0•10

B

Battery - 0•10
 charging - 9•4
 check - 1•14
 removal, installation, inspection and maintenance - 9•3
Bleeding brake system - 7•14
Bodywork - 8•1 et seq
 air ducts - 8•5
 fairing - 8•3
 mirrors - 8•3
 mudguard - 8•6
 seat cowling - 8•2
 seats - 8•2
 trim panels - 8•5
Brake fluid - 0•12, 1•2, 1•20
Brake hoses inspection and renewal - 1•24, 7•13
Brake lever - 6•6
 switch - 9•10
Brake light
 bulb renewal - 9•7
 check - 9•5
 switches check and replacement - 9•10

Brake pedal removal and installation - 6•3
Brakes, wheels and tyres - 7•1 et seq
 bleeding - 7•14
 calipers - 7•4, 7•9
 check - 1•12
 discs - 7•5, 7•11
 hoses, pipes and unions - 7•13
 master cylinder - 7•6, 7•11
 pads - 7•3, 7•8
 sprocket coupling bearing - 7•19
 tyres - 7•20
 wheels - 7•14, 7•15, 7•16, 7•17
Buying spare parts - 0•8

C

Cables - 2•30, 4•12, 4•14, 9•11
 check - 1•14
 lubrication - 1•10
Calipers
 removal, overhaul and installation
  front - 7•4
  rear - 7•9
 seals renewal - 1•24
Camshaft drive gear assembly removal, inspection and installation - 2•22
Camshafts removal, inspection and installation - 2•13
Carburettor - 4•4
 disassembly, cleaning and inspection - 4•6
 reassembly and float height check - 4•10
 removal and installation - 4•4
 separation and joining - 4•8
 synchronisation - 1•15
Chain - 0•13, 1•2
 check, adjustment and lubrication - 1•6
 removal, cleaning and installation - 6•23
Charging system leakage and output test - 9•22
Chemicals and Lubricants - REF•21

Choke cable
 check - 1•14
 removal and installation - 4•14
Clutch
 cable replacement - 2•30
 check and adjustment - 1•8
 lever - 6•6
 removal, inspection and installation - 2•25
 switch check and replacement - 9•17
Compression check - 1•24
Connecting rods
 bearings - 2•40, 2•42
 removal, inspection and installation - 2•41
Contents - 0•2
Conversion factors - REF•20
Coolant - 0•13, 1•2
Cooling system - 3•1 et seq
 check - 1•12
 draining, flushing and refilling - 1•23
 fan and switch - 3•3
 hoses - 3•8
 pressure cap - 3•2
 radiator - 3•6
 reservoir - 3•2
 temperature gauge and sender - 3•4, 9•13
 thermostat and thermostat housing - 3•4
 water pump - 3•6
Crankcase halves
 inspection and servicing - 2•40
 separation and reassembly - 2•37
Crankshaft removal, inspection and installation - 2•46
Cylinder block removal, inspection and installation - 2•2
Cylinder bores inspection and servicing - 2•40
Cylinder compression check - 1•24
Cylinder head
 disassembly, inspection and reassembly - 2•18
 removal and installation - 2•16

Index

D

Daily (pre-ride) checks - 0•11 et seq
  brake fluid - 0•12, 1•2
  chain - 0•13, 1•2
  coolant - 0•13, 1•2
  engine/transmission oil - 0•11, 1•2
  final drive - 0•13, 1•2
  fuel - 0•14
  legal and safety checks - 0•14
  lighting - 0•14
  pressures - 0•14
  safety - 0•14
  signalling - 0•14
  steering - 0•13
  suspension - 0•13
  tread depth - 0•14
  tyre pressures - 0•14
Dimensions - REF•1
Diode check and replacement - 9•18
Discs inspection, removal and installation
  front - 7•5
  rear - 7•11
Drive chain
  check, adjustment and lubrication - 1•6
  removal, cleaning and installation - 6•23

E

Electrical system - 9•1 et seq
  alternator - 9•26
  battery - 9•3, 9•4
  brake lever switch - 9•10
  brake light - 9•5, 9•7, 9•10
  charging system - 9•25
  clutch switch - 9•17
  diode - 9•18
  fault finding - 9•3
  fuses - 9•4
  handlebar switches - 9•15
  headlight - 9•5, 9•6
  horn - 9•18
  ignition (main) switch - 9•15
  indicators - 9•5, 9•8, 9•9
  instruments - 9•5, 9•11, 9•12, 9•13
  lighting system - 9•5
  neutral switch - 9•16
  oil pressure switch - 9•14
  regulator/rectifier - 9•28
  sidelight - 9•6
  sidestand switch - 9•16
  speedometer - 9•11, 9•12,
  starter motor - 9•19, 9•20
  starter relay - 9•18
  tachometer - 9•13
  tail light - 9•5, 9•8
  temperature gauge - 9•13
  turn signal lights - 9•5, 9•8, 9•9
  warning light bulbs - 9•13
  warning lights - 9•5
Electricity - 0•10
Engine oil pressure
  check - 1•25
  switch check, removal and installation - 9•14

Engine, clutch and transmission - 2•1 et seq
  camshafts - 2•13, 2•22
  clutch - 2•25, 2•30
  connecting rod bearings - 2•40, 2•41, 2•42
  crankcase halves - 2•37, 2•40
  crankshaft - 2•46
  cylinder block - 2•2
  cylinder bores - 2•40
  cylinder head - 2•16, 2•18
  gearchange mechanism - 2•34
  idle/reduction gear - 2•24
  main bearings - 2•40, 2•47
  oil cooler - 2•12
  oil pump - 2•32
  oil sump, oil strainer and pressure relief valve - 2•31
  piston rings - 2•45
  pistons - 2•44
  removal and installation - 2•1
  running-in procedure - 2•57
  selector drum and forks - 2•54
  starter clutch - 2•24
  start-up after overhaul - 2•56
  transmission shafts - 2•48, 2•49
  valve cover - 2•12
  valves/valve seats/valve guides - 2•18
Engine/transmission oil - 0•11, 1•2
  change - 1•10, 1•13
Exhaust system removal and installation - 4•14

F

Fairing panels removal and installation - 8•3
Fan check and replacement - 3•3
Fault finding - REF•30 et seq
  braking problems - REF•37
  clutch problems - REF•34
  driveline noise - REF•35
  electrical problems - 9•3, REF•38
  engine doesn't start or is difficult to start - REF•31
  engine noise - REF•35
  exhaust smoke - REF•36
  frame and suspension noise - REF•36
  gearchanging problems - REF•35
  oil pressure warning light comes on - REF•36
  overheating - REF•34
  poor handling or stability - REF•37
  poor running at low speed - REF•32
  poor running or no power at high speed - REF•33
Fault Finding Equipment - REF•39 et seq
Filter
  air - 1•11, 1•23, 4•3
  fuel - 1•14
  oil - 1•13
Final drive - 0•13, 1•2, 1•6, 6•23
Fire - 0•10
Float height check - 4•10
Footrests removal and installation - 6•3
Forks - 6•19
  disassembly, inspection and reassembly - 6•8
  oil - 1•2, 1•25
  removal and installation - 6•6

Frame, suspension and final drive - 6•1 et seq
  brake lever - 6•6
  brake pedal - 6•3
  chain - 6•23
  clutch lever - 6•6
  footrests - 6•3
  forks - 6•6, 6•8, 6•19
  frame - 6•2
  gearchange lever - 6•3
  handlebars - 6•4
  handlebar levers - 6•6
  shock absorber - 6•17, 6•20
  sidestand - 6•4
  sprockets - 6•23, 6•25
  steering head bearings - 6•16
  steering stem - 6•15
  suspension - 6•18, 6•19
  swingarm - 6•20, 6•22
Fuel - 0•14
Fuel and exhaust systems - 4•1 et seq
  air filter housing - 4•3
  carburettor - 4•4, 4•6, 4•8, 4•10
  choke cable - 4•14
  exhaust system - 4•14
  fuel pump - 4•15
  fuel tank and fuel tap - 4•2, 4•3
  idle fuel/air mixture adjustment - 4•4
  throttle cables - 4•12
Fuel filter cleaning - 1•14
Fuel hoses renewal - 1•24
Fuel pump check, removal and installation - 4•15
Fuel system check - 1•14
Fuel tank
  cleaning and repair - 4•3
  removal and installation - 4•2
Fuel tap removal and installation - 4•2
Fumes - 0•10
Fuses check and renewal - 9•4

G

Gearchange lever removal and installation - 6•3
Gearchange mechanism removal, inspection and installation - 2•34
Glossary - REF•43 et seq
Grey imports - 0•8

H

Handlebar switches
  check - 9•15
  switches removal and installation - 9•16
Handlebars and levers removal and installation - 6•4
Headlight
  aim check and adjustment - 1•17
  assembly removal and installation - 9•6
  bulb renewal - 9•6
  check - 9•5
Horn check and replacement - 9•18
HT coils check, removal and installation - 5•3

Index

I

Identification numbers - 0•8
Idle fuel/air mixture adjustment - 4•4
Idle speed check and adjustment - 1•7
Idle/reduction gear removal, inspection and installation - 2•24
Ignition (main) switch check, removal and installation - 9•15
Ignition system - 5•1 et seq
   check - 5•2
   control unit - 5•5
   HT coils - 5•3
   pulse generator coil - 5•4
   throttle position sensor - 5•6
   timing - 5•5
Imports - 0•8
Indicators - 0•14
   assemblies removal and installation - 9•9
   bulbs renewal - 9•8
   circuit check - 9•8
   lights check - 9•5
Instruments
   check and replacement - 9•12
   removal and installation - 9•11
Instrument lights
   bulbs renewal - 9•13
   check - 9•5
Introduction - 0•4

L

Legal and safety checks - 0•14
Levers removal and installation - 6•6
Lighting - 0•14
   check - 9•5
Lubricants - REF•21

M

Main bearings - 2•40
   shell selection - 2•47
Maintenance schedule - 1•3
Master cylinder
   removal, overhaul and installation
     front - 7•6
     rear - 7•11
   seals renewal - 1•24
Mirrors removal and installation - 8•3
Model years - 0•8
MOT Test Checks - REF•22 et seq
Mudguard removal and installation - 8•6

N

Neutral switch check, removal and installation - 9•16
Nuts and bolts tightness check - 1•20

O

Oil
   chain - 1•2
   engine/transmission - 0•11, 1•2, 1•10, 1•13, 1•25
   forks - 1•2, 1•25
Oil cooler removal and installation - 2•12
Oil filter change - 1•13

Oil pressure
   check - 1•25
   switch check, removal and installation - 9•14
Oil pump removal, inspection and installation - 2•32
Oil sump removal, inspection and installation - 2•31

P

Pads
   renewal
     front - 7•3
     rear - 7•8
   wear check - 1•8
Parts - 0•8
Piston rings inspection and installation - 2•45
Pistons removal, inspection and installation - 2•44
Pivot points lubrication - 1•9
Pressure cap check - 3•2
Pressure relief valve removal, inspection and installation - 2•31
Pulse generator coil assembly check, removal and installation - 5•4

R

Radiator removal and installation - 3•6
Regulator/rectifier check and replacement - 9•28
Reservoir removal and installation - 3•2
Routine maintenance and servicing - 1•1 et seq
   air filter - 1•11, 1•23
   battery - 1•14
   brake hoses - 1•24
   brake system - 1•12
   brake fluid - 1•20
   cables - 1•10
   caliper seals - 1•24
   carburettors - 1•15
   chain and sprockets - 1•6
   choke cable - 1•14
   clutch - 1•8
   cooling system - 1•12, 1•23
   cylinder compression - 1•24
   drive chain and sprockets - 1•6
   engine/transmission oil and oil filter - 1•10, 1•13
   fork oil - 1•25
   fuel hoses - 1•24
   fuel system - 1•14
   headlight aim - 1•17
   idle speed - 1•7
   maintenance schedule - 1•3
   master cylinder seals - 1•24
   nuts and bolts - 1•20
   oil pressure - 1•25
   pads - 1•8
   pivot points - 1•9
   sidestand - 1•17
   spark plugs - 1•8, 1•15
   stand - 1•9
   steering head bearings - 1•18, 1•25
   suspension - 1•17
   swingarm and suspension linkage bearings - 1•25
   throttle cables - 1•14
   tyres - 1•11
   valve clearances - 1•20
   wheel bearings - 1•20
   wheels - 1•11
Running-in procedure - 2•57

S

Safety first! - 0•10, 0•14
Seals - 1•24
Seats
   cowling removal and installation - 8•2
   removal and installation - 8•2
Selector drum and forks removal, inspection and installation - 2•54
Shock absorber - 6•20
   removal, inspection and installation - 6•17
Sidelight bulb renewal - 9•6
Sidestand
   check - 1•17
   removal and installation - 6•4
   switch check and replacement - 9•16
Signalling - 0•14
Spare parts - 0•8
Spark plugs
   gaps check and adjustment - 1•8
   renewal - 1•15
Speedometer
   cable removal and installation - 9•11
   check and replacement - 9•12
Sprockets
   check and renewal - 1•6, 6•23
   coupling/rubber damper bearing - 7•19
   check and renewal - 6•25
Stand, lever pivots and cables lubrication - 1•9
Starter clutch and idle/reduction gear removal, inspection and installation - 2•24
Starter motor
   disassembly, inspection and reassembly - 9•20
   removal and installation - 9•19
Starter relay check and replacement - 9•18
Start-up after overhaul - 2•56
Steering - 0•13
Steering head bearings
   check and adjustment - 1•18
   inspection and renewal - 6•16
   lubrication - 1•25
Steering stem removal and installation - 6•15
Storage - REF•27 et seq
Sump removal, inspection and installation - 2•31
Suspension - 0•13
   adjustments - 6•19
   check - 1•17
Suspension linkage
   bearings lubrication - 1•25
   removal, inspection and installation - 6•18
Swingarm
   bearings lubrication - 1•25
   inspection and bearing renewal - 6•22
   removal and installation - 6•20
Switches - 3•3, 9•10, 9•14, 9•15, 9•16, 9•17

Index

T

Tachometer check and replacement - 9•13
Tail light
  assembly removal and installation - 9•8
  bulb renewal - 9•7
  check - 9•5
Technical Terms Explained - REF•43 *et seq*
Temperature gauge - 9•13
  check and replacement - 3•4
Thermostat and thermostat housing
  removal, check and installation - 3•4
Throttle cables
  check - 1•14
  removal and installation - 4•12
Throttle position sensor check and
  replacement - 5•6
Timing general information and
  check - 5•5
Tools and Workshop Tips - REF•2 *et seq*
Transmission shafts
  disassembly, inspection and
    reassembly - 2•49
  removal and installation - 2•48
Trim panels removal and installation - 8•5
Turn signals - 0•14
  assemblies removal and installation - 9•9
  bulbs renewal - 9•8
  circuit check - 9•8
  lights check - 9•5
Tyres - 0•14
  check - 1•11
  depth - 0•14
  general information and fitting - 7•20
  pressures - 0•14

U

Unofficial (grey) imports - 0•8

V

Valve clearances check and
  adjustment - 1•20
Valve cover removal and installation - 2•12
Valves/valve seats/valve guides - 2•18

W

Warning lights
  bulbs renewal - 9•13
  check - 9•5
Water pump check, removal and
  installation - 3•6
Weights - REF•1
Wheel bearings
  check - 1•20
  renewal - 7•17
Wheels
  alignment check - 7•14
  general check - 1•11
  inspection and repair - 7•14
  removal and installation
    front - 7•15
    rear - 7•16
Wiring diagams - 9•30 *et seq*
Workshop Tips - REF•2 *et seq*

Haynes Motorcycle Manuals – The Complete List

Title	Book No
BMW 2-valve Twins (70 - 96)	♦ 0249
BMW K100 & 75 2-valve Models (83 - 96)	♦ 1373
BMW R850, 1100 & 1150 4-valve Twins (93 - 04)	♦ 3466
BSA Bantam (48 - 71)	0117
BSA Unit Singles (58 - 72)	0127
BSA Pre-unit Singles (54 - 61)	0326
BSA A7 & A10 Twins (47 - 62)	0121
BSA A50 & A65 Twins (62 - 73)	0155
DUCATI 600, 750 & 900 2-valve V-Twins (91 - 96)	♦ 3290
Ducati MK III & Desmo Singles (69 - 76)	◊ 0445
Ducati 748, 916 & 996 4-valve V-Twins (94 - 01)	♦ 3756
GILERA Runner, DNA, Stalker & Ice (97 - 04)	4163
HARLEY-DAVIDSON Sportsters (70 - 03)	♦ 2534
Harley-Davidson Big Twins (73 - on)	2536
Harley-Davidson Twin Cam 88 (99 - 03)	♦ 2478
HONDA NB, ND, NP & NS50 Melody (81 - 85)	◊ 0622
Honda NE/NB50 Vision & SA50 Vision Met-in (85 - 95)	◊ 1278
Honda MB, MBX, MT & MTX50 (80 - 93)	0731
Honda C50, C70 & C90 (67 - 99)	0324
Honda XR80R & XR100R (85 - 04)	2218
Honda XL/XR 80, 100, 125, 185 & 200 2-valve Models (78 - 87)	0566
Honda H100 & H100S Singles (80 - 92)	◊ 0734
Honda CB/CD125T & CM125C Twins (77 - 88)	◊ 0571
Honda CG125 (76 - 00)	◊ 0433
Honda NS125 (86 - 93)	◊ 3056
Honda MBX/MTX125 & MTX200 (83 - 93)	◊ 1132
Honda CD/CM185 200T & CM250C 2-valve Twins (77 - 85)	0572
Honda XL/XR 250 & 500 (78 - 84)	0567
Honda XR250L, XR250R & XR400R (86 - 03)	2219
Honda CB250 & CB400N Super Dreams (78 - 84)	◊ 0540
Honda CR Motocross Bikes (86 - 01)	2222
Honda CBR400RR Fours (88 - 99)	◊ ♦ 3552
Honda VFR400 (NC30) & RVF400 (NC35) V-Fours (89 - 98)	◊ ♦ 3496
Honda CB500 (93 - 01)	◊ ♦ 3753
Honda CB400 & CB550 Fours (73 - 77)	0262
Honda CX/GL500 & 650 V-Twins (78 - 86)	0442
Honda CBX550 Four (82 - 86)	◊ 0940
Honda XL600R & XR600R (83 - 00)	2183
Honda XL600/650V Transalp & XRV750 Africa Twin (87 - 02)	◊ ♦ 3919
Honda CBR600F1 & 1000F Fours (87 - 96)	♦ 1730
Honda CBR600F2 & F3 Fours (91 - 98)	♦ 2070
Honda CBR600F4 (99 - 02)	♦ 3911
Honda CB600F Hornet (98 - 02)	◊ ♦ 3915
Honda CB650 sohc Fours (78 - 84)	0665
Honda NTV600 Revere, NTV650 & NT650V Deauville (88 - 01)	◊ ♦ 3243
Honda Shadow VT600 & 750 (USA) (88 - 03)	2312
Honda CB750 sohc Four (69 - 79)	0131
Honda V45/65 Sabre & Magna (82 - 88)	0820
Honda VFR750 & 700 V-Fours (86 - 97)	♦ 2101
Honda VFR800 V-Fours (97 - 01)	♦ 3703
Honda CB750 & CB900 dohc Fours (78 - 84)	0535
Honda VTR1000 (FireStorm, Super Hawk) & XL1000V (Varadero) (97 - 00)	♦ 3744
Honda CBR900RR FireBlade (92 - 99)	♦ 2161
Honda CBR900RR FireBlade (00 - 03)	♦ 4060
Honda CBR1100XX Super Blackbird (97 - 02)	♦ 3901
Honda ST1100 Pan European V-Fours (90 - 01)	♦ 3384
Honda Shadow VT1100 (USA) (85 - 98)	2313
Honda GL1000 Gold Wing (75 - 79)	0309
Honda GL1100 Gold Wing (79 - 81)	0669
Honda Gold Wing 1200 (USA) (84 - 87)	2199

Title	Book No
Honda Gold Wing 1500 (USA) (88 - 00)	2225
KAWASAKI AE/AR 50 & 80 (81 - 95)	1007
Kawasaki KC, KE & KH100 (75 - 99)	1371
Kawasaki KMX125 & 200 (86 - 02)	◊ 3046
Kawasaki 250, 350 & 400 Triples (72 - 79)	0134
Kawasaki 400 & 440 Twins (74 - 81)	0281
Kawasaki 400, 500 & 550 Fours (79 - 91)	0910
Kawasaki EN450 & 500 Twins (Ltd/Vulcan) (85 - 04)	2053
Kawasaki EX & ER500 (GPZ500S & ER-5) Twins (87 - 99)	♦ 2052
Kawasaki ZX600 (Ninja ZX-6, ZZ-R600) Fours (90 - 00)	♦ 2146
Kawasaki ZX-6R Ninja Fours (95 - 02)	♦ 3541
Kawasaki ZX600 (GPZ600R, GPX600R, Ninja 600R & RX) & ZX750 (GPX750R, Ninja 750R) Fours (85 - 97)	♦ 1780
Kawasaki 650 Four (76 - 78)	0373
Kawasaki Vulcan 700/750 & 800 (85 - 01)	♦ 2457
Kawasaki 750 Air-cooled Fours (80 - 91)	0574
Kawasaki ZR550 & 750 Zephyr Fours (90 - 97)	♦ 3382
Kawasaki ZX750 (Ninja ZX-7 & ZXR750) Fours (89 - 96)	♦ 2054
Kawasaki Ninja ZX-7R & ZX-9R (ZX750P, ZX900B/C/D/E) (94 - 00)	♦ 3721
Kawasaki 900 & 1000 Fours (73 - 77)	0222
Kawasaki ZX900, 1000 & 1100 Liquid-cooled Fours (83 - 97)	♦ 1681
MOTO GUZZI 750, 850 & 1000 V-Twins (74 - 78)	0339
MZ ETZ Models (81 - 95)	◊ 1680
NORTON 500, 600, 650 & 750 Twins (57 - 70)	0187
Norton Commando (68 - 77)	0125
PEUGEOT Speedfight, Trekker & Vivacity (96 - 02)	◊ 3920
PIAGGIO (Vespa) Scooters (91 - 03)	3492
SUZUKI GT, ZR & TS50 (77 - 90)	◊ 0799
Suzuki TS50X (84 - 00)	◊ 1599
Suzuki 100, 125, 185 & 250 Air-cooled Trail bikes (79 - 89)	0797
Suzuki GP100 & 125 Singles (78 - 93)	◊ 0576
Suzuki GS, GN, GZ & DR125 Singles (82 - 99)	◊ 0888
Suzuki 250 & 350 Twins (68 - 78)	0120
Suzuki GT250X7, GT200X5 & SB200 Twins (78 - 83)	◊ 0469
Suzuki GS/GSX250, 400 & 450 Twins (79 - 85)	0736
Suzuki GS500 Twin (89 - 02)	♦ 3238
Suzuki GS550 (77 - 82) & GS750 Fours (76 - 79)	0363
Suzuki GS/GSX550 4-valve Fours (83 - 88)	1133
Suzuki SV650 (99 - 02)	♦ 3912
Suzuki GSX-R600 & 750 (96 - 00)	♦ 3553e
Suzuki GSX-R600 (01 - 02), GSX-R750 (00 - 02) & GSX-R1000 (01 - 02)	♦ 3986e
Suzuki GSF600 & 1200 Bandit Fours (95 - 04)	♦ 3367
Suzuki GS850 Fours (78 - 88)	0536
Suzuki GS1000 Four (77 - 79)	0484
Suzuki GSX-R750, GSX-R1100 (85 - 92), GSX600F, GSX750F, GSX1100F (Katana) Fours (88 - 96)	♦ 2055
Suzuki GSX600/750F & GSX750 (98 - 02)	♦ 3987
Suzuki GS/GSX1000, 1100 & 1150 4-valve Fours (79 - 88)	0737
Suzuki TL1000S/R & DL1000 V-Strom (97 - 04)	♦ 4083
Suzuki GSX1300R Hayabusa (99 - 04)	♦ 4184
TRIUMPH Tiger Cub & Terrier (52 - 68)	0414
Triumph 350 & 500 Unit Twins (58 - 73)	0137
Triumph Pre-Unit Twins (47 - 62)	0251
Triumph 650 & 750 2-valve Unit Twins (63 - 83)	0122
Triumph Trident & BSA Rocket 3 (69 - 75)	0136
Triumph Fuel Injected Triples (97 - 00)	3755
Triumph Triples & Fours (carburettor engines) (91 - 99)	♦ 2162
VESPA P/PX125, 150 & 200 Scooters (78 - 03)	0707
Vespa Scooters (59 - 78)	0126
YAMAHA DT50 & 80 Trail Bikes (78 - 95)	◊ 0800
Yamaha T50 & 80 Townmate (83 - 95)	◊ 1247
Yamaha YB100 Singles (73 - 91)	◊ 0474

Title	Book No
Yamaha RS/RXS100 & 125 Singles (74 - 95)	0331
Yamaha RD & DT125LC (82 - 87)	◊ 0887
Yamaha TZR125 (87 - 93) & DT125R (88 - 02)	◊ 1655
Yamaha TY50, 80, 125 & 175 (74 - 84)	◊ 0464
Yamaha XT & SR125 (82 - 02)	◊ 1021
Yamaha Trail Bikes (81 - 00)	2350
Yamaha 250 & 350 Twins (70 - 79)	0040
Yamaha XS250, 360 & 400 sohc Twins (75 - 84)	0378
Yamaha RD250 & 350LC Twins (80 - 82)	0803
Yamaha RD350 YPVS Twins (83 - 95)	1158
Yamaha RD400 Twin (75 - 79)	0333
Yamaha XT, TT & SR500 Singles (75 - 83)	0342
Yamaha XZ550 Vision V-Twins (82 - 85)	0821
Yamaha FJ, FZ, XJ & YX600 Radian (84 - 92)	2100
Yamaha XJ600S (Diversion, Seca II) & XJ600N Fours (92 - 03)	♦ 2145
Yamaha YZF600R Thundercat & FZS600 Fazer (96 - 03)	♦ 3702
Yamaha YZF-R6 (98 - 02)	♦ 3900
Yamaha 650 Twins (70 - 83)	0341
Yamaha XJ650 & 750 Fours (80 - 84)	0738
Yamaha XS750 & 850 Triples (76 - 85)	0340
Yamaha TDM850, TRX850 & XTZ750 (89 - 99)	◊ ♦ 3540
Yamaha YZF750R & YZF1000R Thunderace (93 - 00)	♦ 3720
Yamaha FZR600, 750 & 1000 Fours (87 - 96)	♦ 2056
Yamaha XV (Virago) V-Twins (81 - 03)	♦ 0802
Yamaha XVS650 & 1100 Dragstar/V-Star (97 - 04)	♦ 4195
Yamaha XJ900F Fours (83 - 94)	♦ 3239
Yamaha XJ900S Diversion (94 - 01)	♦ 3739
Yamaha YZF-R1 (98 - 01)	♦ 3754
Yamaha FJ1100 & 1200 Fours (84 - 96)	♦ 2057
Yamaha XJR1200 & 1300 (95 - 03)	♦ 3981
Yamaha V-Max (85 - 03)	♦ 4072
ATVs	
Honda ATC70, 90, 110, 185 & 200 (71 - 85)	0565
Honda TRX300 Shaft Drive ATVs (88 - 00)	2125
Honda TRX300EX & TRX400EX ATVs (93 - 04)	2318
Honda Foreman 400 and 450 ATVs (95 - 02)	2465
Kawasaki Bayou 220/250/300 & Prairie 300 ATVs (86 - 03)	2351
Polaris ATVs (85 - 97)	2302
Polaris ATVs (98 - 03)	2508
Yamaha YFS200 Blaster ATV (88 - 98)	2317
Yamaha YFB250 Timberwolf ATVs (92 - 00)	2217
Yamaha YFM350 & YFM400 (ER and Big Bear) ATVs (87 - 03)	2126
Yamaha Banshee and Warrior ATVs (87 - 03)	2314
ATV Basics	10450
TECHBOOK SERIES	
Motorcycle Basics TechBook (2nd Edition)	3515
Motorcycle Electrical TechBook (3rd Edition)	3471
Motorcycle Fuel Systems TechBook	3514
Motorcycle Maintenance TechBook	4071
Motorcycle Workshop Practice TechBook (2nd Edition)	3470
GENERAL MANUALS	
Twist and Go (automatic transmission) Scooters Service and Repair Manual	4082

◊ = not available in the USA   ♦ = Superbike

The manuals on this page are available through good motorcycle dealers and accessory shops. In case of difficulty, contact: Haynes Publishing
(UK) **+44 1963 442030**   (USA) **+1 805 498 6703**
(FR) **+33 1 47 17 66 29**   (SV) **+46 18 124016**
(Australia/New Zealand) **+61 3 9763 8100**

Preserving Our Motoring Heritage

The Model J Duesenberg Derham Tourster. Only eight of these magnificent cars were ever built – this is the only example to be found outside the United States of America

Almost every car you've ever loved, loathed or desired is gathered under one roof at the Haynes Motor Museum. Over 300 immaculately presented cars and motorbikes represent every aspect of our motoring heritage, from elegant reminders of bygone days, such as the superb Model J Duesenberg to curiosities like the bug-eyed BMW Isetta. There are also many old friends and flames. Perhaps you remember the 1959 Ford Popular that you did your courting in? The magnificent 'Red Collection' is a spectacle of classic sports cars including AC, Alfa Romeo, Austin Healey, Ferrari, Lamborghini, Maserati, MG, Riley, Porsche and Triumph.

A Perfect Day Out

Each and every vehicle at the Haynes Motor Museum has played its part in the history and culture of Motoring. Today, they make a wonderful spectacle and a great day out for all the family. Bring the kids, bring Mum and Dad, but above all bring your camera to capture those golden memories for ever. You will also find an impressive array of motoring memorabilia, a comfortable 70 seat video cinema and one of the most extensive transport book shops in Britain. The Pit Stop Cafe serves everything from a cup of tea to wholesome, home-made meals or, if you prefer, you can enjoy the large picnic area nestled in the beautiful rural surroundings of Somerset.

John Haynes O.B.E., Founder and Chairman of the museum at the wheel of a Haynes Light 12.

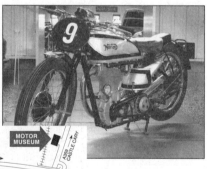

The 1936 490cc sohc-engined International Norton – well known for its racing success

The Museum is situated on the A359 Yeovil to Frome road at Sparkford, just off the A303 in Somerset. It is about 40 miles south of Bristol, and 25 minutes drive from the M5 intersection at Taunton.
Open 9.30am - 5.30pm (10.00am - 4.00pm Winter) 7 days a week, *except Christmas Day, Boxing Day and New Years Day*
Special rates available for schools, coach parties and outings Charitable Trust No. 292048